D1380816

Ecologies of Knowledge

SUNY Series in Science, Technology, and Society
Sal Restivo and Jennifer Croissant, Editors

Ecologies of Knowledge

Work and Politics in Science and Technology

Susan Leigh Star, Editor

STATE UNIVERSITY OF NEW YORK PRESS

Published by
State University of New York Press, Albany

© 1995 State University of New York

All rights reserved

Printed in the United States of America

No part of this book may be used or reproduced
in any manner whatsoever without written permission.
No part of this book may be stored in a retrieval system
or transmitted in any form or by any means including
electronic, electrostatic, magnetic tape, mechanical,
photocopying, recording, or otherwise without the
prior permission in writing of the publisher

For information, address State University of New York
Press, State University Plaza, Albany, N.Y., 12246

Production by E. Moore
Marketing by Dana E. Yanulavich

Library of Congress Cataloging-in-Publication Data

Ecologies of knowledge : work and politics in science and technology /
Susan Leigh Star, editor.
p. cm. — (SUNY series in science, technology, and society)
Includes bibliographical references and index.
ISBN 0-7914-2565-7 (alk. paper). — ISBN 0-7914-2566-5 (pbk. :
alk. paper)
1. Science—Social aspects. 2. Technology—Social aspects.
3. Social ecology. I. Star, Susan Leigh, 1954– . II. Series.
Q175.5.E28 1995
306.4'5—dc20 94-33447
CIP

10 9 8 7 6 5 4 3 2 1

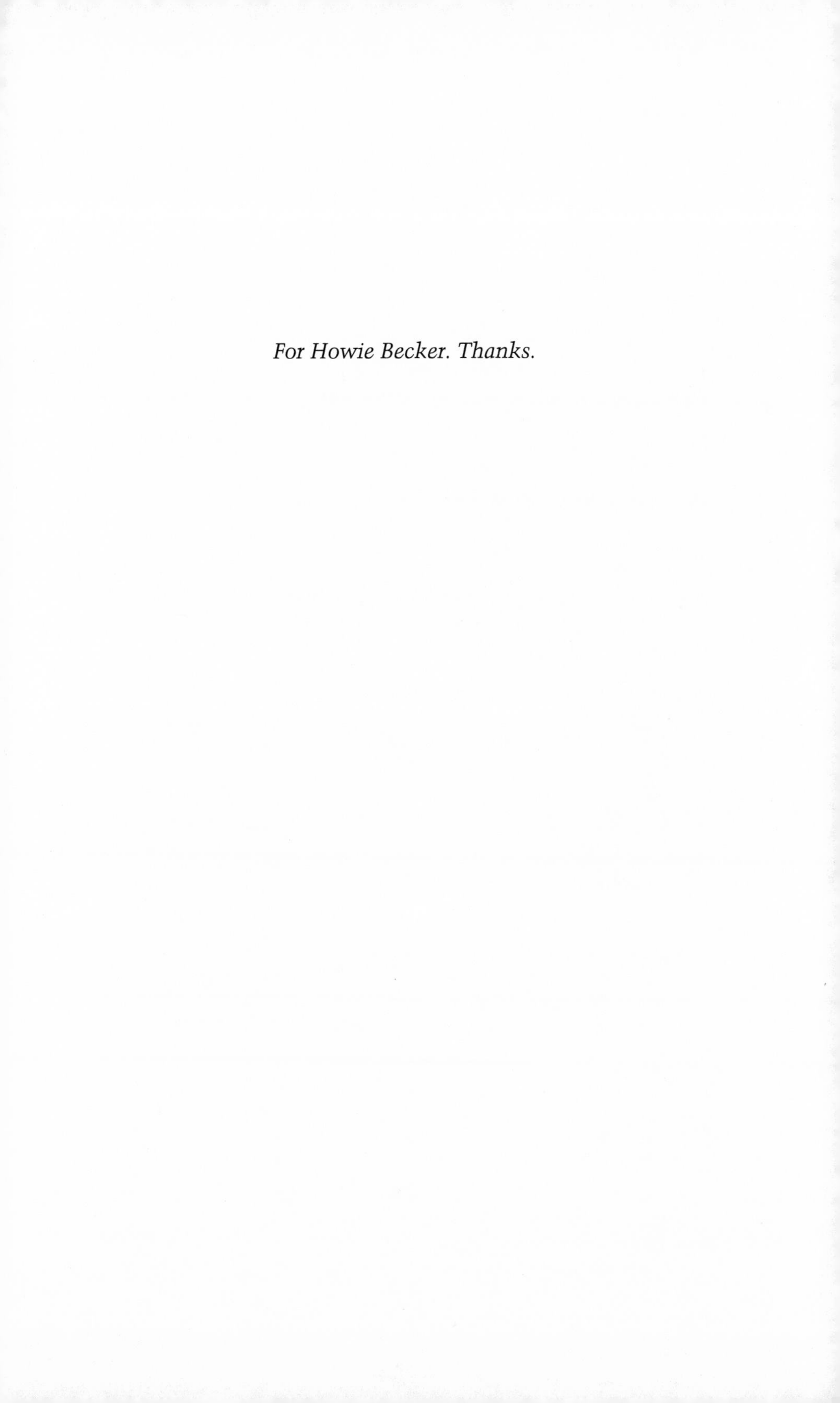

For Howie Becker. Thanks.

Contents

Acknowledgments

Howard Becker first suggested the special issue of *Social Problems* in which several of these papers appeared, and encouraged me to edit the volume. I would like to acknowledge his help and ongoing, generous commitment to dialogue between people in different areas of specialization within sociology. To him this book is affectionately dedicated. Joseph Schneider, who was editor in chief of *Social Problems* at that time, was a patient editor and became a friend in the process. Series editor and contributor Sal Restivo provided much help and encouragement throughout the long editing process. Betty Barlow, Delores Jean Hill, Gina Manning, and Eva Ridnour of the Sociology Department, University of Illinois, did a lot of fiddly typing and organizing at a very busy time of the semester—thanks. The Centre de Sociologie de l'Innovation of the École Nationale Sup. des Mines, Paris, and The Institute for Research on Learning provided space and congenial atmospheres in which to begin and finish bringing the volume together. NSF grant Number EVS 83-62, then the Ethics and Values Program, Division of Biological and Behavioral Sciences, provided support for analysis of the ethical and political issues, some of which are discussed in the introduction and in my chapter. Adele Clarke provided perceptive and extensive comments on the revised introduction. Marc Berg and Stefan Timmermans made helpful critiques of the introduction. Geof Bowker gave ongoing encouragement for the volume's completion. My grateful thanks to all of these good friends.

FORMAL ACKNOWLEDGMENTS

Earlier versions of several of these papers appeared in *Social Problems* 35(3)(June 1988). © Society for the Study of Social Problems. They are reprinted with permission from University of California Press:

"The Mobilization of Support for Computerization: The Role of Computerization Movements," adds an epilogue.

Sal Restivo's "Modern Science as a Social Problem" has been incorporated into an extended essay with Jennifer Croissant.

"Mixing Humans and Nonhumans Together: The Sociology of a Door-Closer." (An extended version of the argument can be found in Latour, Bruno. "Where are the Missing Masses? The Sociology of a Few Mundane Artefacts," in Wiebe E. Bijker and John Law, eds. *Shaping Technology/Building Society*. Cambridge, MA: MIT Press, 1992.)

"Engineering and Sociology in a Military Aircraft Project: A Network Analysis of Technological Change" (incorporates minor changes).

"The Molecular Biological Bandwagon in Cancer Research: Where Social Worlds Meet" is extensively revised here as "Ecologies of Action: Recombining Genes, Molecularizing Cancer, and Transforming Biology."

In addition, the following permissions are gratefully acknowledged:

An earlier version of "The Politics of Formal Representations: Wizards, Gurus and Organizational Complexity" appeared in *Fundamenta Scientiae* under the title "Layered Space, Formal Representations and Long-Distance Control: The Politics of Information," © 1990, *Fundamenta Scientiae* , reprinted with permission.

"Representation, Cognition and Self: What Hope for an Integration of Psychology and Sociology?" by Steve Woolgar first appeared in S. Fuller, M. De Mey, T. Shinn, and S. Woolgar, eds., *The Cognitive Turn*. Dordrecht: Kluwer Academic Publishers, 1989. © Kluwer Academic Publishers. Reprinted with permission of the publisher.

"Research Materials and Reproductive Science in the United States, 1910–1940," Adele E. Clarke. First appeared in Geison, Gerald, ed. *Physiology in the American Context*. Bethesda, MD: American Physiological Society. © American Physiological Society. Reprinted with "Epilogue: Studies of Research Materials (Re)Visited," courtesy of American Physiological Society.

"Laboratory Space and the Technological Complex: An Investigation of Topical Contextures," first appeared in *Science in Context*, 4 (1991), 51–78. © Cambridge University Press. Reprinted with permission of the publisher.

SUSAN LEIGH STAR

Introduction

In 1967 Howard Becker wrote an article that became a clarion call to sociology, his presidential address to the Society for the Study of Social Problems. It was entitled "Whose Side Are We On?" and reminded sociologists that pretensions to value neutrality were themselves value laden. He argued that we must choose to recognize that all perceptions are located in a hierarchy of credibility. In other words, people consider the source of any statement or perception, and discount those produced by lower-status people.

Whose side are we on in social studies of science and technology? What hierarchies of credibility are we tacitly or explicitly assigning? And what language can we invent to investigate these questions honestly?

We could do well to borrow from Patricia Hill Collins's (1986) essay on black feminist thought and its contributions to the structure of sociological knowledge. She argues that African American women's radical explorations of the meaning of self-definition and valuation, the interlocking nature of oppression, and the importance of redefining culture constitute a challenge to sociology's basic beliefs about itself. The challenge takes the form of asking sociology to learn from "the outsider within"—the double glasses of insider and outsider, articulating the tension of both being a sociologist and being excluded by its frame of reference.

The papers in this volume are all attempts to frame the question of whose side we are on by examining science as a radically contextual, problematic venture with a very complicated social mandate, if any. Our purpose here is more than polemics; rather than valorizing or denigrating science as a monolith, we are taking an ecological view of work and politics. And we, too, are "outsiders within," as Woolgar's essay in this volume argues—both strangers and intimates in the world of science. Our work challenges the moral order of science and technology making—and in turn places us in a complex, often tense moral position.

Ecologies of knowledge[1] here means trying to understand the systemic properties of science by analogy with an ecosystem, and equally important, all the components that constitute the system. This is not a functional (or functionalist) approach, with a closed-system organic metaphor at its core. As Sal Restivo notes in his description of science as a social problem, we want to approach science as a set of linked interdependencies inseparable from "personal troubles, public issues and social change agendas," not a social structure with one or more *dysfunctional parts.* Science and technology become monsters when they are exiled from these sorts of questions (Law 1991; Haraway 1992b; Star 1991a; Clarke, in press b), just as other symbolic monsters have been borne from the exile of women's strength from the collective conscience, or in the demonizing of people of color. In Michele Wallace's words: "The absence of black images in the reflection of the social mirror, which such programmatic texts (from *Dick and Jane,* to Disney movies, to *The Weekly Reader*) invariably contract, could and did produce the void and the dread of racial questions . . ."(1990). In our exorcism, we simply want to see scientists and technologists as ordinary, as citizens, neither villains nor heroes.

Each of the authors in this volume calls in different ways for an ecological analysis, including a restoration of the exiled aspects of science. Thus by *ecological* we mean refusing social/natural or social/technical dichotomies and inventing systematic and dialectical units of analysis. I think this reflects the dissatisfaction with conventional ways of approaching organizational scale and units of analysis, a dissatisfaction brought on no doubt in part by our respondents (scientists and technologists), who themselves are continually plagued by these questions. Restivo and Croissant, in this volume, examine a large-scale set of relationships, as between science and all other institutions and social arrangements. Kling and Iacono put the context back into analyses of technological innovation, noting that only such an approach can overcome simpleminded technological determinism or technocracy. Star makes a similar point for the class of "artifacts" called formalisms, or formal representations, and their relationship to organizational complexity. Fujimura looks at the ecology of the workplace in a Hughesian sense, and thus goes beyond the simple adoption of new technologies as a determining factor in scientific social change. Law and Callon, Woolgar, and Latour each break the traditional boundaries of what can be included in an analysis of technology and social organization, recommending a broader, more democratic kind of analysis that is both moral and deeply ecological. In this, Law and Callon call attention to the local expertise of the scientist/engineers who, in their eyes, make no distinction between *technae* and *politaea* (Winner 1986). Woolgar refuses another "great divide" in so doing—that between individual psyche and collective repertoires of behavior. Clarke challenges us to examine the stuff of science—the material substrate—instead of ignoring it in the service of idealist theories. Lynch's work extends this point to a revaluation of the idea of "place" in science, arguing instead for distributed, material " topical contextures" in which to examine scientific genre. Latour uses the device of analyzing an ordinary, low-tech system—the door—and its surprising complexity to point to the inseparability of the technical and the social.

Our key questions here are those of general political theory and of feminist and third world liberation movements: *Cui bono*? Who is doing the dishes? Where is the garbage going? What is the material basis for practice? Who owns the means of knowledge production? The approach begins in a very plain way with respect to science and technology by first taking it "off the pedestal" (Chubin and Chu 1989)—by treating science as just something that people do together. Some of this means looking at science and technology as the occasion for people to do political work—not necessarily by other means, but fairly directly. Science as a job, science as practice, technology as the means for social movements and political stances, and science itself as a social problem—collectively, these articles take science/technology as the occasion for understanding the political and relational aspects of what we call knowledge. This introduction situates these papers with respect to science and technology studies (STS), and gives a sense of the problems they are addressing in social theory.

Most work in STS has not been seen as general social theory or as contributing in a fundamental way to social science theory. While historians, philosophers, and even computer scientists show a great deal of interest in the new sociology/anthropology of science, most social scientists view it as a kind of luxury, an arcane corner of the discipline offering only specialized insights. One purpose of this book is to demonstrate that social studies of science and technology are addressing a set of questions central to all social science. In this selection of science studies research, readers will see that science and technology are the vehicles for analyzing some very old questions. How do people come to believe what they believe about nature and social order? What are the relationships between work practices and social change? What is the trajectory of social innovations? Who uses them and for what purposes? As people from different worlds meet, how do they find a common language in which to conduct their joint work? How can we study people's work critically, yet as ethnographers or historians respect the categories and meanings they generate in the process? Finally, what are the boundaries between organism and environment; how fixed are they; how can we know them; and are they meaningful a priori?

Perhaps because learning another scientific language is a prerequisite for doing the kind of social studies of science/technology described herein, scholars in science studies tend to be an omnivorous bunch. We read history and philosophy of science as well as scientific tracts in the substantive fields we study, feminist theory, and *Science* magazine. We often borrow models from those writings as well as from other areas of social science. We work across national boundaries in informal groups clustered around an analytic topic: the use of metaphor, for example, or the extent of technological determinism or infrastructural change. The field is small (although growing), lively, and filled with debates, cross-fertilization, and often surprising collaborations. For example, sociologist Michael Lynch, who is interested in visual representation (and whose paper on the topic appears in this volume), has collaborated with art historian Samuel Edgerton on analyzing the pictures astronomers create (Lynch and Edgerton 1988). Steve Woolgar became the project manager of an industrial computer development firm in order better to understand the process of technology construction. Diana Forsythe (1992, 1993), a cultural anthropologist by training, has worked for many years in collaboration with computer scientists building medical expert systems, both critiquing the notion of expertise and acting as a designer. Because the field is so interdisciplinary, the term *science studies* often replaces the disciplinary-specific label, such as *sociology of science.*

HISTORICAL REVIEW

In the United States, most sociology of science before the late 1970s was dominated by the work of Robert K. Merton and his associates (see, for example, Merton 1973; Zuckerman 1979) and by a group of researchers conducting bibliographic citation analysis (from which work came the concept of "invisible colleges"; see Crane 1972; Mullins 1973). While there had been much criticism of functionalist sociology on a number of fronts, particularly from symbolic interactionism and Marxism, the sociology of science received scant attention. Symbolic interactionists, for example, had studied work and occupations, medicine, deviance, gender, the family, urban life, and education, critiquing functionalism in each of these areas. Yet they had produced only a few scattered monographs and articles on science (for example, Glaser 1964; Strauss and Rainwater 1962; Marcson 1960; Becker and Carper 1956; Bucher 1962) and had undertaken no large research programs in this area.

Meanwhile, in Europe in the early to mid-1970s,[2] a group of researchers began a series of studies to demonstrate, contra Merton, that science was not "disinterested, communistic, and universal." Many of them had also been deeply influenced by Kuhn's (1970) *Structure of Scientific Revolutions*, a book that had questioned the cumulative nature of science and raised the issue of the incomparability of scientific viewpoints or paradigms. They were concerned to show that science was not neutral, that the outcomes and content of science as well as access to it as a profession were determined by structural commitments, political positions, and other institutional considerations. MacKenzie's (1981) work on the interrelationships of statistics and eugenics is a good example of these efforts. These researchers were also concerned to demonstrate the constructed nature of science and its view of nature. Thus, they were strongly antipositivist. Some of this work was done at the University of Edinburgh, and the "interests" model became especially associated with the "Edinburgh School" (see, for example, Barnes 1977). Other important centers with overlapping approaches were in Paris and Amsterdam.

In 1979, Bruno Latour and Steve Woolgar published *Laboratory Life*, probably the book from the field that is most well known outside of science studies. The book was an ethnographic study of a scientific laboratory, and its purpose was to document the creation of a scientific fact. Using a variety of techniques from anthropology, semiotics, and ethnomethodology, they traced the birth of a biological fact in the context of lab work. They concentrated on a process they called "deletion of modalities," a progressive stripping away of contextual information about production, with the end result being a fact bare of its own biographical information. The book was an immediate success and was one of the factors helping spawn a series of laboratory studies and descriptions of act-making, often ethnomethodological in approach.

The combination of fieldwork and antipositivism was familiar to American symbolic interactionists, who welcomed the chance to apply these techniques to science and to learn from colleagues in Europe. A number of collaborations ensued among researchers in America, England, France, and the Netherlands pursuing these viewpoints. (See Fujimura et al. 1987; Clarke and Gerson 1992; Star 1992b; Clarke 1990; 1991 for reviews of this work.) Among our common interests and beliefs was the necessity of "opening up the black box" in order to demystify science and technology; that is, to analyze the process of production as well as the product. Methodological directives for those of us working in the interactionist tradition were familiar: Understand the language and meanings of your respondents, link them with institutional patterns and commitments, and, as Everett Hughes once said, remember that "it could have been otherwise." Many of our colleagues in Europe held similar views, albeit from very different traditions: Do not accept the current constructed environment as the only possibility; try to understand the processes of inscription, construction, and persuasion entailed in producing any narrative, text, or artifact; try to understand these processes over a long period of time (some of this work is represented in Law 1986a; Bijker et al. 1987; Callon et al. 1986).[3]

There were and are many other groups throughout the world studying science or technology: policy makers, historians, analysts of the impact of technology (particularly computing and automation), and the number is growing rapidly. Another important development began as programs of social science research gelled into Science, Technology, and Society programs at a number of technical institutions and regular universities. New undergraduate and graduate STS programs began to spring up, both within traditional departments and as interdisciplinary programs. Early STS programs represented an amalgam of interests: ethics and values in science and engineering, studies of social impacts of technology, and history of science and technology. They were often an academic home for science criticism, that is, studies that demonstrated scientific bias (racism, sexism, classism) or danger resulting from scientific and technological research and development (nuclear and toxic wastes, recombinant DNA, technological disasters). Criticism of the sacrosanct institution of science and explication of the constructed nature of nature have remained core problems in science studies, and there is currently lively debate about the role of activism in the field, as Restivo and Croissant indicate in their paper for this volume.

QUESTIONS OF ORGANIZATIONAL SCALE

Questions of organizational scale have always plagued (or some might say, graced) social science. Is social change individual or aggregate? How can we understand the relationship between social facts and individual experience?

These questions appear in science as it is interlaced with presumptions about the nature of scientific inquiry. If researchers accept that nature is simply "out there" waiting to be discovered, they may append to that belief the idea that "anyone can do it, geniuses better than the rest of us." There is nothing that logically ties these two together, but much of the received mythology about science involves great men (*sic*), great moments, great labs, and great accidents of Nature revealing herself. This combination of individualism, positivism, and elitism conspires to confuse the question of the appropriate level of organizational scale at which to conduct inquiry. So, the secondary literature on science is littered with psychologism, reified "societies" that act in mysterious ways on believers, and participant histories that claim exclusive centrality for powerful, rich institutions and people.

Against this trend is a lively debate, partially represented in the pages of this book, about the right unit of analysis for studying science. In escaping from the nasty things mentioned in the previous paragraph, sociologists and anthropologists of science have invented, borrowed, or transformed units of analysis from other parts of the discipline or from science itself: bandwagons, social movements, political economy and large-scale work organization, units of action and activity that cross human/nonhuman boundaries, the taken-for-granted truth about Nature that reflects old and widespread conventions (and superstitions).

How is the little black box of the computer, the test tube, or the door-closer joined with phenomena at larger scales of organization? This is a fundamental question about science and technology, but it appears whenever one questions the nature of local social arrangements as articulated with those at a distance or with considerably more power and purview. All of the articles herein attempt to answer this question ecologically and propose several modes for doing so.

At the largest scale of organization, questions are raised here about the utility and role of science or technology in maintaining or changing the *status quo*. This is asked not simply in terms of technological determinism, but in terms of larger scale issues, a central one being: Can there be a revolutionary science/technology in the absence of revolutionary social change in other spheres? To the extent that one believes in the interpenetration of spheres and science as a social institution of its historical time and place, the answer must be no. This puts the question of political commitment squarely at the center of science studies. For one thing, it is difficult to escape examining oneself as a scientist while engaged in studying scientists full time. Truly revolutionary science or technology thus means full-scale revolution. The sociology of science might allow us better to understand what that might mean.

METHODOLOGICAL ISSUES

Interwoven with questions of scale and politics are questions of method. Scientists are very challenging respondents. For one thing, all scientists share with us concerns about reliability and validity of data, robustness of findings, and the meaning of those findings. I never met a scientist who had not thought about the issues raised in this introduction. As a group of respondents, scientists are particularly difficult and rewarding because they have often thought rigorously about the issues we are investigating, and about which we are ourselves uneasy (Woolgar, this volume). So the work of our respondents blends with our own. The meaning of participant observation in this case can begin radically to change.

There are several kinds of work to be juggled in doing research in STS, each of them methodologically challenging. First, there is the map and language that the scientists themselves use in their work. Second, there is our mapping of the work practices and organization. Third, we create maps of the communications between domains. Fourth, a complex "nested" map is generated that shows who answers to whom, and why. It is at this level that questions of unit of analysis, or scope, often show up in force. Reconciling the different maps is a nontrivial methodological problem, again common across many domains of social science and political life.

WHY I AM NOT A NAZI: REALISM AND RELATIVISM IN SCIENCE AND TECHNOLOGY STUDIES

One of the curious things about being in the STS field is that one is immediately plunged into philosophical debates about realism and relativism. Briefly, realism is the position that "there really is a there out there, and it's true in some absolute sense." Relativism holds that truths are relative to a place, time, or person (often a historical situation or geographic/cultural location). All researchers in science studies have had the experience of being challenged about the "underlying truth" of science: What about the scientific method? What about truth? What about the laws of gravity?

I have been involved in science and technology studies for about fifteen years, first as a science critic, then as a historical sociologist and ethnographer. I have given over one hundred talks on various aspects of the sociology of science and technology. In almost every presentation, I have been asked some version of what I now call the "there there" question: But are you saying it's *all* socially constructed? Doesn't that mean anything could be true? Isn't there anything out there? Are you saying that scientists are making it all up? Are you saying germs don't really make you sick, or gravity doesn't really make things fall down?

It is indicative of the central place of science in mainstream Western belief systems that *merely to imply* that the acquisition of scientific knowledge is work, not revelation, seems to involve the kind of radical idealism (if not radical autism) alluded to above. But this is not necessarily the case.

To say, as Hughes did about social order, that "it could have been otherwise" is not to say that it *is*. And to say that the conditions of nature or science are the result of collective enterprise that includes humans and nonhumans is not to imply that the merest whim on the part of an individual could overturn them. Rather, as social scientists, let's ask: Under what conditions do such questions about reality routinely get raised?

First, the term *socially constructed* is a reformist term, inserted into titles in sociology/anthropology of knowledge and science. Its initial purpose was to demonstrate that the reports of science that had been stripped of production history were missing important historical and situated accounts; second, to restore accounts of the actual work and its organization to those reports. Furthermore, if one takes "society" as the scientific problem, then the image of a society "out there" structuring an experience that is then entered into the canon of research doesn't make any sense either.

I call the idea that cumulative collective action is flimsy the "mere society" argument. It is paralleled by simplistic perceptions about, for example, socialization and gender. The argument goes something like this: "So, she's been socialized as a girl. Well, let's just let her into the corridors of power and de-socialize her, and then everything will be ok." Such a statement depends on a trivial and reified conception of both socialization and gender. Whatever bundle of actions, past and present, we might think of as "socialization" here is far more complex and durable than most of us realized in the early days of feminism. Similarly, the notion of "institutionalized racism" has been crucial in understanding that racism is not simply a matter of people not being nice to each other, nor necessarily to be found in a single set of micro-interactions—rather, it is a web of racist discourse and practices that extends through and informs all human practice—and cannot be simply transcended (hooks 1990). The durable bundle of actions and experience that comprise "science" has a similar sturdy complexity. This complexity does not defy its ontological status as "created," however. The constructivist or relativist schools in science studies (and I will not explicate the subtle differences between them here) have often been accused of flimsiness or mentalism on grounds that deeply confuse epistemology (how do you know what you know) with ontology (how are you what you are).

Scholars in science studies have disagreed about this issue and will continue to do so for some time to come. Yet a thread runs throughout the work of the groups represented in this issue: Let's replace the either/or dichotomy of constructed versus real with more useful concepts. Concepts such as workplace ecology, *irréductions* (Latour 1987), sociological imagination, networks and translations, and boundary objects (Star and Griesemer 1989) are important here. Wimsatt's (1980) concept of "robustness" has similarly been an important replacement for more restrictive concepts of reliability and validity. He, borrowing from biologist Richard Levins, defines it as "the intersection of independent lies," or more sociologically, the durability of collective action despite the fragility of any one instance.

During the 1980s there were scores of articles and books addressing this class of questions in science and technology studies. They have important links as well with earlier work in other parts of sociology and anthropology. For example, the debate in the 1960s about labeling deviance asked whether some things aren't *really* sick (or unnatural). The sociology of art has been concerned with the question of whether some things aren't *really* just beautiful (in a timeless or transcendental fashion).[4] The enduring concern with ethnocentrism in anthropology has recently exploded in debates about the place of the anthropologist and whether the knowledge constructed by anthropology is rightly seen as a jointly created fiction. In sociology and anthropology of medicine, the debate occurs as a question about whether one can differentiate physiological disease from "illness behavior"—aren't some things *really* just germs?

But the analytic trick in each of these cases is to raise the concept of "really" to the status of rigorous, reflexive inquiry and ask: *Under what conditions does the question get raised?*

One of the difficult things about trying to analyze an institution as central as science is that one challenges the received views of things for audiences and respondents. In giving talks that defend the above position, I have sometimes been called a Nazi, or parallels have been drawn between the social construction of science and Nazi science. It took me a while to figure out what people were talking about in these accusations, since being a Nazi is anathema to me.

If one takes the point of view that fascism requires a kind of situation ethics and requires that one redefine the situation according to opportunism or a kind of distorted view of science and nature, then any attempt to make relative any situations (especially natural or scientific ones) becomes morally threatening. This is so because one antidote to fascist ideology is to affirm an overriding value in human life, a universal value that cannot be distorted by the monstrosities informed by local, parochial ideologies of racism and genocide. Ethicists often base their arguments on this presumption. The worst thing for an ethicist is to hear arguments that plead "special circumstances"—the name of the game is finding good universals.[5] Yet this criticism of relativists as Nazis shows another kind of confusion, which again relies on a separation of the social and the natural and a separation between the conditions of production and the product. If the relative ontological status of a phenomenon is inextricably embedded in the conditions of production, then it's not a question of an analyst legitimating genocide or situation ethics. Rather, the question on a meta-level becomes: How can we make a revolution that will be ontologically and epistemologically pluralist yet morally responsible? Can we be both pluralist and constructivist, hold strong values and leave room for sovereign constructions of viewpoints? These are not new questions, either; both the French and American Revolutions were fueled by them. I would claim that there is stronger evidence for Nazism arising from ignorance of the conditions of production of knowledge than from exploring the relative configurations of these conditions in different times and places; more oppression from the appeal to absolute natural law than from negotiations about findings. While I'm not implying here that science studies is the best weapon against totalitarianism, the fact that this question arises so frequently in so many different contexts is to me indicative of the fact that science has been such an inviolate institution, certainly in academia.

CURRENT INTELLECTUAL DEVELOPMENTS IN STS

Taking on science as a social construction grew beyond either interest explanations or laboratory ethnographies by the end of the 1980s. Science and technology studies (STS) has over the past several years worked hard at two central intellectual currents, both of which are at the core of an ecological analysis of science (or perhaps, in some sense prior to it). The first is the establishment of science as materially based (see especially Clarke and Fujimura 1992a and 1992b; Haraway 1989; Clarke, this volume; Lynch, this volume); the second is science as a form of practice (see especially Star 1989a; Pickering 1992).

It is remarkable for how long accounts of science (in history, philosophy, and sociology/anthropology of science) neglected to notice that much of the activity we call science consists of people manipulating materials , including specimens, media and cultures, breeding colonies, and display items. This material culture of science is important not just as another form of exoticism, but for the ways in which it is constituent of scientific findings and constraining of the ways we perceive scientific meaning.

OF HUMANS AND NON-HUMANS

One of the issues that appears in different ways in the papers in this volume is the issue of "where to draw the line" in analyzing science and technology. Traditional studies usually drew the line at the edge of the black box, whatever it might be: the computer, the laboratory, the closed scientific work group. The argument in this volume is that science studies in the past have left out some of the most important actors, the "nonhuman" ones. Many of the new sociologists of science are engaged in a kind of democratization of this analysis, as the papers here demonstrate. If one adopts an ecological position, then one should include all elements of the ecosphere: bugs, germs, computers, wires, animal colonies, and buildings, as well as scientists, administrators, and clients or consumers (see Clarke and Fujimura 1992b and Latour 1987 for an analysis of this). The advantages of such an analysis are that the increased heterogeneity accounts for more of the phenomena observed; one does not draw an arbitrary line between organism and environment, one can empirically "track" lines of action without stopping at species, mechanical or linguistic boundaries, and especially without invoking a reified conception of society.

On the other hand, this kind of analysis presents some serious ethical problems—on both sides (Singleton and Michael 1993). For many years feminists, radical ecologists, and pantheists have recommended a kind of analysis that does not exclude anything from the natural world. The exclusion of animals, the biological environment, and other parts of the natural context has been one of the major sources of alienation under patriarchy, late capitalism, or religions that are antinature (Griffin 1978; Merchant 1980; Harding 1991). Restoring the natural world to the research context would be an ethical and political advance. On yet another hand, the papers by Kling and Iacono and that by Star are written from within a research context in which it is not humans who have been privileged at the expense of nonhumans, but vice versa. It is computers and automation that have occupied a privileged position vis-à-vis human beings, often because of the inadequate social analysis held by computer movement advocates. An ethical social problems position in this case most likely involves checking the power of nonhumans and their advocates and seeing that humans understand it contextually, not democratizing the nonhuman position. Thus where Latour is concerned to restore ecology from one side, Kling and Iacono and Star are concerned to restore it from another.

What are the moral values invoked by such analyses? I think that there are no simple answers. The dividing lines should not really be between advocates of humans and advocates of nonhumans, but between ecologists and reductionists. In furthering the cause of an ecologically responsible, socially and philosophically sophisticated analysis of science and technology, we need to confront head-on questions of scale, of boundary drawing, and of mystifying science and technology, *as well as* questions of race, sex, and class. To do that, recursively and reflexively, we need an ecological approach.

A recent collection of papers in the sociology of science highlighted this debate in the field (Pickering 1992). Collins and Yearley's paper in that volume accuses Latour and Callon of playing "epistemological chicken" in the interests of advancing the nonhuman analysis at the expense of the human. Callon and Latour respond with a defense of their position, claiming it as a heuristic analytic device that pushes the boundaries of science studies beyond reified sociological categories. Fujimura's paper in the volume, taking a symbolic interactionist/pragmatist and feminist perspective, rejoins with the claim that neither side has comprehended the human stakes involved, and that when the debate is phrased as humanists versus poststructuralists, once again concerns of all women and men of color, as well as other minorities, are ignored.

The debate between the British and the French, on the one hand, and Fujimura's claim that from an American pragmatist perspective the issues are misframed are important for the ecological analyses presented here. If we take ecological to mean treating a situation (an organization or a country or interactions and actions) in its entirety looking for relationships, and eschewing either reductionist analyses or those that draw false boundaries between organism and environment, then indeed the human/nonhuman question is reframed. The axes within the ecological space are four:

1. continuity versus discontinuity
2. pluralism versus elitism
3. work practice versus reified theory
4. relativity versus absolutism

On the left-hand, radical side go continuity, pluralism, and relative ecologies of work practice; the reactionary side is discontinuity (or divides great and small), elitism (or pretensions to a single voice), reified theory (or deletions of the work in representations of it), and absolutism (or "there really is a little bit of determinism").

A central fight within American sociology, and subsequently within sociology of science as discussed above, has been against functionalism, that school of thought which sees a closed-world, top-down, organismlike social order that draws its imperative from an imputed physiology-writ-large. The sociological field in America is mined with "hot spots" that come from the scars of these historical battles. From the pragmatist side are the words *consensus, boundary maintaining, natural,* and *obvious.*

The fights between the British and the French resonate along these axes in three ways. First, the relativism of both schools satisfies the pragmatist concern with pluralism. They both imply that there are frames of meaning, definitions of situations, different perspectives based on experience. Due to the long history of fighting that pseudosingular voice, such pluralism became the salient relativist dimension for pragmatists/interactionists in STS. The fact that it also has profound resonance along the axis of nature/society, people/things, and so on doesn't carry as much historical weight, for the reason that pragmatists never believed in that divide in the first place. John Dewey and Arthur Bentley, for example, spent their entire careers fighting these notions in analytic philosophy; they appear as similar fights in the work of symbolic interactionist sociologists influenced by them, such as Howard Becker and Anselm Strauss. So the work of Collins and Pinch (1982) on the different valuations placed on parapsychology and psychology resonates with that commitment to pluralism, too, seeming to restore the voice of the underdog to scientific debates and balance it out—to even out the "hierarchy of credibility" discussed above.

The French actor-network theories, and their emphasis on the inclusion of nonhumans (see Latour, this volume), find a match in the pragmatist concerns about continuity and process. Because the mandate of the pragmatist research program since the 1920s has been to "follow the actors," it is not surprising that there have developed strong ties between French researchers and American symbolic interactionists.

A central tenet of the pragmatist work in STS has been to think of scientists as people who are doing a certain kind of work.

Among other things, science is a job. It's a very interesting one, because it turns out that even *calling* it a job invokes the wrath of American functionalists, most philosophers, many deans and administrators, and most computer scientists. Simultaneously, this vision of science as work invokes the appreciation and support of many historians.

As a symbolic interactionist, I agree with Callon and Latour and Lynch and Woolgar that new methods that will lead us across traditionally accepted boundaries are crucial, and those that will help manage the "rich confusion" of things and people are absolutely critical for science studies at this time. Because of my pragmatist concerns with work, I would add work itself to the rich confusion, in the form of activity, practice, and/or work organization. I would also like to emphasize a neglected dimension in the worlds of STS and science/technology: the great divide between formal and empirical.

Work, Formalisms, and Divides. One of the most confusing actants in this complex ecology is the work of scientists and technologists who create formalisms,[6] including those working with information technologies. The impact of STS in these spheres is not so much limited by concerns with relativism/realism—indeed, here Latour's early point about scientists not being naive realists is absolutely true. But many scientists I've known *are* naive formalists, especially those in information technology and computer science. One thing about computing technology is that it allows one, paradoxically, to use a very concrete thing to manipulate representations that are quite formal. The great divide that computer science itself then produces is between the formal and the empirical—this is reproduced many times across the sciences, including social sciences. On the one side are formalists who believe that computers are embodied mathematical theories—theories come to life. (I do not exaggerate here; if anything, I am understating the case.) On the other side are engineers who argue that *only* making things (e.g., machines, programs, new speed records) really counts for technological advance. And on a third (much, much smaller) side are a few brave souls who argue (mostly for reasons of safety or ethics) that empirical studies must join together inseparably with formal models. They are as yet few in number, and have suffered enormous academic stigmatization for their stance. Some details of the debate in computer science itself can be found in Newell and Card (1985, 1987) and Carroll and Campbell (1986).

These pieces and Fetzer (1988) give a sense of the vituperative flavor of the debate. Star (1992b) reviews the work of people in several disciplines working in this "third force" (see also Star, in press b).

The formal/empirical canyon is a complex and compelling great divide (see Bowker et al. 1993 for a discussion of the issues), one that is only beginning to receive attention in STS. And here are the stakes. A formalist would argue that when building air traffic control systems, human fallibility and bias is such that we are virtually killing people to rely on it. The computations are so many and so intricately interconnected that only a machine can be smart enough (statistically, and there's the rub) to do them fast enough before the (now much more complex) airplanes fall out of the sky. The empiricist argues that all such mathematical systems are unprovably fallible (using strong formal allies like Gödel and Schrodinger), too big or too dangerous to test, and that we'd better stop the reliance on formal testing or we will all (literally) be blown sky high by Star Wars or a series of high-tech accidents.

A significant divide indeed, since it captures all life forms in its technological threads. . . .

The formal/empirical divide is also richly represented in social sciences. On the one side, in American social science especially, there are formalist fundamentalists who believe that life "really is" a mathematical model, and that empirical data are incidental at best to its representation. Only quantitative truth matters. On the other are empiricist fundamentalists who sneer at numbers, algorithms, or other sorts of formal models.[7] Star's and Woolgar's papers in this volume explore ways to refuse this great divide of formal versus empirical in computer science, psychology, and in social theory more generally.

As both a scientist and a citizen, I have a great stake in closing this divide. I do believe that the stakes are as high as the empiricists in computer science claim, although I don't think they are as centralized as many of them envision. Rather, I think the consequences of maintaining the formal/empirical divide are highly distributed, and reside as much in forms of bureaucracy, education, and exchange as in bombs and air traffic control systems.

The Status of Matter and the Absolute

I have on occasion taught Latour's *Science in Action* (1987) to scientists. The book's central tenet is that "nature" is nowhere to be found apart from the web of work and inquiry constituting the relations of science. To my initial surprise, the class discussions often became theological in nature. It seems my students who are in the sciences aren't afraid to use words like *God* or *soul*. However, among social scientists such discussions seem to be completely taboo. It's easier to talk about sex or excrement or almost anything than to talk about one's deepest spiritual or metaphysical beliefs. (I suppose in the United States at least it's because religious fundamentalists conjure up images of antiscience, antifeminist fanatics.)

There are a set of questions in STS of science that resemble metaphysical and theological questions throughout the ages. Beyond the questions of humans and nonhumans, and of formal and empirical knowledge, many of the deeper issues in the debate seem to reflect divergent opinions about the status of matter itself. The questions go something like this: A couple of years ago I gave a talk at UCLA about my work in collaborating with computer scientists in artificial intelligence. I talked among other things about the primacy of distributed (cross-personal, organizational, or community) cognition. Harry Collins was also there, and as "devil's advocate" asked me a question: "Agreeing that cognition is social, isn't there a limit to that? How do you explain, for example, that you could wake up in the morning speaking English, go alone into a locked room, and come out at night still speaking English? Doesn't that imply that there's some cognition only in the head?" At the time, I didn't have much of an answer for him, but in thinking through the human/nonhuman debate for this volume, I will venture a bit of translation.

First of all, when speaking of the brain in that fashion, one is implicitly speaking about matter, about physicality. Traditionally, people have had a difficult time modeling or speaking of "brain" and "thought" without invoking one of the original great divides, that between mind and matter. If we want to refuse that great divide, we must thus carefully examine how we think about matter itself. This is perhaps the microarchitecture of an ecological analysis. If we are to go beyond the current debates, STS researchers must confront this basic dichotomy and not allow in "a little bit of determinism" here and "a little bit of realism" there.

Think of matter as composed of arrangements of space and time—some very, very fast, as with light, some very, very slow, as with rocks. In an Einsteinian/quantum mechanics fashion, this matter has no absolute speed or rhythm. Rather, its rhythm and speed derive from its context. The rearranging of space-time configurations is a constant, never-stopping process, although some speeds are too slow for us to perceive as anything but stopped.

Another way in which this rearrangement works is as a relative location. Analytically, it is extremely useful to think of human beings as *locations* in space-time. We are relatively localized for many bodily functions and for some kinds of tasks we perform alone. But for many other kinds of tasks we are highly distributed—remembering, for example (Middleton and Edwards 1990). So much of our memory is in other people, libraries, and our homes. But we are used to rather carelessly localizing what we mean by a person as bounded by one's skin. Pragmatist philosopher Arthur Bentley cautions against the philosophical contradictions this brings about in his brilliant essay, "The Human Skin: Philosophy's Last Line of Defense" (1975 [1954]). The skin may be a boundary, but it also can be seen as a borderland, a living entity, and as part of the system of person-environment. Where the skin is, indeed, under some conditions becomes a very interesting question. But as an unthinking, linelike division between inside and outside, where "self" is on the inside, it makes no sense philosophically. Parts of our selves extend beyond the skin in every imaginable way, convenient as it is to bound ourselves that way in conversational shorthand. Our memories are in families and libraries as well as inside our skins; our perceptions are extended and fragmented by technologies of every sort.

All the matter in our body can be thought of this way, including the brain. In 1896, John Dewey wrote a critique of reflex psychology that still stands today (1981 [1896]). He noted that the common image of psychology was that a stimulus would happen, "go in" to the brain, stop there, be processed, and something would come back out. This was complete nonsense, said Dewey. It doesn't "go in" through nothingness . . .there is an event that changes the air, interacts with skin, with nerves. It is continuous, and there is never a time when it "stops." The arc is a convenient notation for a dualist, reductionist psychology, however, and makes certain things amenable to quantification.

I reiterate Dewey's critique with respect to cognition and the individual, and recommend it to researchers in STS. Learning English is a series of continuous events, of changes, rearrangements in the space-time of your body. Once the process gets going it keeps on going, given constant interactions with other people and all kinds of humans and nonhumans in the world. I don't know enough about death to know whether or in exactly what forms it might keep going afterwards, except that the ongoing actions we leave embedded in the world constitute one such action; for example, the books we write may be read after our deaths.

So the alone person in the room speaking English before and after is one case of a time of aloneness (a typically very short time, otherwise it would be solitary confinement, the ultimate penalty in most cultures). The ecological image for the aloneness of that learning is analogous to holding your breath—you still need oxygen, but you take your lungs "out of play," or put them on hold, for a moment. So the alone person is aside in the sense of not being together with others. Aloneness seen this way is not a vindication for mentalism or for the primacy of the individual; it's just a special case of relocating.[8]

There are historical and contemporary neurophysiologists who view the brain this way, as well as the brain-body-environment. The images have a wide range. Some have seen the brain and cognitive functioning as a "re-entrant, emergent process," where that which is sensed keeps circulating in the brain forever. Others find that we can't process sound without hearing what is before and after it—perception is entirely relative with respect to context. Once something is perceived, the action of perception continues indefinitely, changing and being changed by other events near it, sometimes resonating, sometimes clotting up or clumping up, sometimes fading into background noise.

To think this way we must vastly complexify the way we think and talk about matter. The brain is not a lump of meat with a few electric channels strung though it. The body/brain of any one person is a location of dense rearrangements, nested in like others. When we use the shorthand "individual" or "individual cognition," we are thus only pointing to a *density*.

Thus, in speaking of aggregates of peopled, material ecologies (including in them things, built environments, the natural world), we have a basis of resolution of the realism/relativism dichotomy and of the formal/empirical divide. An ecological analysis refuses to ground beliefs, including scientific beliefs, in something outside of this webbing location.

Moral Implications. What are the moral implications of this view? Scientists certainly don't hesitate to address such concerns in bringing up these issues, and neither should we in STS. Sociology and economics began as moral ventures, addressing this class of problems with respect to the place of human beings on earth, and of the prescriptive nature of our relationships with each other. From this evolved a long investigation of those relationships, in the middle of which we seem to have forgotten that the whole purpose of the enterprise was to speak responsibly, in a disciplined and collective fashion, to Very Big Questions. What is a moral order? What are values and passions, and how are they arranged? What do we owe tradition, and what do we owe innovation? In the case of STS, this includes the question: What is matter?

Activity theorist Yrjö Engeström (1990) argues that technology occurs when joint activity between two actors is articulated. It does not preexist such action, and as the tool occurs, it comes to form part of the subsequent material conditions mediating further action. Material conditions are not only such things as stuff, money, climate, and bodies, but also refer to *durable arrangements* that have consequences on the trajectory of action as material conditions (e.g., Butler 1990).

In order for this not to be read as an idealist statement, one must firmly resist great divides between individual and environment, between technology and knowledge, and between language and thought. *The emphasis is on use and consequences, not antecedents.* And the result is an inversion of everyday thought. It requires resistance to the functionalist defaults of presumption. So ethnomethodologists say that "constraints are also resources"—or feminists observe that structurelessness can be tyrannical (Freeman 1975)—or pragmatists that "things perceived as real are real in their consequences" (Thomas and Thomas, 1970 [1917]). Let me say it another way: If you think of matter (including people) as a space-time arrangement in the way I've described it, and you also then think of rearrangements or reconfigurations in those arrangements as having consequences, you can easily come to a very political but nonreductionist and nonpositivist account of moral orders. There are some very slow moving, very large-scale, quite general standing arrangements and settled questions constituting multiple moral orders that, taken ecologically, constitute what we think of as "societies."

I think Collins and Yearley have a quite legitimate fear that including nonhumans in an undifferentiated way threatens our moral order (and in particular our moral order as social scientists). The very real image behind the passion in their critique of Callon and Latour is, well, does a cat have just as much right as a human being? Are we going to anthropomorphize machines in a nonchalant way so as to render our moral critiques worthless? Aren't we either like silly pantheists on the one hand running around in meadows all day worshiping daisies, or like grim mechanists on the other, giving primacy to machines and their attendant constraints?

I think we need to say that such criticisms occur in the presence of deeply anti-ecological power structures in academia and government, and they are the "third force" or silent partner in the argument coming from Collins and Yearley against Callon and Latour. They seem to be saying, If you let anything into the analysis, and give away the important differences between humans and nonhumans, you are throwing away our birthright as discussants of moral order—our "birthright as social scientists." The debate becomes confounded at precisely this juncture. Latour and Callon insist that they are not "leveling" between humans and nonhumans, merely including. From the pragmatist point of view I must ask: Under what conditions, and for whom, does this inclusion imply leveling, nihilism, and claims to a lost birthright?

One of those conditions does have to do with giving up a privileged position, as indeed Latour states both in the Pickering volume and in this volume. And perhaps this is another place where pragmatist feminists make a contribution. We didn't come into the academy, or science, or social science, as insiders. We come from a numerically small, "holdout" tradition in American sociology that has had pluralism as its deepest commitment, as that from which all sorts of relativisms derive. We were among the first generation with more than a token number of women in our positions. Intellectually, our tradition never thought it had a better way of looking at things than our respondents; our whole social science was fashioned on the premise that THEY were making the news, and we were only reporting it. As long as we were talking about drug addicts or prostitutes, such sentiments seemed acceptable for many liberals. But now that we're talking about scientists, we threaten some basic theological commitments. We honestly believe that there are no positions that are epistemologically superior to any others. But I do at the same time argue with and try to overthrow those I don't agree with! Relativism in this sense does *not* imply neutrality—rather, it implies forswearing claims to absolute epistemological authority. This is quite different from abandoning moral commitments.

WE'RE SCIENTISTS TOO

This brings me to two last notes about the relationship between scientists and STS. First, I have said above that I am a scientist. I notice that the way we talk (I, too) about scientists is usually as "them". Why? We share the same fate, use their results, enter into dialogue with each other, and even go to each others' conferences and try to live in each others' departments.[9] Others have noticed this in other ways—in STS the reflexivists, and in anthropology the so-called "new ethnographers." Fujimura (1992a) has written an analysis of this movement from the points of view of pragmatism, feminism, and antiracism, emphasizing the importance of multivocality (pluralism) that is grounded in experience, practice, and identities.

I think the reflexivists and new ethnographers are moving in an important direction, which is to see ourselves as cocreators of our scientific narratives, and to place ourselves actively in those texts (see Denzin 1989 for a methodological explication of this in social science). But it's a difficult undertaking. In analyzing reflexively, they/we are addressing, among other things, the conditions of their-our own work. And that is dangerous in several ways. The first is simply in overestimating the impact of academic politics. Fujimura (1992b) notes that such reflexivity can become "academic politics" veiled as moral reform. Another "silent partner" in this argument, which she makes explicit, is feminism. For those of us who "grew up" intellectually in feminism, it is a little bewildering to encounter in science studies the attitude that gender is one of those boring, reified categories like "society" itself, that must be endured but not mined (see Dugdale and Fujimura, in prep). Feminist theory is not about civil liberties, or rather, only in the sense that it completely redefines the notion of civil liberties. It began in its recent Western incarnation as a movement about exclusion—"we need more women in x "—or about barriers. But the central, exciting parts of feminist theory concern exactly those issues raised here. We have an example of a personal, often private, pervasively and somewhat unevenly distributed phenomenon: the oppression of women.

Just documenting this was not an easy task. However, when done to the exclusion of other kinds of analysis, especially in the academy, it can indeed be boring. But the exciting part of feminism has been to invent ways to *see* this phenomenon: to understand the role of invisible work, to articulate the asymmetries between listening and speaking, between hearing and agreeing, to form a political reform program that was simultaneously collective and private (consciousness raising and action), to struggle with extant categories and define the subtle ways in which coalition becomes co-optation.

There are two major parts of feminist scholarship and activism that have been largely absent from science studies: community and spirituality. For feminists, the community aspect was especially important for the early years of theory formation as well as for mobilizing social movements. The result was a highly interdisciplinary development, women's studies. At first, feminist scholarship was so beleaguered, so new, and so dependent on emerging community that people from all fields were welcome. So the analysis of poets was equal to if not greater than that of scientists; the experience of an eighty-year-old woman as important for the critique as that of a twenty-five-year-old (although this certainly not without its own struggle), and there was an incredible heterogeneity in the sorts of analysis brought to bear on issues. There was an openness in figuring out the *questions* that was very important for maintaining the scope and vision of the movement as these questions shifted into scholarly arguments. This was coupled with an important inclusion and participation of feminist theologians—Mary Daly, Carol Christ, Rosemary Ruether, Nelle Morton, and others—who were not afraid to tackle questions of God/dess, the Absolute, power, imagination, and so on. These were taken up seriously as questions, discussed and debated; feminist scholars read all sorts of writing. We attempted simultaneously to redefine knowledge, politics, family, race, community, nature, and spirituality. We created experimental forms of writing and of worship, and seriously rethought our commitments to the most basic of questions.

There were, of course, large parts of the women's movement that converted these questions into silly mysticism, or called them silly mysticism. And there were scholars who went about the business of converting gender into just one more variable in a positivist pantheon; "Women in *X*" became a favorite career-building strategy for some. But for many others the scope of the questions remained undiminished. I think that even for women who thought "feminist spirituality" was nonsense and a detraction from bigger issues, such as those of class and capitalism, the dialogue about ultimate questions remained alive, and long-term survivors from both sides came to appreciate the power of the questioning as the "daily-ness of moral, ethical and political conundrums became clearer to us"[10] (see Anzaldúa 1987).

What does it mean, then, to include yourself in debates about such things? In the feminist movement, answers emerged in the form of communities, committees, commitments, and combats, as well as a new field of scholarship. When we say "self-inclusion" in STS, what does it mean? We are tackling one of the most important, widespread institutions of our time—science and technology. And if this doesn't mean social reform of some sort, what, by our own analysis, *does* it mean? (By social reform here, by the way, I include reforming the division of labor in the academy—more on that below.) Where are our communities and commitments? What sort of moral order are we building or trying to reform? I think the problematic of including ourselves in the analysis can be viewed as a problem of work organization, too. Why not just include it as another kind of observation? Why not adopt the French stance of including everything democratically (which to an American most emphatically also does *not* mean without difference)?

One of the most important philosophical tenets of pragmatism is that it is the consequences of an action that constitute logic and belief, not putative antecedents (which are anyway unknowable). So from my point of view, we can't know about the consequences of including ourselves in the analysis until we try. But reflexivity does not work at arm's length. Rather, it implies radical change quite close to home (even if not *in* the home), and the consequences of working for that are serious.

At the same time, for us as feminists, differences such as between human and nonhuman, race and ethnicity are not abstract. The *lived*, experienced differences as embodied in specific locales and moments, and communities, are central (Star, in press a).

STUDYING NONHUMANS: A PROFESSIONALIZATION MOVEMENT IN STS?

With respect to nonhumans, there are similarly large questions that may be viewed politically and pragmatically. Some people have full-time jobs observing or tending nonhumans. We call these people scientists, laboratory technicians, data entry clerks, various kinds of technical monitors, some computer scientists, and still-life painters. Some people have full-time jobs observing or tending humans. We call these people baby-sitters, parents, teachers, attendants, some kinds of servants, and old-fashioned sociologists. Perhaps most people have full-time jobs observing a "rich confusion" of humans and nonhumans: some computer scientists, dressmakers and tailors, bus drivers, doctors and nurses, most scientists, police, literary critics, some of the new sociologists of science, and so on. I confuse the points here in order to make a point. As Latour has beautifully argued in *Science in Action* (1987), we *all* have full-time work that is interacting both with humans and nonhumans—the mingling is inescapable.

The constitution of anyone's work is a mixture of human and nonhuman which can be analyzed ecologically. But the nature and quality of that composition will reflect back on the organization of the work itself in important ways. So in this, I consider myself an amateur nonhuman-watcher as compared with my license and mandate as a sociologist to be a people-watcher. To change the ecological mix with respect to my work organization means changing the organization in which I work. It is not merely an exercise of imagination, but a real political risk. And that brings us back to self-in-the-study: it's pretty dangerous to do so. You walk across so many great divides that your feet develop webs. . . .

But let us consider the movement to include nonhumans as a case of amateurs in a line of work trying to professionalize. We have as a group rather casually (even as passionate amateurs, compared with professional scientists) observed nonhumans. Some of us are making ready to go from amateur to professional status. In so doing, we should draw on lessons from other professionalizing groups, within and outside of academia.

And like all professionalizing movements, we face obstacles in the form of those who already claim the full-time territory in our own organizations (in this case, the university). But you could look at all forms of moving from low-percentage nonhuman watching to high-percentage in these terms—for example, the early years of computer science and the automation of various kinds of work.

What happens in a professionalization movement? I'll be brief: To establish a profession you need several things. First is a license and mandate from those whom you serve (Hughes 1971a). You need to have research to legitimate the movement and professional societies to secure it. You need to seize the means of evaluation for yourself. And you need to be in control of methods. Typically, resistance from the old full-timers who control the domain comes in the form of ridicule, gatekeeping of positions and other resources, the invocation of various great divides, and so on. The key to success in professionalization movements is to develop good infrastructure—training programs, methods, technologies—and to offer a new kind of service. I think, being a pragmatist and a little reflective, that we might even be able to improve on the nasty aspects of professionalization—for example, use the phenomenon of professionalization in order to understand the reaction to our proposed new organizational order, and eschew the elitism and secrecy that too often go with professions. In our studies of science, we try to include the voices of people not traditionally heard in accounts of science: lab technicians, sponsors, administrators, spouses (usually wives), and consumers. In our studies of science we are also trying to observe and account for the nonhumans not traditionally heard, in ways not often practiced by those in power. Following Abbott (1988), as well, we can speak of a *system of professions*[11] rather than stand-alone entities. Sciences and technologies cannot be separated analytically from professional governments, from medicine, or from any other profession.

Let's adopt a stance about this: Rather than simply creating a professionalization movement with respect to nonhumans, and thus becoming scientists in a way that often makes us uneasy, let us change the way science is organized. And it is here that the authors in this volume have the most to say. Perhaps one way to characterize their voices is as a reclaiming of the term *network* from some of its unfortunate discontinuous connotations and affiliations. A web is composed of filaments, and a seamless web should be an oxymoronic term. There's no empty space in a seamless web, but our image of *network* is that it is filaments with space between. For this reason I prefer ecology. Let's use networks-without-voids for an ecological analysis. And in that ecology, let us be epistemologically democratic, including toward our own work organization.

There's so much to be done.

INFORMATION TECHNOLOGY: HOPE AND FUN, HYPE AND DANGER

I live in the summers in Silicon Valley, California. My closest friends for years have been "hackers," many of whom occupy highly paid positions as systems programmers in computer design firms. In the 1980s they rode the remarkable growth boom of The Valley, switching jobs frequently for higher paid and more scientifically interesting situations. By the beginning of the 1990s, the recession had cut deeply into California life. Companies like Apple and IBM are firing vast numbers of people. Like so many others in the United States, even the top programmers and systems developers in Silicon Valley these days are often "working scared."

Outside work, my friends join progressive organizations, working for feminist, gay rights, and antiracist causes, and embrace ecology as both movement and fundamental ideology. They opposed Star Wars in the '80s and continue to work with organizations such as Computer Professionals for Social Responsibility. They are all avid readers of science fiction, and some are active participants in science fiction fandom. In more than the economic sense, they are working scared. As much as any group I know, they grasp the double-edged significance of the technologies they are making and how embedded they are in life in the United States these days—the visionary possibilities and the fun, on the one hand, and the dangers and violence on the other. All the computers in the world will not clean up the terrible pollution in California. We know now that dreams of telecommuting as panacea were flawed in many ways (see Kling and Iacono, this volume). The "neat toys" that my friends create for peaceful, medical, or leisure purposes may come to form the substrate for surveillance or weaponry—or for world peace. They continue to dream and to work and to hope, living with the contradictions as do we all.

My good friends—the Tuesday dinner crew—have taught me more than all the laboratory studies in the world about the links between politics and playing and working in science and technology. I would like to thank them for that (and for the good food, Linnea), with the hope that we will all live to see our better fantasies realized, and not our worst nightmares.

The Cost of Opposing Hype

In boom times there is a terrible cost to opposing a bandwagon, the mirror image of Fujimura's analysis of the blandishments and career advancement prospects for joining one (1987). Silencing and isolation are among them. This summer I was interviewed by *Science* magazine about some studies I had done of computer systems currently being designed for scientists. I had explained some of the difficulties of responsible design of such systems, and that many scientists had told me that they did not like or use such computer systems, even though they liked our carefully designed system. I spoke to the reporter of potential problems, and of my fears that the system I was evaluating might increase structural disparities between rich and poor labs, even while I hoped that it would level the differences. The resulting story was published with no mention of my part of the study or my comments; an apologetic e-mail note to me from the author recently said that there had been no room, and that perhaps some of my words would appear in a later article on computing and the humanities. The whole tone of the article was celebratory of such systems, full of visionary language of the sort criticized by Kling and Iacono in their article in this volume, and by the women of the Women, Information Technology and Scholarship group at the University of Illinois (Taylor et al. 1993).

In an atmosphere of expansion and bandwagons, there is little room for complexity and caution. Perhaps science fiction, poetry, and film *are* the best possible subversive vehicles for countering the tsunami of computer hype—for the sociological imagination (Star, in press b). I wrote the following poem after reading an adulatory article on virtual reality.

The Net

"we are part of the records we keep"
—Gayatri Spivak

i

network

and the word flares trumpets
shining webs
connect me
dissolving time and space

network

soaked with information
all there is to know
the little wire
next to my bed

network

the net
work
feeding the teenage son
waiting in line for the pass
word
shaking the numbness from the shoulders
the arms

ii

my best friend lives two thousand miles away
and every day
my fingertips bleed distilled intimacy

trapped Pavlovas
dance, I curse, dance
bring her to me

the bandwidth of her smell

iii

years ago I lay twisted below the terminal
the keyboard my only hope for work
for continuity

my stubborn shoulders
my ruined spine
my aching arms

suspended above my head soft green letters
reflect back:
Chapter One
no one can see you
Chapter Two
your body is filtered here
Chapter Three
you are not alone

iv

oh seductive metaphor
network flung over reality
filaments spun from the body
connections of magic
extend
extend
extend

who will see the spaces between?

the thread trails in front of me
imagine a network with no spaces between
fat as air
as talk

this morning in the cold Illinois winter sun
an old man, or perhaps not so old
made his way in front of a bus his aluminum canes inviting
spider thoughts
a slow, a pregnant spider
the bus lumbering stopped

and in the warm cafe I read of networks and cyborgs
the clean highways of data
the swift sure knowing
that comes with power

who will smell the factory will measure the crossroads
will lift his heavy coat from his shoulders
will he sit before
the terminal

v

it's too late for romance
the chestnut tree blooms no more
the corn and pigs in this vast flat place
travel the network too
their genes secure in stock indexes

it's too late for bitterness

but still there is a space
in the net

a choice of cyborgs
oh brave new world
for the courage to choose the mundane

the rough wool of a winter coat
draped over an old back
a smell, a feel of her hair
the unfamiliar intimacy of the dancing letters

literacy
or survival

vi

am I the only one who strokes the scars
the Frankenstein neck
who wonders
when the stitches will come
out

CONCLUSION

Each paper in this volume addresses issues common across a variety of domains in social science and for wider questions of social order. Fujimura extends Hughes's concept of the ecology of the workplace to study the creation of standardized "packages" in scientific work and in biotechnology research. Woolgar raises the notion of the nature of moral order in the doing of our own work, and the threats posed by crossing traditional boundaries. Lynch defines genre and representation with respect to a context of textures, conventions, and work practices. Star treats formalisms as a form of technical artifact, and raises ethical and methodological questions about their impact in very large systems.

Kling and Iacono, drawing on literature from social, professional, and scientific movements, look at the symbolic and political uses of computers to create "computer-based social movements." Several of the papers (theirs, Restivo and Croissant's, and Star's) ask: Are science and technology themselves social problems? All eschew scientific or technological determinism and look at how science and technologies are being used in the service of various social movements and structures. All find a conservative effect. Neither the so-called computer revolution nor the scientific revolution has been very revolutionary for many of our lives, certainly not in the sense of moral and political order. Restivo uses this observation to propose a return to Mills's sociological imagination, and to ask important questions about widely-held implicit assumptions, including those held by researchers in STS.

Law and Callon place the analysis of technological innovation and its attendant failure or success at the heart of social change as well. They concern themselves with fidelity to the categories of the actors; this includes a methodological demand for "heterogeneity." We should understand scientists and engineers on their own terms, as engineers of the social as well as of the technical. The failure of the group they studied to create an "obligatory point of passage" within their organization raises a larger question: What does it take to become a gatekeeper? What kinds of autonomy can be negotiated in technological design and implementation? Bruno Latour's (a.k.a. Jim Johnson's) piece, for all its humor, presents a serious philosophical challenge: Where are the boundaries of technical phenomena? What happens when we draw analytic boundaries in unconventional places or when we take the methodological mandate of "following the actors" to include everything in the site chosen, including mundane things? The processes of inscription and pre-inscription that he describes are important tools for understanding technological determinism and social change.

Adele Clarke as well follows all of the actors in the setting of reproductive research, describing a regime of materiality embedded in the very stuff of scientific work. The mundane materials of physiological work become the key pivots upon which findings revolve; only by understanding this previously ignored concrete "substrate" may we get at abstract theoretical developments.

Many great divides have been rejected, crossed, and perhaps will be closed by STS research. I hope that this volume will contribute to that process—that our hopes and our fun, not our fears and violence, will inform the closure.

NOTES

1. After writing this piece Adele Clarke brought to my attention an article by Charles Rosenberg (1979b) with a similar title—and I trust a complementary approach!

2. I am loosely grouping apples and oranges here for the purpose of creating a coherent narrative. There were functionalists and antifunctionalists in both Europe and America, of course, and many different schools of thought represented by this brief description. I do not have space here to detail the contributions of American ethnomethodology, French history and philosophy of science, or Marxist and anarchist analyses, among others. See Bowker and Latour (1987); Lynch, this volume; Croissant and Restivo, this volume.

3. I am grateful to Françoise Bastide, Geof Bowker, and Bruno Latour for discussions of these issues.

4. I am grateful to Howie Becker for pointing this out.

5. I should note that it is precisely this quest, and these terms, that some feminist ethicists are attempting to revolutionize. See Addelson (1991).

6. That is, artifacts such as formal models, algorithms, formulae, quantitative simulations, and so forth.

7. It is important to note here that formal models need not be mathematical. Rather, they are an attempt to explicate the interacting constraints of a set of conditions. In that sense, the work of the American pragmatist sociologists such as Hughes, Blumer, Becker, and Strauss, and their students, has some extremely formal elements. However, this has been combined with a reliance on empirical data—the pragmatist attempt at refusing this particular great divide. (See Glaser and Strauss 1967 for the first volley in this particular version of the war.)

8. We have a paucity of language for this which does not recommit some version of the great divides. The only somewhat adequate language for this I have heard is theological analysis, which would refute the distinction between the immanence and transcendence of God/Goddess.

9. Here I must note, with varying degrees of exhaustion and success, but that is another story.

10. Thanks to Adele Clarke for these words.

11. It is important to note that *system* here does not mean a functional system, or a closed system of any sort. Rather, *system* means a set of interrelated contingencies, whose scope is unknown and that is quite open, both analytically and with respect to other contingencies entering in. *Ecology*, again, in this sense, is not a closed ecosystem, but rather a term that emphasizes the open interdependence of ongoing processes. It is not, in other words, a functionalist description.

PART 1: POLITICS AND PROBLEMS

JENNIFER CROISSANT
SAL RESTIVO

1

Science, Social Problems, and Progressive Thought: Essays on the Tyranny of Science

The following papers are presented here as a two-part argument on science and progress. The first specifically situates science as a social problem in contemporary society. In it, Restivo argues that modern science is a social problem, and that much, but not all, social theory and especially the sociology of science have overlooked this in favor of less critical agendas. The second paper traces critical approaches to science and technology through three major traditions in progressive thought: Marxism, feminism, and anarchism. At issue is the ambivalence of progressive thinkers toward the apparent (but as we will note, arguable) instrumental successes of scientific work within the context of an internationalized regimelike structure that is eroding cultural and environmental diversity and sustainability, and contributing to continued oppression based on class, race, and gender. Notes are provided for each paper. We begin with a brief prelude.

PRELUDE: TECHNOSCIENCE OR TYRANNOSCIENCE REX?

"Science Imitates Art Imitating Science." So goes the title for a box in a *Science* magazine Research News report on molecular paleontology (Morrell 1992). Against the explicit backdrop of Michael Crichton's "sci-fi thriller," the real excitement of scientific discovery and the development of techniques[1] for amplification of prehistoric DNA is dulled by the shadow of terror in the speculative fiction of *Jurassic Park* (Crichton 1990). The real connections between University of California entomologist George Poinar and Crichton are part of the novel, and part of the news report. The potential (fictional or not) reconstruction of organisms might be chilling to an imaginative reader: "So if a big green flesh-eater goes cruising past your bedroom window one of these dark nights, you'll know just who to blame: Michael Crichton and George Poinar."

But the potential for critique is "safely enclosed" in a box, just as the fictional characters assume their re-created dinosaurs will remain "safely enclosed" in their cages. And the questions of responsibility are, in the *Science* review, separated from the contexts of power and the cultural significations that drive research. Alan Grant, the paleontologist in Crichton's novel, is asked as the novel closes (with the animal cages open): "Please, señor, who is in charge?"

And he replies: "Nobody."

Dinosaurs, their evolutionary biology and paleontology, whether macroscopic or molecular, hold great fascination as teratological versions of human origin/apocalypse myths.[2] They serve, like Crichton's novel, as examples of speculative or science fiction (SF), "where possible worlds are constantly reinvented in the contest for very real, present worlds" (Haraway 1989:5). Like primate studies, the debate in paleontology is about what has been, what is, and what might be. What follows is a review of *Jurassic Park* as one of our potential futures and a science fiction lesson in the social relations of science.

The dinosaurs that roamed the earth 235 to 65 million year ago emerged without the help of human beings. In fact, humans would not come into the planetary picture until long after the age of the dinosaurs. Now, in *Jurassic Park*, the human dynamics of science, technology, money, and greed have given new meaning to the old zoological park and circus imperative of Bring Them Back Alive. The dinosaurs have been brought back alive and put on display in *Jurassic Park*. The Dr. Frankenstein behind this Mesozoic Disney Land is John Hammond. Hammond is elderly, eccentric, very rich, and madly enthusiastic about dinosaurs. Under the auspices of his Hammond Foundation, and International Genetic Technologies, Inc., of Palo Alto, Hammond leases an island from the government of Costa Rica. The island is perpetually enshrouded in clouds; thus its name, Isla Nublar, Cloud Island. Hammond then brings together a team of scientists and advisors to exploit state-of-the-art genetic engineering technologies to bioengineer dinosaurs from *Velociraptor mongoliensis* to *Tyrannosaurus rex*. Hammond wants to provide the children of the world with a novel piece of education and entertainment. But his overriding intention, revealed to his chief bioengineer, Dr. Wu, is to make money—"A lot of money." Using science and technology "to help mankind" is a terrible idea, Hammond points out; it drags too many bothersome institutions and regulations into the picture. The idea, Hammond says, is to go after money free of government intervention—any government, anywhere.

Given the existence of the appropriate DNA and a cloning technique, Hammond thinks he can realize what he views as a simple idea. But he fails to reckon with the fact that butterflies are more powerful than millionaires, scientists, and even that greatest of all land predators, *T. rex*. At least, mathematical butterflies are. A mathematical butterfly can stir the air in Brazil today by flapping its wings, and this can lead to a storm in Boston next month. This butterfly, the Butterfly Effect, has become the metaphor and image for the currently fashionable mathematics of fractals and chaos theory.

Crichton organizes his story into seven main "iterations," each one prefaced by the prophetic words of his character, mathematician Ian Malcolm. The story unfolds in correspondence with Malcolm's seven fractal phrases: from 1, "At the earliest drawings of the fractal curve, few clues to the underlying mathematical structure will be seen," to 4, "Inevitably, underlying instabilities will begin to appear," to the penultimate fractal nightmare, 6, "System recovery may prove impossible." Fractals—a way of seeing and studying shapes, dimensions, and geometry grounded in the claim that the degree of irregularity in an object remains constant over different viewing scales—were once described as monstrosities. And this is one of the reasons the fractal metaphor works here. For the dinosaurs of *Jurassic Park* are true monstrosities, living things out of their own time.

The Butterfly Effect in *Jurassic Park* begins with some little flaps on Cloud Island. As we proceed through the story at high-tech thriller speed, we follow the little flaps as they move toward more global effects. Dinosaurs escape their pens on the island and some people die. But almost from the first page, the reader is aware that the Butterfly Effect has already caused the beginnings of a storm beyond the island. In the end, we are left to speculate on just how big the storm will be and how far it will spread.

Crichton carries us along fractal step after fractal step through this tale of suspense, thrills, gore, and geometry with a prose that is as compelling and uncomplicated as the bite of the velociraptor. But this book is much more than a clever goulash of fashionable popularized science and technology and best-seller seasonings, just as the movie, however diluted its cautionary tale about science, is more than a record-setting summer thriller.

While a few reviewers and associates have remarked on the cautions about science in *Jurassic Park*, the movie and book are inadequate as science criticism.[3] They illustrate, perhaps, like *Robocop* and other science fiction films, a genre in the politics of despair (Glass 1989). Here we find consumerist criticism, kitsch (Montgomery 1991), and recycled sentiment from prior generations of science horror films. These films presume a helplessness and a distance between those who might actually be able to do something, the scientist-expert-corporate agent and the viewer-citizen. The result is a critique of "bad science" or bad scient*ists* rather than a critique of the system as technocratic and unresponsive. The "new bad future" films may result in lowered expectations and a resignation or passivity toward a future full of violence, corruption, and inhumanity (Glass 1989:48). But "we cannot expect simplistic bourgeois closure on a narrative that cannot yet be closed" (Glass 1989:47). These science fiction futures are being contested now, in laboratories and marketplaces, and on a number of literal battlefields across the globe.

Modern science is, like all of the other institutions of contemporary (post)industrial societies, a social problem. Crichton captures the everyday minutiae of this still widely resisted social reality. But even Crichton is not immune to the myth of pure science. He seems to believe in a Golden Past, where pure scientists shunned business and money. This is what he tells us as narrator in an introduction that sets the documentary tone for the tale ahead. The Golden Past idea emerges again in the middle of the story's climax when the mathematician Ian Malcolm recalls the basic idea of science—an objective, rational view of reality that was new and appropriate five hundred years ago. But science, like all forms of knowledge, is everywhere and always a product of and guardian of ruling powers and ideas. *Jurassic Park* is thus a story for all forms of science, not just modern science. There are lessons here not only for the scientists and engineers driven to do something because it can be done, but also for politicians and citizens. *Jurassic Park* may be in the Middle East, the Balkan republics, Bosnia; and it may be in your own back yard.

MODERN SCIENCE AS A SOCIAL PROBLEM

Sal Restivo

Sociologists of science do not, in general, doubt the value of modern science. They are, implicitly or explicitly, science advocates. Their research tends to affirm, imitate, and justify modern science as a progressive, well-functioning social system and the paradigmatic mode of inquiry. In this essay, I challenge that view. I consider the arguments of C. Wright Mills, Thorstein Veblen, and other critics who have implicated modern science in problems of alienation, dehumanization, ecological deterioration, and nuclear escalation. My objective is to explore this conception of modern science and reflect on its implications for critique and renewal in the sociology of science. In the section that follows, I consider the development and nature of the sociology of science.

Sociology and Social Studies for Science

The characterization that sociologists of science rarely doubt or question the worth of science holds on both sides of the 1970s watershed that separates the "old" and the "new" sociology of science. On the far side of the watershed, the sociology of science is dominated by Robert Merton (1973) and "the Mertonian paradigm." On the near side, the field is part of a hybrid discipline variously referred to as "science studies," "social studies of science," and "science and technology studies." The Mertonian hegemony has been replaced by a diverse and conflictful arena of realists and relativists, strong and weak programmers, and conflict theorists and neofunctionalists (Collins and Restivo 1983). But neither old nor new sociologists of science linger on the human face of science or on issues of class, power, and ideology. Both affirm that science as it is, with all its social trappings (including elitism and competitiveness), "works." The basic goal of the old and new sociology of science is the same: to explain *how* science works.

The old sociologists of science focus on the social system of science itself and exempt scientific *knowledge* from sociological scrutiny. New sociologists of science are more concerned with social context, social construction, and on-site studies of scientific practice (Knorr-Cetina and Mulkay 1983). They have also made scientific knowledge an object of inquiry. These are significant, even revolutionary, departures from the Mertonian tradition in the sociology of science and from traditional philosophy and history of science. But new sociologists of science, with some notable exceptions (e.g., MacKenzie 1986), are busy developing *new accounts* of how science works. They are not challenging or criticizing modern science as a value system, a worldview, and a way of living and working.

There are, of course, some Marxists, conflict theorists, socialists, anarchists, and radical science advocates who do carry out critical analyses of modern science (e.g., Rose and Rose 1976; Arditti et al. 1980). These analyses tend to be generated as part of a general criticism of the modern social order. But even their criticisms are often reined in by the belief that "socialized science," science in a socialist (or communist, or anarchist) society, or some sort of unadulterated science could realize the promise of a science that would benefit humanity. These works notwithstanding, the idea that modern science is at least as much a factor in as a solution for our personal, social, and environmental ills is not defended by very many sociologists of science. C. Wright Mills's (1963:229–30, 417) conception of modern science as a subordinate part of "the wasteful absurdities of capitalism," the military order, and the national state is not ascendant in the sociology of science.

The constructivist and relativist agendas in the new sociology of science have alarmed the guardians of the scientific community. They view them as threats to the integrity and autonomy of science, to the realist assumptions of scientific inquiry, and to the quest for truth and objective knowledge (e.g., Campbell 1987:390–91). But the leading "constructivists" and "relativists" are not antirealists in any simple sense, and many are explicit defenders of the methods and worldview of science. Bruno Latour and Steve Woolgar (1986), for example, disassociate themselves from naive relativism; they do not deny the existence of facts or of reality. Karin Knorr-Cetina (1979:369) also explicitly divorces her constructivist interpretation from an idealist ontology. She does not deny the existence of an independent reality. Latour's (1988d:26–27) "recantation" is even more dramatic "in spite of our critiques—and to be fair, in spite of a few of our early claims,"—the new sociologists of science, he writes, are no more "relativist" than Einstein, "and for the same reason."

By fighting absolute definitions of observations that do not specify the practical work and material networks that give them meaning, we take as seriously as everyone else the construction of reality—indeed, we might be the only ones to take it seriously *enough.* Other sociologists of science explicitly announce that they are "for science". David Bloor (1976:141), for example, long associated with relativism and interest theory, bases his strong program in the sociology of knowledge on the dictum: "only proceed as the other sciences proceed and all will be well." Within this strong program, relativism is not only not a threat to science; it is a basic condition for "good" science, that is, disinterested research (Barnes and Bloor 1982:44–45). Even the high priest of (empirical) relativism, Harry Collins (1985:165–67), views his work as a defense of the authority of science ("the best institution for generating knowledge about the natural world that we have") and of the ultimate (however uncertain and fallible) expertise of scientists.

The conservative, neofunctionalist bias in the new sociology of science reflects the influence of Thomas Kuhn's (1970 [1962]) *The Structure of Scientific Revolutions.* But his reception also suggests that bias was present from the beginning among the most prominent founders of the science studies movement in the 1960s and 1970s. Barry Barnes (1982) has played an important role in championing Kuhn as a significant, even radical, contributor to the new sociology of science and knowledge. Kuhn has even been hailed as a hero in the radical and feminist science studies communities. This is, as I have noted in detail elsewhere, a great curiosity (Restivo 1983a; cf. King 1971 and Hesse 1980:32). Kuhn's work is an asociological, prescriptive defense of science, and is compatible with Merton's sociology of science, especially at the level of values, as both he and Merton have acknowledged. Michael Mulkay (1979) illustrates this point. Like most of the new sociologists of science I criticize, Mulkay can conceive an alternative interpretation of how modern science works but not an alternative to modern science. This helps to explain why he can discuss Karl Marx and the sociology of knowledge without commenting on Marx's distinction between "science" and "human science" (Marx 1956:110–11; 1973: 699ff.). Bloor's (1976:144) closing remarks in *Knowledge and Social Imagery* make quite clear what is at stake here:

> I am more than happy to see sociology resting on the same foundations and assumptions as other sciences. This applies whatever their status and origin. Really sociology has no choice but to rest on these foundations, nor any more appropriate model to adopt. *For that foundation is our culture. Science is our form of knowledge* [my emphasis]. That the sociology of knowledge stands or falls with the other sciences seems to me both eminently desirable as a fate, and highly probable as a prediction.

Bloor not only understands the interdependence of science, culture, and the sociology of knowledge, he approves of it and its forms.

It should be clear by now why so much of what goes by the name of sociology of science, science studies, or science criticism remains fundamentally conservative on the question of the value of science: The most influential authorities on the "sociological" nature of science, notably Kuhn and Bloor, are science advocates. Advocacy in itself is not so much the problem, I want to stress, as the fact that, in the cases I refer to, it interferes with a critical sociology of science. What is missing from science criticism and from the sociology of science is the Millsian blend of structural analysis (sociology in the strong sense), social criticism, epistemological relevance, and an activist orientation toward social change—in brief, the sociological imagination.

There are two basic reasons why sociologists of science, old and new, have been unable or unwilling to follow Mills (1961:8) in linking modern science to the "personal troubles of milieux" and the "public issues of social structure." First, the idea that science "works" and a "science fix" orientation have been amplified by runaway technological "progress." In the heady atmosphere of material plenty, people have been seduced by the icons, myths, and ideologies of modern science. Second, sociologists of science can not afford to alienate the scientists they study by criticizing their ideas and actions, including how their social roles, organizations, and products fit into society. It is precisely this sort of criticism of "our" science, "our" culture, and "our" sociology of science and knowledge that I want to encourage.

Modern science, from such a critical perspective, is a threat to democracy, the quality of human life, and even the very capacity of our planet to support life at all (cf. Feyerabend 1978). Moreover, modern science is a *social problem* because it is part of modern society which itself is a social problem. I turn next to a discussion of what I mean by the term "social problem" and why I consider modern science and modern society social problems.

Seeing Science As a Social Problem

There is a reluctance among students of social problems to include modern science in their analyses, criticisms, and policy studies. Social problems courses and textbooks do not, as a rule, devote space to modern science, although it may receive indirect attention in studies that deal with "technology" (e.g., Mankoff 1972). One reason for this is that modern science is not yet widely appreciated as a social phenomenon in the strong constructivist sense; scientific knowledge itself is a social construction. Another reason is the assumption that science (and especially "pure" science) and technology are separate, relatively independent phenomena. Other more general reasons were identified by Mills (1963:535–36) in his 1943 paper on "The Professional Ideology of Social Pathologists." Mills criticized the situational, institutional case-by-case approach to social problems typical of most introductory textbooks. Mills proposed instead a social structural approach to reveal the interdependencies linking all the activities, organizations, and institutions in a society. This requires theories that address these interdependencies.

My assertion that modern science is a social problem because modern society is a social problem is a cryptic criticism of the situational approach. By "social problem," I want to convey nothing more complicated than the Millsian notion that modern science is implicated in the personal troubles and public issues of our time. The idea that modern society is, again in the Millsian sense, a "social problem" means that concerns about personal troubles, public issues, and social change agendas should focus on a total social structure rather than one or more of its "dysfunctional" parts.

The term *society*, it should be stressed, poses a conceptual problem. In standard usage, it often refers to an imaginary, undifferentiated entity, and it tends to connote cooperation and "democracy." The problem can be readily identified by considering what it means to use the term *state* in place of *society* (Mills 1963:538–40). Our conceptions of "society" have methodological and political implications (Mills 1963:537; Restivo 1991).

The situational approach to the study of society and social problems has two important consequences for the study of science. First, it makes it possible to isolate science from other institutions and classify it with the "healthy" as opposed to the "unhealthy" ones. Second, it means that, even when science is examined critically, the total social structure is unlikely to become the focus of criticism and analysis. While I cannot describe all of the ramifications of a total social structural approach to the critical sociology of modern science, I can, in the course of this essay, provide some of the conceptual resources for such an approach.

Mills's (1963:530–31n) critique of the conception of "social problems" in his time is still relevant. This is not only because some social problems research continues to be guided by the strategies Mills criticized. More importantly, contemporary strategies in social problems research and theory are subject to Mills's argument that they are not "of a sort usable in collective action which proceeds against, rather than well within, *more or less tolerated channels*" [my emphasis]. It is not at all clear, for example, that the fashionable "definitional" or "constructionist" (sometimes constructivist) approach avoids the pitfalls Mills identifies. I will return later in the essay to the "realistic" and "activist" implications of the sociological imagination. It is important, however, to clarify a conceptual problem that cuts across the new sociology of science and contemporary social problems research and theory: the relationship between constructivism and relativism.

Constructivism and Relativism

It is important to understand modern science, including scientific knowledge, as a social construction in order to appreciate it as a social problem in the Millsian sense. But there is some confusion inside and outside of science studies about what the constructivist interpretation of science means. The idea of social problems as social constructs is a key part of the framework of contemporary social problems research and theory (Spector and Kitsuse

1977; Gusfield 1981; Schneider 1985; Best 1987). In their critique of constructivism in social problems research, Woolgar and Pawluch (1985) assume that *constructivist* and *definitional* are synonymous and that they entail relativism as opposed to realism (but see Latour and Woolgar 1979:180; Latour and Woolgar 1986:277). The fact is, however, that the genesis of constructivism in the new sociology of science is closely associated with if not coincident with the sociological realism of ethnographic studies of scientific laboratories. In this context, constructivism is not merely a matter of reality being constructed "by definition." It tends, rather, to be a fashionable way of talking about social structures as the causal forces that generate thoughts and actions, with a stress on the day-to-day, moment-to-moment activities of scientists as they go about producing and reproducing scientific culture. There is thus no necessary connection between constructivism and relativism. Given my earlier discussion of the fact that sociologists of science such as Bloor and Knorr-Cetina are not relativists in any simple sense, and certainly not in any antiscientific sense, it should not be assumed that constructivism in science studies and constructivism in social problems research mean the same thing.

The foregoing should alert the reader to the fact that I proceed according to constructivist principles, but do not adopt any sort of naive relativism. My approach is probably better described as "realistic" rather than "realist" or "relativist" (cf. Hooker 1987). The wedding of constructivism and a realistic worldview does imply that there are things that are true and things that are false, and that some sort of objective knowledge is possible. But constructivism does not leave these ideas untouched; it transforms them into sociological concepts and makes a sociology of objectivity both possible and necessary (Restivo 1993a).

Science in Context

Seeing modern science as a social problem depends on getting behind the facade of ideology and icons in science to the "science machine," and on exposing the cultural roots of science. These are my objectives in the following sections. I begin by examining some of the important reasons for reconceptualizing science and its social relationships. I discuss Mills's theory of the Science Machine and the conceptual problems that need to be resolved in order to pursue it. Then, the important distinction between the autonomy of individuals and structural autonomy is introduced, followed by the rationale for dissolving the traditional boundaries that separate science, technology, and society.

The Science Machine. Mills (1961:16) observed that a variety of troubles and issues are rooted and reflected in the relationships between modern science and other social institutions:

> science seems to many less a creative ethos and a manner of orientation than a set of Science Machines, operated by technicians and controlled by economic and military men who neither embody nor understand science as ethos and orientation.

There are in Mills's (1963:234) conception of the transformation of science into a Science Machine by "military metaphysicians" echoes of Marx's (1956: 110–11; 1973:699ff.) notion of modern science as alienated and of Veblen's (1919:1–55) critique of modern science as a machinelike product of our "matter-of-fact" industrial and technological era. Thinkers with this turn of mind have described modern science as an "instrument of terror," an assault on the natural world, and a tool of greed, war, and violence (Broad 1987:39; Schwartz 1972; Dickson 1984). In order to understand the grounds for such claims, we must distinguish clearly between isolated scientific biographies, methods, findings, experiments, and theories on the one hand and modern science as a social institution on the other. By focusing on modern science as a social institution, we not only can see how it is connected to other social institutions, we also transform biographies, methods, findings, experiments, and theories into social facts.

Scientific activity, rooted in what Bernard Barber (1952:26,52) refers to as the "generic human attribute of empirical rationality," occurs in all societies. But modern science emerged in Western Europe after 1500, became organized around the social role of the scientist, and has grown without interruption for nearly five hundred years. By the mid-nineteenth century, modern science had crystallized as a social institution. Since then it has undergone transformations in scale and power coincident with processes of professionalization and bureaucratization internally and changes in its relationship with the state externally (cf. Restivo and Vanderpool 1974:3).

The origin and development of modern science is inextricably intertwined with the origin and development of modern society. This has been recognized in varying degrees by students of the Scientific Revolution. Weber (1958:24), for example, noted that the technical utilization of scientific knowledge was encouraged by favorable economic conditions. Merton (1970 [1938]:55), in his study of the reciprocal relations between science and society, concluded that arguments for the utility of science arose in a variety of institutional spheres, including religion, economy, and the military. The "reciprocal influences" and "utilitarian" hypotheses formulated by Weber, Merton and others surface again in Ben-David's (1965:15) analysis of the scientist's role. He argues that the Scientific Revolution occurred because certain scientific discoveries convinced people that science had economic value and because certain people were convinced of the intrinsic value of science and were able to gain general acceptance for their view independently of any evidence for the utility of science.

There is an alternative explanation for the Scientific Revolution that helps to resolve some of the difficulties of conjectures about "reciprocal influence," "utility," and "pure motives." The Scientific Revolution was one of an interrelated set of parallel organizational responses within the major institutional spheres of Western Europe from the fifteenth century onwards (including Protestantism in the religious sphere and modern capitalism in the economic sphere) to an underlying set of ecological, demographic, and political economic conditions (Karp and Restivo 1974). This perspective does not readily yield a conception of modern science as an autonomous social system. Modern science is autonomous in a sociological sense to the extent that it is a structurally and functionally differentiated social activity. But the "parallel responses" thesis sets modern science into the very core of the modern state and its technological foundations. This notion requires some further discussion to clarify the distinction between structural autonomy and the autonomy of individuals.

Structural Autonomy. The concept of autonomy has played a key role in research on professions and bureaucracies. In general, students of autonomy in this context tend to focus on the autonomy of *individuals*, and in particular of professionals in bureaucracies (Bledstein 1978:87–88). Some (e.g., Scott 1966) view professions and bureaucracies as institutions but are primarily interested in professionals in bureaucracies (cf. Friedson 1986:166). In some cases, the analysis may shift to the social role. But in neither case is the focus on autonomy as a structural variable, especially at the organizational and institutional levels of analysis (but see Kornhauser 1962; on the dysfunctions of professionalization, see Restivo 1983b:152ff.; Bledstein 1978:94; Friedson 1970; and Brewer 1971). Autonomy in this sense refers to the nature and degree of organizational or institutional demarcation and closure and to the degree to which the boundaries of social activities and systems are distinct, permeable, open, or closed. The more, for example, a system can function independently of the resources of other systems, the more autonomous it is.

The structural sense of autonomy lends sociological meaning to the concepts "internal" and "external." The use of these concepts to refer to the two basic types of factors that can affect science and to the distinction between contextual and noncontextual analysis has been properly criticized in the new sociology of science (e.g., Johnston 1976). But these terms can be usefully applied in the context of analyzing the interaction of social systems that vary in terms of degree of closure, that is, degree of autonomy. Thus, an "internalist" approach would be an appropriate part of the research strategy for studying a relatively autonomous social activity or system. A traditional internalist would likely consider factors such as scientific "ideas" to be independent of social forces. An internalist analysis in my sense would focus on the social structure of the system under study as a determinant of the knowledge produced in that system. Thus autonomy and the internal/external dichotomy can be rendered sociologically meaningful if we conceptualize them in terms of a structural analysis.

The internal/external dichotomy is just one of a number of ideas that new sociologists of science have discarded or transformed conceptually. Another idea that has increasingly posed problems in science and technology studies, an idea that is of central importance for the argument in this essay, is the science, technology, and society triad.

Science, Technology, and Society. The boundaries between the concepts "science," "technology," and "society" in traditional studies of science and technology have been more or less dissolved by some of the leading new sociologists of science. Harry Collins (1985:165), for example, argues that his study of how scientific facts are established dissolves those boundaries two ways: First, it points to the continuity of the networks of social relationships within the scientific professions with networks in society as a whole. Second, it points to the analogy between cultural production in science and all other forms of social and conceptual innovation. The point that needs to be stressed about boundaries and networks in this context, a point relevant to the discussion of structural autonomy, is that they are to varying degrees dynamic and protean.

The relative stability of social boundaries and networks over a long period of time gives rise to systems for which we can determine degrees of autonomy. But even so, we must be alert to changes, including periodic changes, in the character of those boundaries. They may be so fixed that it makes sense to say they define an organization or an institution; but even within that framework, the boundaries may periodically break down. The system may be more closed, more autonomous, at some times than at others. Thus, Latour (1987:174) uses two expressions to refer to two aspects of the activities of scientists and engineers:

> I will use the word *technoscience* from now on, to describe all the elements tied to the scientific contents no matter how dirty, unexpected or foreign they may seem, and the expression "science and technology," in quotation marks, to designate *what is kept of technoscience* once all the trials of responsibility have been settled.

His conclusion is that "the name of the game will be to leave the boundaries open and to close them only when the people we follow close them" (Latour 1987:175; cf. Knorr-Cetina 1981:88ff.).

I think Latour's conclusion means that whether it makes sense to talk about science or technology or technoscience or wider cultural spheres depends on our perspective at any given time and the details of the system we are studying. In some cases, in fact, we may find we are studying a feature of "science" that is so widely diffused across and interdependent with other cultural spheres that we will need to use a new term to describe it.

Pinch and Bijker (1984, 1986; and see Russell 1986) propose an interesting but less radical rationale for eliminating the distinction between science and technology. They argue that technology and science should be treated within the same social construction framework. Obviously, this argument is a contribution to the more radical project Latour is engaged in. Establishing technology as a social construct is to some extent less difficult than showing scientific knowledge to be a social construct but still contributes to the groundwork necessary for seeing technoscience where we have traditionally seen science and technology. In that sense, placing technology in its social context, and treating artifacts as "political," or more generally as social constructs, is relevant to the theory of the Science Machine (cf. Winner 1977, 1985; and see Bijker, Hughes, and Pinch 1986). But the relevance of this strategy tends to be limited because it can be carried out while implicitly or explicitly sustaining the traditional distinction between science and technology (e.g., MacKenzie and Wajcman 1985a; Trescott 1979).

Ruth Schwartz Cowan's (1983) work is a good example of how the contextualization of technology (in this case using the concept of "technological system") can be accomplished while implicitly treating science as a distinct phenomenon. And in the end, Cowan (1983: 215–16) misses the linkages among the work process, technological systems, and social structure. Her proposal for "neutralizing" the sexual connotation of household technology and the "senseless tyranny of spotless shirts and immaculate floors" is not a sociologically viable solution to the social problems of technology (or technology as a social problem). She fails to see the profound and far-reaching structural changes necessary to achieve the goals she sets, and this failure reflects the fact that she does not see the sorts of connections embodied in a concept such as technoscience. There is now good reason on both empirical and conceptual grounds to argue that modern science is of a piece with modern technology and the central values, interests, and structures of the more powerful classes in modern society. As I pursue my social problem thesis, I will continue to focus on "modern science," even though I have now provided a rationale for either dispensing with the term or using it more cautiously. But that rationale needs to be developed further or developed in new directions (not only in this essay but in science studies generally) before we can confidently adopt a new conception of the referent for "modern science."

The idea of the Science Machine and concepts such as technoscience help us to see (1) the connections between science and technology and the fragility of the boundaries that ostensibly separate them; and (2) the connections between science and other social activities. In the next section, I pursue these connections by sketching the emergence of modern science as one of the key ingredients of European expansionism.

The Cultural Roots of Science

By 1500, on the eve of the Scientific Revolution, Europeans were taking command of the world's oceans and beginning to subjugate the cultures of the Americas. William McNeill

(1963:569–70) identifies three "talismans of power" that enabled the Europeans to conquer oceans and cultures:

> (1) a deep-rooted pugnacity and recklessness operating by means of (2) a complex military technology, most notably in naval matters; and (3) a population inured to a variety of diseases which had long been endemic throughout the Old World ecumene.

European militarism of the period had its roots in Bronze Age barbarian societies and the medieval military habits of the merchant classes and certain lesser aristocrats and landowners. It was in this most warlike of the major civilizations that modern science arose.

The maritime supremacy of the Europeans was the basis for the enlarged scope of their militarism beginning in the sixteenth century. Their superiority at sea was the result of deliberately blending science and practice, first in the Italian commercial cities and, ultimately, under the guidance of Prince Henry the Navigator and his successors, in Portugal (cf. Law 1986a). The Scientific Revolution institutionalized this inseparable blend of science and practice, science and technology. Modern science has been primarily a tool of the ruling elites of modern societies from the time of its origin in sixteenth- and seventeenth-century Europe (cf. Noble 1979; Dickson 1979).

In its earliest stages of development modern science was a part of the repertoire of "gentlemen" who were embracing capitalism and seeking to destroy the monopolies of the old landed aristocracy. But by the 1690s in England, the tie between science (and in particular Newtonian science), the culture of the ruling Whig oligarchy, and the established church (in particular the latitudinarian hierarchy) was well established: "The scientific ideology of order and harmony preached from the pulpits complemented the political stability over which [the Whig] oligarchy presided" (Jacob 1988:121-23).

The Scientific Revolution organized the human and cultural capacity for inquiry in ways that stressed laws over necessities, the value of quantity over quality, and strategies of domination and exploitation over strategies based on an awareness of ecological interdependencies. As a product of the commercial, mercantile, and industrial revolutions that transformed Europe and the world between 1400 and 1900, modern science emerged and developed as an alienating and alienated mode of inquiry. It arose as the mental framework of capitalism and the cognitive mode of industrialism (Berman 1984:37; Geller 1964:72). Capital accumulation and industrial products and processes became prominent features of social life and the primary factor in shaping our ways of thought, our science. We learned to think the way modern technological processes act (Veblen 1919:7). Modern science (including scientists and images and symbols of science) came into the world as a *commodity* and has developed in close association with the discipline of the machine (cf. Meiksins 1982). A number of researchers from Karl Marx to David Noble have recognized this, although many of them have implicitly or explicitly distinguished "science," more or less "pure," from "modern science," science adulterated by capitalism and technology.

Given the tenacity of the myth of pure science, it is important to remember that science in every form has always been as much a part of the economic, political, and military fabric of society as modern science is. The "scientific community" did not, as Noble (1979:4) contends for example, have to overcome "Platonic prejudices." There is some legitimate confusion about the relationship between modern science in its formative stages and modern science in its fully institutionalized form. It might be assumed that science was characterized by purity and Platonic prejudices before it became a differentiated part of European social structure through institutionalization and the crystallization of the social role of the scientist. But even if this were the case in the earliest and most diffuse stages of the history of modern science, it is clear that, once it became a major force in European culture, it "took on an immensely practical posture that moved it from an intellectual pursuit to a source for industrialization" (Jacob 1988:259; Musson and Robinson 1969; cf. Carroll 1986; and see McNeill 1982; Postan et al. 1964).

The thesis that modern science in its earliest stages was a purely "intellectual pursuit," however, cannot survive careful scrutiny. As Jacob (1988) notes, early modern science was a tool of "gentlemen capitalists" who were not then a ruling elite but an elite on the road to ruling power. Moreover, seventeenth-century natural philosophers already expressed values of the "world politick" in their efforts to develop a mechanical description of the "world natural": "At every turn, that linkage ensured its integration into the larger culture and made its ideological formulation immediately and directly relevant to those who held, or sought to hold power in society and government" (Jacob 1988:38).

All of the foregoing helps us appreciate the Marxian notion that science was transformed into a productive force distinct from labor and pressed into the service of capital by modern industry. This is the starting point for David Noble's (1977:xxiv) study of science, technology, and the rise of corporate capitalism in *America By Design.* The issue of "science" aside, it is fairly clear that modern science emerged as a means of capital accumulation and thus an economic good and an article of commerce.

We can, as I hinted earlier, trace the roots of modern science to the knowledge-producing activities of earlier cultures. Those activities are everywhere inseparable from military, political, and economic interests and power (cf. Dickson 1984:107). The very foundations of modern science are permeated by a sense of the war-making utility of scientific knowledge, expressed by the most brilliant as well as the most ordinary scientific practitioners. Most of the texts from the formative period of modern science that recommend science also point out its utility for improving the state's capacity for waging war more effectively and destroying life and property more efficiently (Jacob 1988:251–52). One way to illustrate this deep-rooted relationship between science and power is to reflect on the reality that lies behind the myths of "pure" science.

The Myth of Purity

The notion of pure science has two basic referents. One is the production of ideas or knowledge through purely "mental" acts, that is, pure contemplation. The second is the pursuit of "knowledge for its own sake." The idea of pure mental or cognitive creation, of mental acts and events untouched by social facts, has been challenged by constructivist sociologists of science. They argue that all knowledge, including "scientific facts," is indexical, situational, contextual, and opportunistic (Knorr-Cetina 1981; Latour and Woolgar 1979; Zenzen and Restivo 1982; Star 1983). In my most recent work in the sociology of mathematics, I have pressed this idea in the direction of a theory of mind and cognition as social structures (Restivo 1992:130–35). The fact that there are individual scientists driven by higher motives (curiosity, for example) does not mean that their social roles are not serving social interests (cf., Merton 1968: 661–63). Similarly, we do not have to deny that individual scientists may be motivated by a desire to "understand" the world "objectively" in order to see the social functions of labeling scientific work "pure" or "objective." The labels *pure* or *basic,* for example, can be used to demonstrate or symbolize a nation's capacity for research or its potential for generating "fundamental" discoveries that may find applications in various areas of social life. The self-applied labels *pure* and *objective* can call attention to and defend the autonomy, solidarity, and professionalism of scientists seeking access to scarce societal resources and independence from external social controls. Pure science can thus be used to intimidate competitors and enemies, project status claims, and establish territories. The scientific research centers Germany established in the early 1900s in Samoa, Argentina, and China were tools of cultural imperialism (Pyenson 1983). The various national research camps and outposts in contemporary Antarctica are labeled "pure science" efforts to mask their functions as informal territorial claims and attempts to estimate the potential value of access to Antarctica's land and resources for military, economic, and political purposes. Individual scientists may be unaware of these functions or otherwise mistaken about just what their social roles are in the Antarctica context. They may be curious about Antarctica, and they may make "fundamental" discoveries. But neither their personal views nor their motives can alter the functions of their camps and outposts in international political economy (cf. Elzinga 1993:138; Elzinga and Bohlin 1989).

It should be noted that a certain amount of trained incompetence is necessary if scientists are going to exhibit ignorance or be mistaken about their social roles. Sharon Traweek's (1984, 1989) anthropological studies of the high-energy physics community illustrate some of the social mechanisms that bring about this trained incompetence. Physicists are trained to value certain emotional qualities (for example, meticulousness, patience, and persistence). As students, they go through a process of intense professional socialization focused on physics and the physics community. The social contexts of their activities are obscured and their conceptions of their social roles dramatically narrow. They are introduced to highly idealized portraits of scientists. Archimedes, Newton, and Einstein are the principal icons of, respectively, ancient, modern, and contemporary science. They are portrayed as men of pure contemplation who have little or no use for the vulgar aspects of science. Their interests in practical affairs are deemphasized or overlooked. They are described as "geniuses," but the development of the idea of genius in conjunction with the concept of intellectual property under capitalism goes unnoticed (Hauser 1974:163; Restivo 1989).

One of the most important social functions of the purity label is to mitigate resistance to and criticism of established interests. The state, for example, grants scientists who adopt the purity label the freedom to pursue their individual research interests so long as what they do keeps them from criticizing or resisting state actions, and especially so long as they do not interfere with the state's efforts to appropriate scientific discoveries and inventions in pursuit of military, economic, or political goals. Even in the most democratic societies, the state can enter the scientist's ivory tower with requests for secrecy and cooperation in the interest of national security and defense. If scientists resist such requests, the state can issue demands and back them up using various controls on the flow of resources for research. In extreme cases, the state can draw on its police powers and the means of violence at its disposal to control scientists.

"Basic" or "pure" science can easily end up focusing on mechanisms instead of causes. As a result, problems can be abstracted from their social contexts, and solutions sought that do not threaten prevailing social arrangements. The focus on basic cellular biology in cancer research, for example, assumes a solution that interrupts the carcinogen process rather than one that rearranges the social order to remove carcinogens from the environment (Ozonoff 1979: 14–16).

To the extent that ideologies of purity stress "science for its own sake" they reinforce the scientist's alienation and obstruct the development and pursuit of interests outside of science, including the realization of the collective interests of scientists as workers. To the extent that they stress the independence of scientific knowledge from social interests, historical and social contexts, and individual subjective experiences, these ideologies help to isolate that knowledge and alienate the knowledge producers from the social processes of production and reproduction in science. To the extent that these ideologies reify the realm of purity, they function as justifications for the authority of ideas and heroic figures in the sciences as well as of texts and teachers and reinforce the principle of authority in everyday life.

The ideology of pure mathematics grows in large part out of ideologies of God and Nature as ultimate authorities. In the end, authority comes to reside in the realm of the purest of the pure sciences, logic. Classical logic, for example, as the intuitionist mathematician L. E. J. Brouwer recognized, is an abstraction from, first, the mathematics of finite sets, and then the mathematics of finite subsets. These mundane origins were forgotten when logic was elevated to a position prior to and beyond all mathematics. The substitute God logic (the Durkheimian spirit cannot be missed here) was then applied to the mathematics of infinite sets without any justification. Reified realms of purity such as logic can be functional equivalents of God, and can serve as moral imperatives and constraints that in one way or another bind us to established professional and state interests and reinforce obedience at the expense of criticism and rebellion in our relationships within established institutions.

Feminists have examined the problem of science and authority from another angle, that of male domination. In the next section, I explore the social problem of gender and science.

Gender and Science

Earlier, I discussed the role of ruling elites in the Scientific Revolution. The analysis of gender and science underscores the masculine stamp they put on modern science (Easlea 1983). In human terms, this meant treating women (along with blacks, wage laborers, and nature) as commodities in the modern world system (Merchant 1980:288). As Keller (1985:143) reminds us, "modern science evolved in, and helped to shape a particular kind of social and political context," including an ideology of gender. Arguments about science being gender neutral or value free are inevitably based on tearing individual scientific sentences or statements out of the social fabric in which they are conceived, produced, and used. Individual facts, or strings of facts, are then exhibited as science. But in order to understand feminist critiques of science, science must be seen as a social activity and a social institution.

An alternative science should not be conceived in terms of alternative scientific laws or techniques but rather in terms of alternative institutions and societies. This is clear even for equity issues that may at first appear to pose no threats to science. But the achievement of equal opportunity or comparable worth for women in science depends on such factors as reducing gender stereotyping and gendered divisions of labor. It may even, as Sandra Harding (1986:82) argues, "require the complete elimination of sexism, classism, and racism in the societies that produce science." Harding also challenges the widely held view that "the feminist charge of masculine bias" leaves physics, chemistry, and the scientific worldview "untouched (and untouchable)." She points out the apparent contradiction between building a "successor science" and deconstructing science as we know it. Her argument is that we need to pursue both goals (Harding 1986: 246).

I have reviewed some of the features of the feminist challenge that are consistent with the perspective on modern science being sketched in this essay. There are certain limitations of the feminist challenge, however, limitations common to other arenas of science criticism. These limitations reflect the difficulty of loosening the grip of the myth of pure science. Even feminist science studies and feminist science criticism that is pursued with the greatest degree of institutional independence rely on conservative theorists for authoritative accounts of the sociology and history of science. I have already explained why neither Bloor nor Kuhn can be considered a critic of science. Bloor's approach requires adopting the "proven methods of science" and ignoring their social trappings. And Kuhn is first and foremost a traditional internalist historian of science and a firm believer in scientific progress.

Harding's (1986:250) misunderstanding of Kuhn is illustrated by the fact that she finds it ironic that his *Structure of Scientific Revolutions*, which she reads as undermining "the notions of science central to the Vienna Circle," was originally published as part of the International Encyclopedia of Unified Science project. But there is no irony in this association, as Kuhn himself has amply documented in the 1970 "Postscript" to his study and elsewhere (Kuhn 1970, 1983).

Another example of the hold of traditional scientific ideology on feminist science studies is Harding's (1986:248–50) reference to feminist scientists as "the new heirs of Archimedes as we interpret his legacy for our age." Archimedes is lauded for his "inventiveness in creating a new kind of theorizing." This is Archimedes as icon of science, not Archimedes as military engineer.

While Harding is at least willing to consider (rhetorically at least) the idea of "a radically different science," Evelyn Fox Keller (1985:177–78) explicitly divorces herself from efforts to reject science or develop a "new" science. She refuses to follow this line of feminist inquiry because, in her own words, "I am a scientist."

Her reasoning is compelling. Rejecting objectivity as a masculine ideal lends the feminists' collective voice to an "enemy chorus"; and it "dooms women to residing outside of the realpolitik modern culture" (Keller 1982:593). Keller (1985:178) rejects the call for a new science on the grounds that it destroys the positive features of modern science. Her goal is to reclaim those positive features from *within* science and to renounce the features that make it a masculine project. But renouncing the "division of emotional and intellectual labor that maintains science as a male preserve" means, from my perspective, renouncing everything associated with the culture of modern science. In practical terms, it is necessary to correct the gender inequalities in contemporary science. But, as Elizabeth Fee (1983:24–25) argues, we also have to "push the epistemological critique of science to the point where we can begin to construct a clear vision of alternate ways of creating knowledge."

The sociological perspective, in the strong structural sense I argue for, is not a prominent feature of feminist science studies and criticism. This does not mean that the feminists do not draw attention to problems of social structure. They do not, however, do so in ways that transform epistemology from a philosophical to a sociological project. This makes it difficult for them to transcend the ideology of pure science. A sociological theory of knowledge must replace epistemology before we can begin to construct alternate ways of inquiry.

Science and Progress

To the extent that we have learned to think the way the machines around us act, so have we learned to see progress where the values of machines reign. The Scientific Revolution made modern science, rationality, and progress synonymous. By transforming "science" into a Science Machine, it turned rationality into a logic of unrelenting and unthinking machines in motion. *Progress* then became the label for "more and more" of whatever the Science Machine produced. As a result, the very things we take to be signs or measures of progress are coupled with the social problems that make us doubt whether there has in fact been any progress at all. Drug addiction, alcoholism, and "nervous breakdowns" are not considered signs of progress; profitable industries are, no matter what they produce. But drugs and alcohol can serve as lubricants, and mental health establishments as reservicing factories that help to keep the human machines in those profitable industries from completely breaking down (Camilleri 1976:42; cf. Berman 1984:7–8). We measure the short-term progress of our economy in terms of Gross National (now Domestic) Product. But like other measures of progress, GN(D)P does not measure the human, social, and environmental costs and risks of producing goods and services. We "progress," then, as Theodore Roszak (1973:426) has put it, "only toward technocratic elitism, affluent alienation, environmental blight, nuclear suicide." C. Wright Mills (1963:238) drew attention to the "highly rational moral insensibility" of our era, raised to higher and more efficient levels by the "brisk generals and gentle scientists" who are planning the Third World War: "These actions are not necessarily sadistic; they are merely businesslike; they are not emotional at all; they are efficient, rational, technically clean cut. They are inhuman acts because they are impersonal."

It may seem paradoxical to argue that modern science (allied with technology and progress) is a social problem because it is impersonal. After all, impersonal, machinelike truths and measures are supposed to guarantee that what we do is scientific and progressive. But it is precisely this notion of validation through proof-machines, logic-machines, language-machines, and number-machines that we must challenge in order to see the world of Science Machines and false progress described by Mills and others. The sociological imagination offers us a way out of this machine morass.

Modern Science and the Sociological Imagination

The core features of Mills's (1961) sociological imagination are: (1) the distinction between personal *troubles* and public *issues*; (2) a focus on the intersection between biography and history in society; and (3) a concern with questions about social structure, the place of societies in history, and the varieties of men and women who have prevailed and are coming to prevail in society. This perspective draws attention to new questions for the sociology of science: What do scientists produce, and how do they produce it; What resources do they use and use up; What material by-products and wastes do they produce; What good is what they produce, in what social contexts is it valued, and who values it; What are the personal and social consequences of their work and work habits; What costs, risks, and benefits does their work lead to for individuals, intimate relationships, communities, classes, genders, and the ecological foundations of social life; What is the relationship between scientists and their various publics, clients, audiences, patrons, colleagues, and friends and acquaintances; How do they relate to their intimates and especially to (their) children; What is their relationship as workers to the owners of the means of scientific production; What are their self-images, and how do they fit into the communities they live in; What kinds of teachers, mentors, and educators are they; What are their goals, visions, and motives? The collective hagiography that portrays scientists as "ingenious," "creative," and "benefactors of humanity" does not tell us what sorts of people scientists are or what sorts of social worlds they are helping to build.

Normal sociologists of science in normal society have concluded that normal science is efficient, productive, and progressive (Kuhn 1970; cf. Restivo 1983a). But normal science is a factor in the production and reproduction of a society burdened by widespread environmental, social, and personal stresses. Normal sociologies of science cannot help us see, let alone prevail in, a world of Science Machines and Cheerful Robots (Mills 1961). Even the sort of Millsian perspective on science, society, and sociology of science that informs my views may prove too limited for the task of critique and renewal. Fundamental categories of experience must be examined, challenged, and changed to even begin to address the social problems of science and society. The dichotomy between "nature" and "culture," for example, has fostered a dominative, exploitative orientation to nature, women, workers, and the underclasses in general. A fascination with spectacular discoveries, inventions, and applications in the physical sciences and with "genius" has blinded people to alienation in scientific work and in the lives of scientists. Inside and outside of sociology proper (especially in the United States) there has been resistance to unadulterated structural analysis. Individualistic and voluntaristic assumptions and perspectives have obstructed the development and diffusion of sociological conceptions of self, mind, cognition, and knowledge.

The full implications of sociology as a Copernican revolution that has moved the group, the collectivity, and social structure to the center of the social universe have yet to be realized in sociology. This revolution has transformed the individual from a being of "soul" and "free will" to a set of social relations and a vehicle for thought collectives (Durkheim 1961 [1912]; Gumplowicz 1905; Fleck 1979 [1935]). This idea does not subordinate the individual to society. Rather, by giving us a better understanding of what an individual, a person, "really is," it helps us to recognize the liberating as well as oppressive nature of the variety of social formations human beings can be socialized in.

Sociologists have generally traced their origins to ideologues of modern industrial society, notably Saint-Simon and Auguste Comte. The Marxist origins of sociological thinking have not been ignored, but (again, especially in the United States) they have not received the attention they deserve. More importantly, the working class and anarchist origins of sociology have been ignored (Thompson 1980; Godwin 1971; Kropotkin 1970 [1927]). So have the origins of sociological thinking among women scholars and writers, and especially among feminists such as Harriet Martineau (Spender 1983). These "oversights" have prevented the development of a sociology and a sociology of science infused with values, interests, and goals that would permit, indeed provoke, critical analyses of science and society. In particular, norms of skepticism and criticism have not been unleashed so that they could act on our deepest, our "unshakable," beliefs and assumptions. Recognizing the diverse origins of sociology depends on recognizing the distinction between the history of sociology as the history of a discipline and profession on the one hand, and as a way of looking at the world on the other. That distinction can help us to identify plural origins of science in general, and identify alternative, unrealized possibilities for the Scientific Revolution of the Galilean and Newtonian ages. That there is such an alternative in the history of science is illustrated by Merchant's (1980) study of women, ecology, and the Scientific Revolution.

The rebirth of the sociological imagination would help transform and clarify some fundamental but still cloudy issues in the sociology of science. The norm of disinterestedness, for example, is usually interpreted in psychologistic ("spiritualized") terms. To interpret it structurally means seeing its implications in terms of social interests. That is, disinterestedness means that commitments to specific social institutions are either dissolved or diffused. This is not a rarified notion. In practical terms, it means that we are in a better position to understand the world around us and ourselves to the extent that we put aside specific commitments to and interests in, for example, the national state, religions, and the bureaucracy of science. The more generalized and diffuse our interests, the more disinterested we are and, by definition, the more objective our statements about the world and ourselves are. Objectivity, then, is a social process and always a matter of degree (Restivo 1983b; Restivo and Loughlin 1987).

Thus, the sociological imagination is not neutral or relativist (in any naive or radical sense) on the question of truth. Mills (1963:611) argued that the social role of the intellectual involved a politics of truth, an absorption "in the attempt to know what is real and what is unreal." But he was not a naive realist. He argued in opposition to Hans Speier, Talcott Parsons, Robert MacIver, and Robert Merton that the sociology of knowledge was relevant for epistemology. Since our experience of nature and reality is mediated through social life, social studies have consequences for norms of "truth and validity" (Mills 1963:427–60). While many researchers who share Mills's political and intellectual concerns conceive of scientific objectivity in a social vacuum (e.g., Harris 1987:13–16), he continually stressed the social structural roots of logic and even of mind (Mills 1963:423–38). While he did not have the advantage of our current knowledge about the social processes of inquiry, he clearly appreciated the idea of a sociology of objectivity I have alluded to here and developed at length elsewhere (Restivo 1983b:147–51).

The sociological imagination is not, in Mills's hands, an abstract exercise. It is implicitly and explicitly a call to arms. It is not something to exercise in a political vacuum. It is true that Mills often spoke and wrote as a reformist rather than a revolutionary. But his proposals on social problems and social change challenged and continue to challenge prevailing social arrangements in fundamental ways. What sorts of rearrangements are necessary, for example, to transform intellectuals from hired hands to peers of the powerful or, more radically, to make intellectual work and politics coincident; What sorts of changes are necessary to develop "a free and knowledgeable public"? Mills addressed these problems and sought for solutions in conventional forms of democratic reform. In fact, such changes, like the changes feminist science critics seek, require much more far-reaching social transformations than usually imagined.

What sorts of social formations foster disinterestedness and objectivity? That is, under what conditions can inquiry proceed unburdened as much as possible by mundane interests and commitments and within the most expansive network of information and knowledge possible? Based on the preceding conjectures, my answer is: social formations in which the person has primacy, social formations that are diversified, cooperative, egalitarian, nonauthoritarian, participatory. The person has primacy in such social formations in an anarchist sense. That is, people are neither mere parts of social systems nor isolated individuals. Their potential for developing a multitude of mental, physical, and emotional dimensions of self is recognized and nourished; it is not surrendered to the authority of one or a few parts of the self, or to external real and imagined authorities. Social formations that allow for this sort of primacy offer the most fertile environments for inquiry because they do not, by definition, demand allegiances to specific institutional interests or subordination to specific authorities. The values people rally around, for example, are very general. We are more likely to learn things that will promote individual liberty, enhance community life, and cultivate healthy environments in such social formations. There is no hope for evading the endemic conflicts, tensions, and contradictions of the human condition. The sociological imagination should not be viewed as a pathway to utopia. Rather, it should be seen as a guide to social change—and not only to social change on a grand scale, but also to marginal improvements in the conditions under which we live and inquire.

SCIENCE AND PROGRESSIVE THOUGHT

Jennifer Croissant and Sal Restivo

Science is widely assumed to be a successful and valuable enterprise. But critics who recognize the need to challenge current institutions and societies should also recognize that modern science is a social problem.[4] It is a machinelike product of industrial and technological society, and indeed the mental framework and cognitive mode of industrial capitalism (Berman 1984: 37; Geller 1964: 72). Its consequences have certainly not all been benign, and the negative consequences should not be brushed off as "mere side effects." Foucault's lectures and interviews provide a subtle perspective on the relations of power and truth, and point to a major source of a continuing ambivalence toward science: "What makes power hold good, what makes it acceptable, is simply the fact that it doesn't only weigh on us as a force that says no, but that it traverses and produces things, it induces pleasure, forms knowledge, produces discourse" (Foucault 1984:61).

In this essay we explore critical thinking and ambivalence about science in three progressive traditions: Marxism, anarchism, and feminism.[5] Our objective is to draw attention to sources of this ambivalence and to show that it reflects a fundamental bankruptcy in modern science as a social institution. We frame this inquiry in a context of structural and intellectual crises brought about by transformations in the relations of global capital, generally identified as "postmodern" (cf. Jameson 1984). Each of the three progressive traditions has complex and contradictory relations to each other and to these transformations, which we will use to highlight our concerns with the regressive tendencies of science as institutionalized inquiry and Science as authoritarian icon.

It makes no more sense for progressives to support what is essentially state-science than it would for them to support state-church, state-justice, or any of the other core institutions of modern industrial societies. In fact, Science, Reason, Logic, Truth, and Objectivity, along with Obligation, Duty, Morality, and God are all tyrannical abstractions that have been used singly and collectively to intimidate and restrict humans, and to attack the foundations for liberating human and cultural development. We will be using capital letters to signify the iconic or symbolic use of science and related concepts as powerful abstractions that suppress dissent and constrain discourse. We recognize the complexity and multiplicity of scientific practices, but at the same time we see and experience the hegemonic power of science as system, institution, and icon. This re-radicalizes the proposition that "Science *is* Culture" current in social studies of science (e.g., Latour 1988a; Haraway 1991b:230), but grounds it in an older tradition in radical science studies (e.g., *Radical Science Journal* or *Science as Culture*). This essay is thus a polemic against capitalized science (in all possible senses of that term). We also suggest how to diffuse the trap question: What are the alternatives to Science for critical thinkers?

Modern science (including the scientific role and images, symbols, and organizations of science) came into the world as a commodity, and it has developed in close association with the discipline of the machine. One of the widely ignored consequences of looking to impersonal machinelike truths and measures (relying on proof-, logic-, language-, and number- machines as validating mechanisms) to guarantee knowledge has been to transform inquirers and thinkers into machines behaving in a purported value-free, value-neutral wonderland. The tyranny of science means in part that while it is possible and easy to criticize and oppose "distorted" versions of science (science corrupted by capital, politics, and sexism), it is impossible and futile to criticize or oppose "true," "pure," or "unadulterated" science. Claims about "the scientific method" and "pure science" are convenient myths, ideologies, and rhetorical constructs (a tyranny of methodolatry, in Mary Daly's [1985:11] terms) that justify objectification and alienation, and mask the constructed nature of the *necessary* statements scientists formulate as "laws of nature."

The tyranny of Method makes it an obstruction to inquiry. No one has made this clearer than Paul Feyerabend. His slogan, "Anything Goes" is the conclusion of a thorough search across the history of science for *the* scientific method, a fruitless search in the end (Feyerabend 1975, 1978). Similarly, Foucault's (1972) search for a final cause of the emergence of "man" was thwarted by the multiplicity of medical and social practices, interests, and authorities framing discourse.

Given how science is practiced rather than how it is characterized in myth and ideology, Feyerabend argues that "reason cannot be universal and unreason cannot be excluded." This is the grounds for "Anything Goes." The idea of an unadulterated science (as institution or method) is certainly not yet dead. It has become increasingly vulnerable thanks to a half-century of research in the sociology of science and nearly a quarter-century of research in the interdisciplinary field of social studies of science (cf. Cozzens and Gieryn 1988; Knorr and Mulkay 1983; Latour and Woolgar 1979; Collins and Restivo 1983).

What is not yet clearly and unequivocally resolved is: What does it mean to criticize science, and what can it possibly mean to prophesy a new science? Marx and Foucault challenged intellectuals to work toward changing rather than exclusively studying the world. And the questions plaguing progressive traditions in the era of postindustrial societies and postmodern theories are about what knowledge about the world is needed and what the bases for that knowledge must be to effect the desired changes. Today, these questions are raised in vital and volatile forms in debates and discussions about feminism and science. But they are a central feature of the history of progressive thought. Marxists, anarchists, socialists, feminists, and other progressive thinkers over the past 150 years or so have struggled with these questions in one way or another. Progressives tend to recognize that science is social relations and that it is a problematic activity in modern societies; and on the other hand they feel that they need to be "scientific" or "rational" in order to carry out their intellectual and social change agendas, and to retain some level of legitimacy in the arenas of public (and especially intellectual) discourse.

The concept of science as social relations leads to a recognition that science in modern societies is in the service of capital, patriarchy, and authority. We understand the complexity of these concepts, and of their relationship to modern science. Nonetheless, it is important that we not let this complexity veil the extent to which modern science is implicated in the social and environmental problems of our times, including alienation, dehumanization, ecological degradation, and nuclear, chemical, and biological hazards and warfare. It has played a role in increasing anomie, alienation, environmental disasters, the commodification of individuals and social relationships, and the spread of authoritarianism. In its current institutionalized form, scientific inquiry requires the control and co-optation of intellectual labor at several levels, and is inextricably linked to the agendas of the state and capital.

In recent years, scientific work has itself become the object of degradation as industrial forces have moved to complete the rationalization of the knowledge production process. This is part of the general process of reducing mental labor that is an inevitable strategy of late capitalism. The process is well underway in the university research centers we and some of our colleagues have been studying. Research centers bring industrial, governmental (including military), and academic interests into direct contact, and serve as crucibles within which to begin the industrialization of universities, including laboratories, classrooms, and graduate programs, in earnest. Let us be clear that what is new about this is its intensity and breadth of application.

If, in fact, information is rapidly becoming the new "industrial base" for first world nations (Reich 1991) we can expect to see the further degradation of science as an institution, and yet with further invocations of Science as a justification for repressive policies and authoritarian tendencies in the social relations of multinational capital and states. We do not wish to refer to some golden era of science (cf. Merton 1973) for judging the relative degradation of scientific institutions. We wish to inquire as to what progressives might make of such transformations. For example, what does "Science," or technoscience, have to do with the rapid commodification of biotechnology and the erosion of peer review, public accountability, and regulation as concerns about property rights and international competitiveness in this field arise (Lewontin 1992)?

Marxism: Bourgeois versus Human Science

Karl Marx (1956:110–11; 1973:699ff.) introduced the potential for Marxist ambivalence about science by distinguishing between bourgeois science and human science. He did not spend much time clarifying this distinction. But a couple of points are clear. First, he recognized perhaps more clearly than any other thinker of his age that science was social relations. More importantly, he understood that scientists themselves were social relations. He thus also understood that modern science (as a social institution) was a product of and an ingredient of modern capitalist society. If there is any question about this, it is dispelled by Marx's introduction of the concept "human science." The new social order Marx blurrily envisaged would give rise to a new form of science; science-as-it-is would be negated, and a new science—dealienated, integrated (but not Unified), wholistic, and global—would emerge. Following Marx, it is relatively easy to conclude today that modern science is part of the hegemonic ideology of modern capitalism, and an integral part of the relations of production. For example, David Dickson notes that "one can detect a formal similarity between the calculus as a mathematical language, and the forms of representation required by the capitalist labour production." In the seventeenth century, algebra "provided an abstract language in which commodity transactions could be readily calculated." Later, the calculus helped to establish a quantitative relationship between process and product, exactly what is "required by capital for the full articulation of, and control over, the links between the labour process and the commodity" (Dickson 1979:23–24).

Science and technology in contemporary society are, for at least some Marxist scholars, systems out of democratic control, controlled by industrial and military moguls, and threatening not only to whatever shards of democracy exist but to the very existence of life on this planet (Rose and Rose 1976; Dickson 1988). It is also increasingly clear that it is misleading at best to continue to separate science and technology (Latour 1987; Wright 1992). This separation has as one of its consequences the allocation of blame for many of our social and environmental ills to technology (and engineers) while science is held aloft as an exemplar of Platonic purity. Perhaps this purity might be better understood as an indicator of alienation (Sohn-Rethel 1975).

Marxist critiques of the effects of contemporary science and technology (or "technoscience") grow primarily out of studies of the workplace and the relations of production, and of the ideology of science as a justification for capitalist enterprises. Critical analyses of the historical roots of the factory, the rise of Taylorism and scientific management, the introduction of new technologies in new industries in the context of manipulations of the world labor market, unemployment, and the degradation of work all take us to the threshold of tyrannical science (Noble 1977, 1986; Hacker 1990; Arditti et al. 1980; Perrolle 1987; Hales 1974).

The technosciences (including the social sciences) serve capital. They have provided rhetorical ammunition for justifying social programming and policies and militarism. Technoscientific expertise has been used as "cultural capital" to justify imperialism and the patronization of weaker nations by stronger ones (Adas 1989; Nandy 1988; Aronowitz 1988). And yet none of this has been able once and for all to exorcise the spectre of Science the Good from Marxist scholarship.

The Marxists (here including Marx, with due respect for his objections) over and over drew attention to the facts that those who control the means of material production control the means of mental production, that new theories and technologies are grounded in social conditions, that the commodity form penetrates all areas of society, that the imperialist expansions and conquests of Capital require the development of the technosciences, and that—finally—the technosciences and capital are inextricably linked. And yet, in this same tradition, we find a tendency to exempt science and technology from the basic principles and perspectives of Marxism, to conceive of the content of technoscience discoveries as independent of social relations, and to identify bourgeois philosophies as ideological but not the empirical sciences. The bottom line seems to be, then, that the empirical sciences, based on observation and experiment, are not ideologically mediated.

The sources of these contradictory and ambivalent views should be obvious. The material successes of the natural sciences and the recalcitrance of the material world demand of Marxists as realistic (and perhaps in some cases, Realist) thinkers that they accept the findings of science and the methods of science that led to those findings. The dynamics of power and knowledge are simultaneous processes of coercive relations and the making of useful "goods." What many Marxists have missed is the fact that these findings and methods are inseparable from a variety of social relations. The associations of science that make it problematic for

Marxists include a domineering orientation to nature and humans, and the alienation of humans from each other and from the natural world as objects of study, sexist social relations, and gendered knowledge. Within this set of problematic social relations, science is especially troublesome for feminists because it embodies the contradictory impulses of liberalism within the debates on realism, knowledge, and postmodernism.

Feminisms: Criticizing Science without Losing Your Voice

While Mary Wollstonecraft relied on evidence and the emerging discourses of rationality and science in her arguments for the equality of women, her daughter Mary Shelley created *Frankenstein*, the paradigmatic tale of male appropriation of reproduction and creativity and the terror of masculinized inquiry. The most radical feminist agendas reject science, as method and institution and as patriarchal discourse. Science requires that we objectify nature in the natural sciences and persons in the social sciences. For ecofeminists, scientific inquiry is the rape of nature and has produced not "knowledge" but rampant degradation of environments and peoples. The symbolic imagery of "penetrating arguments," "seminal theories," and "unveiling nature" is described and decried by feminist authors (Keller 1985, 1982; Merchant 1980). Because women and nature are associated historically, the domination of nature demanded by science mirrors, justifies, and reinforces the domination of women (Mies 1990; Easlea 1983). For radical feminists, the scientific method is by its very nature in the service of Capital, Authority, and Patriarchy. Scientific technologies under the influence of masculinist ideologies entail the exploitation of the natural world, women, and other powerless people (Rothschild 1983; Wajcman 1991).

One of the problems with the radical feminist critique of science is that it leaves them without a voice, that is, without a legitimate institutional discourse. This is why, in part, many feminists, like other progressives, devote considerable energy to conceiving and devising "radical sciences," in this case "feminist sciences." Sometimes the agenda is to show that what women do in the area of knowledge production is "science" that has simply not been legitimated by patriarchal institutions (Harding 1991; Code 1991). Occasionally, women (or minorities) are granted (or grant themselves) a privileged epistemological status because of their experiences as objects and subjects of domination, exploitation, and oppression (Harding 1986; Hartsock 1983), a familiar perspective in Marxism. The argument then is that their increased participation in science will "fix" and "improve" science (cf. Longino 1990).

Science becomes a question mark for feminists because there is sexism in science as in other social worlds, and because whatever progress has fallen out from developments in technoscience has not fallen equally across the lives of men and women. The conclusion that a sexist/gendered society will produce a sexist/gendered science is not adhered to by all feminists, but it is a sensible conclusion from the sociological principles of Marxist thought. Still, the conclusion is not transparent. To make it transparent, we need to clarify some points about the phrase, *science and society*.

Society can be looked at as a set of more or less well-defined institutional sectors. The more well-defined the boundaries of a given sector, the more sense it makes to conceive of it as "autonomous." But autonomy does not entail "separation." It simply identifies the degree to which a given sector of the society is dependent on or independent of the resources of other sectors, and the extent to which it interacts with other sectors through communication and exchange linkages. Thus, even the most autonomous of institutions is in and of the society it is part of, and conditioned by that society. Its members are initially socialized in families, schools, and religious organizations; this prepares and allows them to be socialized later in life in the technoscience professions. The social structures and value systems of those professions thus reproduce the social structure and value system dominant in the wider society. The social institution of science is thus an arena like any other institution in which power relations between social groups and conflicts in relation to value systems are reflected and acted out.

It is clear that science and the roots of liberalism were effective legitimating and liberating ideals, relative to the strictures and structures of feudal Christianity and to the confines of women's lives. Scientific knowledge has served as an ally for feminist agendas. In the writing and lives of thinkers from J. S. Mill and Harriet Martineau to Mary Wollstonecraft, and later, Charlotte Perkins Gilman, the eighteenth- and nineteenth-century roots of women's liberatory activities in the West relied on notions of rationality and the emerging rhetoric of science and proof to argue for the equality and emancipation of women.

In both the early debates of modern feminism and its current discourse, science is called into service on pro- and antifeminist agendas (Schiebinger 1989; Poovey 1988; Helsinger et al. 1989). Feminist scholars who have argued that science is sexist and gendered find that science is used to justify the social arrangements in contemporary society (Hubbard 1990). Haraway (1989, 1991a) has documented the efforts of sociobiologists to argue from primatology that woman's "place" and the nuclear family are "natural" and "inevitable," and that their support and continuity are in the best interest of the human species. Norms of sexuality that justify violence, rape, and male "promiscuity," and myths of female passivity are reified in sociobiological agendas. New studies of the heritability of intelligence and personality are called on to find (and justify) novel treatments for the "ills" of drugs and poverty in modern society, and to explain away the underrepresentation of women in science, mathematics, and engineering. Much of the science of "woman" fails to adequately foster the emancipation of women, whether in the now discredited phrenological enterprises of the nineteenth century, or in present theories of sexual difference. And this is despite numerous attempts to develop feminist biologies and explicit challenges to the existing frameworks (cf. Tuana 1989; Hubbard and Lowe 1983; Brighton Women and Science Group 1980; Martin 1987; Fausto-Sterling 1992).

Recent expressions of concern about gender and science come from "liberal" traditions concerned with equity and employment. The quality and availability of work for women in the technosciences has long been questioned. Women have not had access to the professions in the same way men have; and when they have gained access, the tendency has been to marginalize them. Ethnic groups face similar problems. Many of those concerned with these issues assume that scientific practice is "objective" and that it is matters of equity and civil rights that need to be addressed. The idea is that "science" would "work" and produce objective knowledge and unambiguous progress without alienating and damaging women's lives if society were working properly. As forms of feminist empiricism, these agendas leave untouched the core ideas of science as pure and disinterested (Harding 1986). Further, these perspectives beg questions of whether or not, and if so, why, women and feminists might be interested in or capable of doing or organizing science any differently (Ginzberg 1989; Fee 1981).

The ambivalence about science is a variation on the theme (and tension) of a feminist separatist politics where we find the general problem of voice. Feminist activists have worked assiduously to give women a voice. The problem is that the language of science which is most likely to give them a voice may be inherently oppressive and exploitative, a danger to women and to democratic and socialist principles. However, few feminist critics of science support Audre Lorde's claims about the limits of the master's tools in dismantling the master's house (1981).

The reactions against postpositivist critiques of science are similar to other feminist reactions to postmodernist agendas. Radical critiques of knowledge leave an inquirer and activist at a loss in the face of the apparent necessity for *real* knowledge regarding causes, effects, and facts to be wielded in the name of emancipation (cf. Hartsock 1987). Demands for realistic assessment of science's instrumental effectiveness and for certain knowledge (and Truth) are indicative of the hegemonic power that Science as Icon has as a legitimating idea (Wright 1992).

Postpositivist critiques of science, whether in sociology or philosophy of science and knowledge, across feminist, Marxist, and other progressive traditions, are tenuous in the face of unbridled enthusiasm for technoscience's instrumental effects. These effects are not unequivocal successes. Numerous critics echo Haraway's rejection of "epistemological anarchism" (1991b: 80). But anarchism, whether of the most limited epistemological varieties or the most radical social imaginings, does not necessarily "justify not taking a stand on the nature of things" and becomes "little use to women trying to build a shared politics" (Haraway 1991b:79).

The possibilities of partial knowledge coming from situated knowers and standpoint epistemologies illustrate an emerging sophistication about the social roots of epistemologies (Harding 1991; Hartsock 1983; Haraway 1991a; Code 1991; Smith 1990). The realist/relativist dilemmas are not merely intellectual conundra, but are central to the debates about the possibility of social change and the grounds for action. They manifest the contradictory impulses of individualistic liberalism and the promises of the Enlightenment, and the problems of identity for feminists (Hekman 1990) destabilized by the postmodern deconstruction of the natural category of woman. These contradictions have been reinforced by the fracture of feminist movements along racial, class, and international lines (Alcoff 1988). To date, except perhaps in healing and medicine, the need for legitimate discourse has generally outweighed hesitations about the foundations of science.

The question remains about what the grounds for reliable knowledge are. Certainly, establishing such grounds requires refining the distinctions between the institutionalized inquiry we call Science, and other processes for devising reliable, sustainable knowledge about ourselves and our worlds. It is not so much a problem that the works of Foucault, Derrida, and other postmodern scholars have left those desiring social change without an "absolute" ground for action, rather, it has become a matter of considering what grounds are available, and under what conditions (Alcoff 1988). "Although we recognize that our Archimedean point may be historically contingent, it is nonetheless real and we stand on it as we move the world" (di Leonardo 1991: 30). The multiple visions of internationalized progressive, feminist consciousnesses arising from the paradoxes of occupying multiple, and contradictory, social roles invites heterogeneous knowledge. These multiplicities, however, have not moved far from their roots in the social and health sciences. Such an agenda and approach toward knowledge requires a radical re-vision and perhaps also a reengagement of the other human senses in the social relations of inquiry, knowledge, and authority. And science, sociology, and sociology of science can finally be "*for* only those groups represented in it" (Loughlin 1993:18).

Anarchism: Social Chaos or Social Theory?

Anarchist thought emerged in a nineteenth-century milieu in which the sociological perspective was crystallizing, and was itself considered one of the sociological sciences (by Peter Kropotkin [1970/1927: 191], for example). But the anarchists had trouble, along with other progressives, making and sustaining the distinctions among science as a social institution, science as a set of statements about regularities in the world, and science as a symbol for the best form(s) of inquiry. Kropotkin recognized that the state and capitalism are inseparable, and that state-justice, state-church, and state-army are inextricably linked in a network of insurance for the landlords, warriors, judges, and priests of modern society. But he was unable to see that state-science could be added to the list of capitalist institutions and that the social roles of scientists and scholars could be added to his list of exploiters and oppressors. Already in Kropotkin's time, the hegemonic ideology of pure science had contaminated anarchist thought as it had contaminated all of the emerging social sciences (including, of course, Marxism).

Kropotkin's defense of mutual aid and individual liberty was grounded in a commitment to the scientific method. By scientific method, he meant the inductive method of the natural sciences. The goal of anarchism, as a sociological science, was to use this method to reveal the future of humanity in its progress toward "liberty, fraternity, and equality." Kropotkin's ambivalence about science is reflected in the fact that while on the one hand he was a founder of scientific (and even scientistic) anarchism, he was also aware of some of the problems with using the natural scientists as role models for anarchists. Most of the scientists of his time, he argued, were either members of the possessing classes and shared their prejudices, or they were actually employed by the state. He did not, however, seem capable of imagining that science and the state were or would become so inextricably intertwined that any convergence between anarchism and science would in the end prove impossible.

The increase in scale of organization is a consequence and cause of the organizational search for "control, discipline, and standardization" (Presthus 1964:555). Bureaucratization rationally extends and deepens this search, and stimulates the development of "personal mortgages" and conservative behavior (Presthus 1964:555). This prevailing logic of the organization is just another manifestation of that authority that every anarchist opposes. But what sort of opposition is this?

In the face of the logic, sheer pervasiveness, and energy of state power, anarchism may seem to offer little more than the possibility of "opposing Goliath" (Horowitz 1964:26):

Anarchism can be no more than a posture. It cannot be a viable political position.

Indeed, Horowitz (1964:59) claims that anarchism has failed in part because it fails to address the problem of bureaucracy and in part because when it does address that problem it becomes enmeshed by it. But anarchism can be viewed more positively as the symbol flag followed by all those on the path to liberty, as opposed to the path to authority. Tucker (1964:173), for example, sees this in terms of the parting of the ways by Marx (state socialism) and Warren and Proudhon (anarchism).

The logic of bureaucracy is not, as might be suggested by Horowitz's argument that anarchism can only be a posture, invulnerable. Indeed, as Horowitz (1964:58–59) is quick to point out, the inertial extension and concentration of bureaucratic logic has made it problematic as an organizational form.

Other anarchists, such as Bakunin (1970/1916; 1980) and Proudhon (1977), although equally enamored of science, were at the same time somewhat more sensitive to the fact that it could be a dangerous institution, one that could divide the world and tyrannize the unlearned masses in the name of Science and Truth. Many anarchists agreed with Proudhon that science was the basis of the unity of humanity, and that society should be organized on its foundations rather than on the foundation of religion or of any authoritative institution. Science and thought, according to Bakunin, must be the "guiding stars" of any social progress. But he was skeptical of arriving at socialist or anarchist convictions *only* by way of science and thinking. He was critical of science because it could be divorced from life, from "the truth of life." Then, its "cold light" would produce only powerless and sterile truths.

On the whole, the anarchists were ambivalent about science because they generally recognized that science was social relations on the one hand, but that on the other it provided important anti-authoritarian ammunition in the conflict with religion and with authoritarian institutions in general. And they struggled with the fact that science seemed to be the source of truths not only about nature but about society. But it also seemed to produce a cold, harsh, violent, distant, and alien truth far removed from the human projects of love, community, and honest, trusting relationships.

We should not conclude this section without mentioning the contributions of Friedrich Nietzsche to the problems before us. Nietzsche needs to be considered here because he was, in an important sense, an anarchist (in spite of his antipathy to the anarchists of his time).[6] He was an uncompromising enemy of the State and an equally uncompromising defender of individual liberty. And like other thinkers we have considered, he was an advocate *and* an opponent of science. But even in his advocacy it is easy to detect the bases of his opposition. For example, he defends mathematics and physics—but not for the usual reasons. Mathematics is good because it helps us to determine our human relations to things, physics because it reflects the honesty that compels us to turn to physics for explanations. But he criticized modern science because even though it had helped to "kill" (anthropologize) God, it was in general rooted in the same motives underlying religion. It was not, he claimed, a way to the goodness and wisdom of God. It has not exhibited the absolute utility claimed for it by Voltaire, or any intimate association with morality and happiness. And it is not immune to evil impulses. Science is dangerous because it has the potential for divesting life of its "rich ambiguity" and turning it into an indoor mathematical diversion. He went so far as to describe science as the stupidest of all interpretations of the world because it deals with the most superficial aspects of life, the most apparent things, only those experiences that can be counted, weighed, seen, and touched. Science might yet, Nietzsche claimed, turn out to be "the great dispenser of pain," a means for making humans cold and stoic.

Unlike other students of science and society of his era, Nietzsche (1974) was able to complement his criticism with a relatively clear vision of an alternative to modern science—what he referred to as "the joyous wisdom" (or the gay science). In place of the scientist, Nietzsche put the thinker. The thinker has inclinations that are strong, evil, defiant, nasty, and malicious in relation to prevailing values. To think is to question and experiment, to try to find out something. Thus, success and failure are equally valued because they are, above all, "answers". And the thinker constantly—day by day and hour after hour—scrutinizes his/her experiences to answer the question, What did I really experience?

Love and passion, laughter, and the complementary roles of the fool and the hero rather than the detached cold light of reason are at the root of the wisdom the thinker seeks. The more emotions we bring to bear on a given problem, Nietzsche claimed, the more eyes we bring to it; and the more emotions and eyes we bring to bear the more complete our conceptions will be and thus the greater our "objectivity." Paul Feyerabend's defense of an anarchistic or dadaistic theory of science has much in common with Nietzsche's joyous wisdom, even though Feyerabend defends a limited and temporary form of anarchism: epistemological anarchism (Restivo 1993b:31–34).

Toward Humane Inquiries

We ourselves are not immune from the ambivalence that we have noted in the three progressive traditions we have sketched. That ambivalence is reflected on a larger scale in the debates about the cultural meanings of science. The defenders—worshipers, advocates, apologists, and ideologues—of science have been heard; they were, and for most of us continue to be, our teachers, mentors, educators, and peers. What types of resources, then, are needed to resolve the tensions within the progressive traditions and in the individual thinkers who create, carry, and change those traditions, caused by ambivalence about science?

The idea that there is somewhere out there a science that is autonomous and free is a pernicious myth. The more we have learned about science as a social and cultural phenomenon, and of scientific knowledge as a social construction, the better we have been able to focus our critique of science. This critique is still widely misunderstood, even within the arenas of criticism themselves, and we would like to try to clarify what the critique is all about. The questions that need to be answered are: What are the bases of the radical critiques of science; What if anything can we do to change scientific institutions and ideas about knowledge in line with the critiques; Does it make sense to talk about replacing science, and if so, what would a "new" or "alternative to" science look like?

The basis for radical critiques of science is the recognition that long prevailing and pervasive ideas about the nature of science are grounded in icons, myths, and ideologies. Archimedes, for example, is a leading icon of ancient science. As an icon, he is a paragon of pure science motives; as a real person, however, he is a military engineer who often serves the interests of his government. Behind every icon, from Euclid to Einstein, is a more or less sinister figure or social role (see, for example, the discussion in Restivo 1989:148–50). It is important to understand the concept of social role because a social role can be consistent with a wide range of personal motives.

Perhaps the most pernicious myth in science is the myth of pure science. In fact, whether we are considering knowledge systems in the ancient world, cultures across time and space, or modern science, we find that in every case the most advanced forms of gathering and using knowledge in a society are closely linked with the centers of power. Just as Platonic knowledge in ancient Greece was tied into the Greek oligarchy, so modern science was institutionalized as the mode of knowing of modern capitalism and the modern nation state. The violence associated with the emergence of modern capitalism and the modern state is at the heart of the modern science generated during the ages of the commercial and industrial revolutions. Will Wright (1992) has made a contribution to expanding the criteria of legitimacy for science to include environmental and social justice in his argument for a "wild," critically reflexive knowledge. He argues that if inquiry does not improve the human socio-natural condition, it fails on epistemological and instrumental grounds; it is thus incoherent and unsustainable. Mary Daly also insists on wild knowledge: undomesticated, asking "unfragmented" questions, ungovernable, and also extreme and prodigious (1978:343– 44). Wild knowledge entails more than linguistic revisions in codes and metaphors. It requires the liberation of human beings and a reorientation to the necessary relations of ecology and the social order.

Perhaps the most important aspect of the ideology of science is that it is (in its allegedly pure form) completely independent of technology; this serves among other things to deflect social criticism from science and to justify the separation of science from concerns about ethics and values. Interestingly, this idea seems to be more readily appreciated in general by third world intellectuals than by the Brahmin scholars of the West and their emulators. Careful study of the history of contemporary Western science has shown both the intimate connection between what we often distinguish as science and technology and also the intimate connection between technoscience research and development and the production, maintenance, and use of the means (and the most advanced means) of violence in society.[7] Not only that, but what we have just written is true in general for the most advanced systems of knowledge in at least every society that has reached a level of complexity that gives rise to a system of social stratification.

For a Critical Inquiry. If religion is the opium of the masses, perhaps we could say that Science is the Valium of the intellectual. It is the sedative, the soporific that engenders a feeble somnambulism rather than active, critical inquiry. What could it possibly mean to argue for a new science? Or to seek to destroy science-as-it-as and to replace it with a better form of science, or a new form of inquiry all together? What will we, if we are indeed Haraway's (1991b) cyborgs, be capable of building and what might we indeed build? The problem is that this is the wrong way to pose this question: it is a tyrannical trap, designed to paralyze the critics who might challenge, or ignore, the hegemony of technoscientific or -scien*tistic* discourse.

Marx was critical of religion, but he did not see any reason or way to go out and destroy religion in particular. Create a new society with new social relationships of the kind I imagine, he claimed—a socialistic or communistic society—and religion will disappear because there will be no need for it, no function for it to fulfill, and no resources to sustain it. This is the way we should approach the critique of any specific institution, including the institution of science. If we create a new society, we will also set the stage for a new form of knowing. This does not mean throwing out "science" as a symbol of our human and cultural capacity to distinguish between truth and falsity, or to learn about how our world works. Every human society generates some form of inquiry that we would recognize as including some of the basic ingredients of what we imagine "ideal," "human," or "humane" science to be; some have been demonstrably more environmentally sustainable and a smaller number supportive of more egalitarian relationships.

The best forms of human inquiry are distinguished by their capacity for criticism (including self-criticism), reflexivity, and meta-inquiry. These are basic epistemic strategies for realistic inquirers, and not for naive realists. Truth and falsity are not determined by the so-called "purity" or alienation of the inquirer, nor by some simple symbolic or linguistic associations between things in the world and terms that refer. They grow out of social relations.

Let us admit that it is difficult to imagine exactly what a new form of inquiry would look like relative to our current understanding of so-called scientific methods. But nonetheless, just as there is something to be gained from imagining new forms of social organization and political economy, however shadowy and ethereal our imagery, so there is something to be gained by imagining what new forms of inquiry might look like.

Imagine, then, a mode of inquiry in which we grant the acceptability of necessary statements and the weight of evidence, but treat claims as nothing more than well founded; and in which we formulate necessary statements rather than laws of nature. In this way, we can begin to imagine how to erase or circumvent the tyranny and hegemony of institutions of Rationality, Logic, Proof, and Method. We can begin, then, to imagine modes of knowing that emerge out of standpoints (local experiences in everyday/everynight lives) rather than out of centers and relations of power (Smith 1990; Hartsock 1984; Haraway 1991b; Harding 1991). We need, for example, to study schools from the perspective of parents and children, rather than from the interests of existing institutional imperatives and the perceived needs of capital and social elites (Smith 1987:187). We need social theory grounded in the perspectives of women, the colonized, the oppressed, not the relations of ruling elites. Technological design must be carried out *in terms of* the perspective and *with* the participation of people as *users*, rather than as passive consumers.

The single most important thing to understand about modern science if we are going to be able to study it critically is that it is a social institution and not an abstract body of knowledge, set of Platonic-like ideas, statements, laws, or facts, or a hall of fame populated with the images of scientists with eyes—to recall Nietzsche—such as no human being has ever possessed, eyes that can see the world unmediated by society and culture. It is equally important that as we become, for a variety of reasons, more aware of the knowledge/power axis in modern science we understand that this is a feature of science and of inquiry throughout history. The charge of envisioning new sciences prior to a new society is an attempt by the hegemonizing interests of the state, military, capitalists, technoscientists, and apologists to defuse and resist potentially explosive visions of radical thinkers.

An alternative progressive science or mode of inquiry can only emerge as the mode of knowing and thinking of an alternative progressive society. Marx offered us a brief and fuzzy view of what such a science might look like when he used the term *human science* in conjunction with his image of a future society. Imagine, then, a social formation in which the person has primacy, in which social relationships are diversified, cooperative, egalitarian, nonauthoritarian, participatory, expressive. The mode of knowing and thinking in such a society would be nonexploitative, nonsexist, nonauthoritarian, and nonelitist. The imperative for progressives, then, is to press forward with their social change agendas. A *nuova scienza* will follow their successes, just as it has the social changes that have gone before.

Harding (1991:173) challenges the science-as-a-social-problem formulation, pushing for specific proposals for the *nuova scienza* and asking what "science and epistemology are to contribute to this project" of fostering humane social forms. There are currently existing prototypes and examples of marginal improvements in achieving goals for a different science and society. The Boston Women's Health Collective is an example of different means and modes for achieving, legitimating, and transmitting useful knowledge. The challenges that AIDS activists present to the National Institutes of Health and the medical community at large for evaluating protocols, and getting onto the research agenda in the first place, indicate participatory promises well beyond "the disease of the month" strategy for managing the national health research agenda. The University of Maryland's rejection of an NIH-sponsored conference on crime and genetics should be seen not as an indication that "the public" is irrational and unprepared to speak about science, but rather that the public is capable of exercising a legitimate voice in and on science.[8] And there are many efforts outside the borders of the United States, however rudimentary or partial, in the participatory design of workplace technologies or public access to scientific information. And participatory discourse on science and ethics exist, however tenuously, as possible models for progressive action.

Epistemology, science, and the philosophy and sociology of science, as professionalized discourses separated from ethics, method, ontology, and metaphysics, and separated from community life, are likely to have little to contribute to new inquiry enterprises. Sociology, feminism, anarchism, or Marxism in bureaucratic or scientistic forms, rather than as imaginative enterprises, will likely be impediments to goals of social and environmental justice. Rather than settling on a craft model of scientific practice (as a number of feminists and progressives do) as a remedy, we need to develop organizations and other modes of supporting and legitimating inquiries.

These are difficult changes to work toward, especially in light of postmodernist theories critical of progress, and the role such theories then play in relation to the status quo (ranging from reactionary to revolutionary). The hope for a better society stumbles across problems of relativism in the service of institutional and capital inertia. Haraway calls it a postmodern "god trick" akin to the "god trick" of the universalizing discourses of modernism (1991b). Relativism is suspicious to a number of feminist scholars, for it seems that as women, people of color, and nations emerging from colonialism begin to speak with authority and out of their experiences, they are told that it's all relative and doesn't much matter anyway (Mascia-Lees et al. 1988). Further, the hopes of progressives are eroded by postmodernism's corrosive nihilism, which induces a cynical paralysis. This can be contrasted with Nietzsche's cautious and constructive use of nihilism, in a manner reminiscent of the way Feyerabend uses anarchism. If linear modes of progress and unambiguous utopias are eliminated as possible futures, we can still remain committed to struggle. We are not reduced to despair and cynicism. And indeed, we can struggle despite and in joyous defiance of our probable futility. Social progress is not a modernist, Enlightenment fantasy, despite the

Enlightenment assumptions of Marx, many feminists, anarchists, and other progressives (Hekman 1990) and despite the end of progress ideologies and utopianism rightly heralded in postmodernist critiques of culture and capital.

Conclusion

Finally, let us consider some questions to pose for and about science and scientists as a methodology for kicking the pedestal out from under them. Wright has asked us to seriously question the efficacy of the instrumental "successes" of science, in the face of widespread environmental degradation and social injustice (1992). Let's not be awed by "great discoveries" or by the "lives of the great scientists." We need to ask the sorts of questions about science and scientists that Restivo proposes in his essay, questions that focus, for example, on what scientists produce, who they produce it for, how they produce it, and with what social, political, economic, and environmental consequences. In the end, we will want to know not what scientists discover or invent, but what sorts of people they are and what sorts of social worlds they associated with. These are not issues of motives and intentions, but of communities and commitments and the assumptions, practices, and agendas by which we create and re-create our social worlds.[9]

NOTES

1. Primarily PCR, polymerase chain reaction, for which patents are held by Cetus corporation.

2. See especially Montgomery (1991) on the dinosaur as kitsch icon and as an "allegory of human history" and a "metaphor of disaster" (13, 18–19). The dinosaur might also be read, like the cyborg (Glass 1989), as a cultural transitional object, making a narrative space for multiple readings by a heterogeneous audience. These multiple readings do not, however, come together into a vision for social change.

3. The editorials in *Technology Review* (Steven J. Marcus, 96(6): 5, August/September 1993) and *Science* (Richard S. Nicholson, 261(5118):143, July 9, 1993) express the professional concerns of engineers and science directly. The first review asks that scientists and engineers "climb off the pedestals," while the second is concerned with the anti-science threat that "The Postmodern Movement" presents. This last is especially curious, especially given the Capital Letters.

4. We use the term *modern science* to mean the institutionalization of inquiry in the West since the scientific and industrial revolutions. We recognize that science is a heterogeneous, changing institution, caught up in contemporary transformations of the social order. However, despite numerous intellectual assaults on the hegemony of science, its rhetorical and ideological power and legitimacy—and its power *to* legitimate—have been extraordinarily resilient.

5. There are a number of other relevant progressive agendas we could have selected. We find these three perennially and preeminently engaged with science as social relations. Others, such as the U.S. civil rights movement or worldwide anticolonialism agendas reflect similar ambivalence about science and technology. This is often made evident in intramovement conflicts among affirmative action, reformist, or assimilative strategies and more revolutionary agendas, or in contradictory uses and critiques of science and scientific information.

6. Classifying Nietzsche with the anarchists may strike some readers as idiosyncratic, but in fact Emma Goldman herself claimed him for the anarchists (she is quoted on this topic in Joll 1993:20).

7. For a discussion of "violence and modern medicine" and of "reductionist science as epistemological violence," see Nandy (1988).

8. See "NIH Kills Genes and Crime Grant," *Science*, April 30, 1993; 260(5108):619.

9. We follow Restivo (1988) and Martin (1979) here.

SUSAN LEIGH STAR

2

The Politics of Formal Representations: Wizards, Gurus, and Organizational Complexity[1]

A map . . . is a document presented in a visual language; and like any ordinary verbal language this embodies a complex set of tacit rules and conventions that have to be learned by practice. Again, like an ordinary language, these visual means of communication necessarily imply the existence of a social community which tacitly accepts these rules and shares an understanding of these conventions.
—Rudwick 1976:151

It is hard to overestimate the power that is gained by concentrating files written in a homogeneous and combinable form.
—Latour 1986: 28

Quantify suffering, you could rule the world.

They cán rule the world while they can persuade us our pain belongs in some order.
—Rich, 1975: 12

PART 1: FORMALISMS, LAYERS, AND LONG-DISTANCE CONTROL

Introduction

Think of space as an arrangement of priorities: Things that are more important are closer to the center; things less important are farther away. The center is always defined with respect to a set of questions.

Here are my questions for defining the space of this essay: How are formal (mathematical, computational, abstract) representations defining the space of our world? What are the moral consequences of using formal representations? *Cui bono?*

These questions represent a journey, personal and collective. I was trained as a qualitative social scientist, in the tradition of grounded theory (Glaser and Strauss 1967; Strauss 1987). For three years I taught in a computer science department, and since 1981 have worked closely with computer scientists (Star 1993, 1992b). The politics of formalism are thus an important part of my professional life. At the same time, most computer scientists are contemptuous of social science, especially qualitative social science and especially sociology (Newell and Card 1985, 1987; Carroll and Campbell 1986; Denning et al. 1989). The values of quantification, to borrow Jean Lave's phrase (1988b), are deeply embedded in that community. Numbers, algorithms, abstractions are valued over qualities, emotions, and deep descriptions. These values are linked with precision, portability, and speed.

Yet engineers and scientists also need social science. As large-scale information systems and engineering projects inevitably create mistakes and disasters (Perrow 1984 ; Vaughan 1989, 1990), engineers and scientists realize that organizational and political factors cannot be separated from issues of safety and reliability. As the United States prepares for an unprecedented investment in a national information infrastructure (Kling and Iacono, this volume), some information systems designers are turning to social scientists for analysis of the social changes and problems that will be generated. So, uneasily, they are turning to us to learn about organizations and to speak for the ethical consequences of the world they are helping design. In so doing, the formal encounters the qualitative.

The primary kinds of work involved in creating formal representations are: abstracting (removing specific properties), quantifying, making hierarchies, classifying and standardizing, and simplifying. These are activities that occupy a large part of computer science. They can be looked upon formally, as rules, or empirically, as forms of work. If they are only regarded as formalisms themselves, many things are lost: the bases for the hierarchical ordering, the choices about what to discard and what to emphasize, whose classification scheme is employed. Elsewhere I have called this "due process" (Gerson and Star 1986), that series of judgments about what to include in a representation. I understand these displacements as work: actions that were taken by people, machines, or other actors. Science and technology, especially information technology, stand at a crossroads. To the extent that they are able to reconcile these formal and empirical descriptions, will they create the possibility of responsible formalisms and formal representations? Goguen (1992: 11-12) names the problem as one between "dry" (formal, logical, reified) and "wet" (ethnographic, situated, empirical) data and their attendant cultures in scientific research, and hints at a direction for solution:

> Dry structures occur "in nature," that is, in ordinary social interaction, in the sense that members of ordinary groups (such as sports fans) organize their talk in ways that correspond to such structures. These structures can be printed in newspapers or shown on television, in a compact geographical form. Nevertheless, the structures are still recognisably *situated*, that is, locally organised, contingent, and *ad hoc*, and they attain a sort of immutability only in retrospect.

This chapter draws on research in several different scientific and technical domains. The central emphasis here is on an ongoing study of the work situations, values, and tradeoffs of computer chip designers (VLSI CAD engineers)[2]. Other data and examples are drawn from my study of a nineteenth-century community of neurophysiologists (Star 1985, 1986, 1989a) and of a group of amateur and professional biologists in the early twentieth century (Star and Griesemer 1989).

In the first section of the paper I lay out several concepts from the sociology of science: immutable mobiles, long-distance control, and re-representation. In the second section, I draw on the study of CAD engineers, as well as some examples from other areas of science, to demonstrate organizational responses to the issues raised in the first section. The final section concludes with a discussion of the ethical and policy implications of the spatial arrangements sketched throughout the paper.

Immutable Mobiles

I begin this paper as a meditation on the model of "visualization and cognition" developed by Bruno Latour (1986). Latour has argued that historical changes in visualization and cognition can and should be tracked via changes in humble manual tasks, rather than, as has traditionally been the case, to a Zeitgeist or "new humanity" or other simple mentalist or reductionist accounts (an analytic stance often taken in accounts of both art and science). In looking historically at the invention of printing, of perspective, and of cartography, he ties their power and success to the "use of eyes and hands."

One primary expression of changes in the way people used their eyes and hands was the creation of "immutable mobiles." These are representations, such as maps, that have the properties of being, in Latour's words, "presentable, readable and combinable" with one another. Such representations also have "optical consistency," that is, visual modularity and standardized interfaces. They are often flattened to make them tractable in combination. They have the important property of conveying information over a distance (displacement) without themselves changing (immutability). Thus, in contrast with a story told from one friend to another that changes with each repetition (like the old children's game "gossip"), immutable mobiles may be taken from one place to another, or sent, without substantial change. Maps, books, sets of specifications transmitted electronically, or readings from a meter submerged into the ground or stationed on Mars are all forms of immutable mobiles.

But no mobiles are completely immutable, as Latour himself has discussed (1987: 241–42). This is because of a "central tension": to be useful, they must be instantiated in, and therefore adapted to, a particular work setting. However, when coupled with a desire for mobility, the need for modular and standardized descriptions subverts the local adaptation. People represent things abstractly in order to send them over distances where they do not know, and cannot control, the local circumstances. When the representations are used, this central tension appears as a tension between representations, which are static and abstract, and work, which is real-time and concrete.

Taking this tension and its attendant tradeoffs into account, we can think of immutable mobiles as traveling along a path of work, where the tensions between mutability and immutability are managed in each situation. This path is a "re-representation path."[3] This is a unit of analysis describing the transformations and use of representations over time that result from this central tension, considered both with respect to technical information content and work organization. Re-representation paths are the story of the trade-offs made along the axis of representations and work.

Becker, in his work on the development of conventions in art worlds, says that such work organization is the foundation for understanding representations. Thus, it is the articulation of various kinds of work which create, for example, a school of painting. This is not aesthetics devoid of contingency, but contingencies that shape aesthetics, ranging from the union-imposed hours of musicians timing a symphony to the large-scale manufacturing of brushes constraining the width of strokes of paint on canvas (1984, 1986). And such work is often invisible to both outsiders and historians, who may come to think of the piece of art as shaped solely by its visual properties, or by an individual artist, devoid of the collective nature of the work implicated in its production.

Latour's (1986:23) final "stage" of advantages for immutable mobiles is the creation of formalisms. "What we call formalism is the acceleration of displacement without transformation." By this he means that information presented in formalisms is the most portable and the most unchanging, precisely because it is both abstract and recoverable. In some sense, one could think of a computer chip as the ultimate "portable formalism," in the sense that it is meant to be built of formalisms and to be portable anywhere for the manipulation of other formalisms in the form of computer programs and digitized information. Latour notes that when immutable mobiles are concentrated they are an impressive means of power and control.

Latour's analysis of the power of regimes of representation is true for both brains and chips. By equating parts of the brain with formal representations, neurophysiologists were able to define functional regions abstractly, and thus create a picture of the brain. An enduring definition of mind-in-brain emerged from this equation (Star 1989a, 1985). Yet to understand the machinations of power Latour describes, we have to look more closely at exactly what is in the cascades of representations. We have to understand how they are put together in light of a vast system of tradeoffs and equivalences.

By concentrating on the management of cascades and regimens, I illuminate a slightly different aspect of Latour's argument. Because representational conventions are the result of regularities in the conduct of humble tasks, tasks faced by teams of people require that work practices be matched or made regular across the domains of design. This "alignment of regimes," to use Latour's term, is a difficult enterprise.

To understand the alignment of regimes, we need to analyze the organizational aspects of managing cascades and concentration of immutable mobiles, rather than individual manual and visual tasks. When people manage very complex representations, the concept of "re-representation path" becomes equally complex: It comes to include mismatches in the division of labor and representational complexity resulting from the complex nature of the work.[4] This includes the fact that hierarchically arranged representations obscure lower or higher layers; communication across the layered boundaries of different task domains is difficult; and large, complex representations require large, complex organizations to manage them. Central to my story here is the constant tension of how to keep visible what keeps slipping away behind layers (bureaucratic or representational) of other things.

In the following section I discuss one type of complex representation especially important for linking space (the ordering of priorities) and formalisms: layered representations.

Layered Representations. One way of trying to control natural complexity and disorder is by stacking things up in hierarchical layers. Much as you would put the piles of paper on your desk into separate piles with important papers on top, scientists and engineers create what I call "layered representations." Unlike maps or photographs, some representations are "sandwiched" on top of each other. The working representation consists of conventional forms which must indicate location both within a layer and between layers. In the case of computer chip design, for example, the representations include both structural and functional aspects of design. There are many examples of this sort of representation in biology, geology, engineering and computer science. For example, in neurophysiology, atlases of the brain have often included representations of brain strata, including cortical layers. These atlases are simultaneously structural and functional. In geology the earth is represented as having neat layers according to geological era of development. In the case of science, these layered representations reduce irregularities in terrain or uncertainties introduced by technology or inaccessibility. In engineering, they are a way to order work and the flow of information, in the artifact and during its development, for maximum efficiency and speed.

Developing a visual representation of something with multiple opaque layers presents a fundamental contradiction: how to make visible what is invisible and must remain invisible for the structural integrity of the phenomenon. This contradiction appears in other circumstances: for example, when things are very far away or very little or scarcely there at all. One way of managing the contradiction is to simplify and abstract the layers, so that they all line up and may be formally manipulated. A probe may be run between layers to demonstrate or assure the alignment.

That is, different parts of the representational process are allocated (or displaced) to different strata of the artifact being constructed; in the end, these strata must be aligned and made to function together in a mutually structuring way. That is, the things constrain and even create each other across strata.

The process of articulating the relationships between layers is a matter of managing both people and things. Because the layers are heterogeneous (e.g., some are abstract—mathematical, logical, or behavioral; some concrete—silicon or flesh), and because they often fall into different task and work domains, the management of information makes considerable demands on the work situation (Jones 1979; Star 1986). The world does not come in neat layers visible to the naive eye. Geological strata simply look like mush when a probe is inserted into the earth to pull them out (Bowker 1988); the neat formal representations of layers one sees in geology and paleontology texts are the results of many decades of abstraction and negotiation in the scientific community. Brains, of course, look similarly mushy to the untrained eye, but even under the microscope seeing the layers is a feat that required decades of training, theorizing and collaboration—all of it laden with contingencies (Star 1989b; 1992b). And in the chip design team, what in the end is an extremely neat, fast bit of silicon and metal engraved with the logical, formal reasoning of a worldwide community of scientists and technicians, appears in the design process to be a battle for terrain and particular viewpoints (Star 1988).

How do work groups manage simultaneously to track their work and produce working artifacts or viable maps of layered phenomena?

Looked at from the point of view of work organization, the information management in, for example, the design of computer chips is similar to that involved in mapping the brain. Both lines of work include creating several kinds of heterogeneous representations which must articulate together to produce a workable artifact. In chip design, this means articulating the logical analysis of circuits carrying information with the layout of these circuits, the actual "etching" of the information on silicon, and the placement of assorted wires, plugs, and power sources. Each stage in this process is accompanied by enormous amounts of documentation and graphical evidence. Mapping the brain meant articulating histological and anatomical information with information garnered from clinical observations, experimental data, and postmortems; it has also meant analyzing and accounting for individual anatomical variation. In both cases, the tools for investigation and visual representation were and are invented simultaneously with the work of design/mapping.

Partial Visibility and Communication. The visual conventions and standardized formulae of the chip design process may be optically consistent at any one layer, but as part of the larger representation are stratified in an inconsistent fashion (see, e.g., Hill and Coelho 1987). This inconsistency derives from several sources. A layered representation is of necessity only partially visible from any one stratum. Any single layer can have the property of either invisibility or partial visibility to another; this is true both in examining an artifact such as a computer chip, and in analyzing the process of design work.

Layers as we are speaking of them here are not simply modular, homogeneous sandwiches; if that were the case there would be few sociologically interesting differences between flat maps and, for example, geological stratigraphic representations. By contrast, layers are heterogeneous in both time and space and with respect to other properties such as abstraction and concreteness.

The flatness and combinability of which Latour speaks is made more uncertain by the layering process, which both complicates and obscures. The ability to superimpose representations is possible only at a fixed temporal/spatial point. These are difficult to find or create, for all the exigencies of scientific work involve juggling many worlds simultaneously, most of them on different timetables (Star 1989b). Thus, a simple geometrical modularity such as Latour describes must always be unstable. Similarly, layering is orthogonal to questions of scale. It is not simply that scale, as Latour notes, is relative for immutable mobiles (1986:12), but is both relative and partially obscured, and distributed as well.

Formal Representations and Long-Distance Control

Recently, sociologists of science and art have been revealing the complex and fascinating organization of work required to create representations where work is long-distance. For example, Law's analysis of the organization of work and technology in the development of Portuguese navigation and cartography required training and managing an array of socio-technical allies over enormous distances (1986a). These ranged from boats and masts to sailors and kings; from wind and waves to stars and note-taking conventions. Star and Griesemer (1989) have discussed the process of creating biological maps, again over large distances, when the information gathered has different meanings to different parties involved. The visual objects thus created occupy a tense but necessarily malleable position between several worlds.

Disciplining Information-Gathering Allies. One of the problems with gathering and collecting information at long distance is one of control and degradation of information. For scientists, as well as for businesses and corporations, accuracy and standardization are important for controlling the information brought back or sent from far afield.

In many fields of inquiry, long distance control of information is achieved through getting agents to use standardization forms for reporting; in some sense this is the heart of bureaucracy.

As methods of collecting information are automated, however, it is not standardized forms that are important as much as standardized interfaces and quantifiable data. At a museum I have studied, for example, information about animal specimens was once collected by amateurs out in the field. They were constrained to record their findings in standardized field notebook forms. They preserved their specimens carefully for transport back to the museum, where each animal was catalogued, measured, weighed, stuffed, and put in a drawer. Nowadays the collecting continues in much the same fashion. But animal specimens may be measured by a pair of digital callipers applied to the body; measurements are read directly into a database where they may easily be manipulated statistically.

Summary of Part 1

I have argued above that the contingencies of scientific work require two things: making sense of the disorder of workplaces and fitting in information coming from a long distance into existing information bases. The fact that scientists work in large, distributed communities and face constant political and sponsorship pressures adds in special ways to the disorder and the need for long-distance control. Some kinds of responses to these situations have been the development of formal representations, including hierarchical layers, standardized forms, and other kinds of immutable mobiles, including formalisms. In the following section I will focus in more detail on the nature of these responses.

PART 2: STRATEGIES FOR ORGANIZATIONALLY CONSISTENT REPRESENTATIONS: WHERE DOES THE MESS GO?

Thus did the imperatives of standardization, interchangeability, and the moving assembly line, reinforced by the logic of scientific management, bring increased control of production by pre-processing away much of the information contained in the final products themselves.
—Beninger, 1986: 298-99

The tension between representations and work forms one of the central axes along which trade-offs occur in both science and engineering. The more general the representations of processes, the less they address those elements unique to a work situation. Because formal representations are so general, and therefore so powerful and so abstract, this tension is especially strong in the use of formal representations. Scientific work can be seen as one long attempt to reduce the tension—a series of trade-offs among, on the one hand, generality, reliability, portability, and integration, and on the other, customization, uniqueness, goodness of fit with local work arrangements, and validity.

In science, the trade-offs result from the need to take account of local variability and still create the most general theory. In engineering, they arise from the strain between customization to a particular situation and standardization to get the benefits of the commodity—in short, the "make or buy" decision.

As large-scale technology development and research programs grow in complexity, become automated, and rely on formal representations, organizations must respond to these tensions.

Deleting the Work

In recent years much sociology of science has been busy documenting the gap between phenomena and representations. John Law (1985), for example, talks of "transforming rats into marks on paper" Latour and Woolgar's (1979) Laboratory Life of the deletion of modalities in descriptions of scientific work. Both of these analyze the gaps between laboratory experiments or scientific research and the ensuing publications (see also Star 1983 and Lynch 1985a, 1985b for similar studies). These gaps become progressively more invisible, or "glossed over" as the work becomes more formal. A narrative of a brain surgery operation, with a tracing of the holes cut in the skull, holds far more visible evidence of contingencies and ad hoc solutions than does a statistical report of successful operations, even though neither of them are addressing work contingencies directly. There is a series of displacements on the road to formalizing, one of which I call "deleting the work." It is in part also this deletion that the empiricist side in the formal/empirical debate is pointing at in stating that, for example, mathematical proofs are ultimately social in nature (see below).

Fujimura (1987) has argued that technical problems become "doable" when certain conditions in the workplace are met, including gathering of outside resources, technical infrastructure, readiness of workers to tackle the problem, and their position in the disciplinary problem structure. She uses the image of aligning a set of transparencies, each with a slightly different window, representing different aspects of the work process and the technical structure of the problem. A doable problem is formed when the windows line up.

In the production of complex representations, the transparencies are themselves mobile, in process, shifting, and n-dimensional. There are ongoing tensions in the articulation of parts of the representations, precisely because they are not docile, optically consistent, flat representations nor purely formal systems. What is sociologically interesting are the complex division of labor and representational processes reflected in the partitioning of the representations and the ongoing attempts to keep the work process "doable." Fujimura emphasizes doability as it arises from the alignment of the local work situation, the larger organizational structure, and the discipline and its problems collectively. The alignment occurs when actors can recognize a clear path to action within their sphere, and look across to other spheres and see a clear road for alliance.

Engineering differs from science along lines of doability because there is a concrete product with a concrete perimeter to encase the layers. Thus, doability obtains both in Fujimura's sense and in the sense of modularized layers that literally "fit" together. Thus, there are two needs in engineering work: a system of alliance across different spheres that together articulate doability, joined with layer-spanning individuals or probing tools that can concretely build things.

The Division of Labor

Layered representations are assembled by teams. These teams are composed of members with disparate skills which must be coordinated. For chip design, for example, this means managers, software tool designers, circuit designers, often those designing other parts of the computer system, and those involved with the physical production of the chip. For brain research, it meant putting together the results of clinical experience and basic research across many medical and scientific specialities. In the museum, amateurs, professionals, and commercial traders cooperated in a vast network to map distributions of flora and fauna. The concerns of the various worlds of work were quite different, as for example, between professional biologists and commercial sellers of animal specimens. The process of crafting and use of such representations is thus also a story of the intersection of many kinds of work—an intersection fraught with inconsistencies, different evaluation criteria and commitments (Star and Griesemer 1989; Star 1989a).

The general problem of mapping and design involved in formal representations requires forcing representations of work processes to become combinable and consistent, however briefly. This must occur despite the formidable above-mentioned problems of distributed information, partial visibility, and heterogeneity of skills and concerns. Teams creating such representations employ a wide repertoire of strategies to overcome these problems and create the equivalent of an "optically consistent" representational language: an "organizationally consistent" representational language that preserves visual and informational continuity.

The push toward formalism under these circumstances is considerable. The more organizational and physical complexity must be aligned, the more necessary it is for formalisms to reduce contingencies, force precise measurement, and become mobile across organizational idiosyncrasies. On the other hand, the work organization is becoming more and more complex, and many people are unwilling to rely on formalisms for testing technologies that may endanger human lives. In many of the mathematical and engineering sciences, particularly computer science, these organizational tensions have given rise to a debate between formalists and empiricists.

Formalism versus Empiricism

Faced with uncertainty and burgeoning complexity, parts of the computer science community have divided along lines of formalism versus empiricism (DeMillo et al., 1979; Fetzer 1988). On the formal side, researchers use statistical and formal modelling techniques to test software, simulate real world situations, and predict reliability. On the other side are those who argue that only real-world testing will hold up to testing, and for reasons quite similar to the line of argument in sociology of science. Scacchi and Bendifallah, for example, argue that the gap between documentation and implementation in software development can only be understood with reference both to the technical specifications and to the dynamics of the work groups involved in development. Thus, software engineering is a "socio-technical" system not amenable to exclusively formal verification (Bendifallah and Scacchi 1987; Scacchi 1984; Scacchi et al. 1986). They argue that one cannot have reliability without analyzing the work processes involved.

The Instability of Formalisms

No one on either side of the formalism/empiricism debate argues that we currently have adequate techniques for measuring work practices. Rather, formalists argue that formalisms serve to correct idiosyncrasies and biases in an individual work site or implementation; empiricists that formal verification prohibits access to work practices.

In terms of current practice, however, formalisms are unstable due to inadequacies of understanding of the work situation, whatever one's position on how they may be remedied. What occurs now is often a cycle, from simulation to formal model then an implementation of a formal model, with many local work-arounds and ad hoc inventions to reinsert the concreteness and specificity that have been left out of the formal model. This reinsertion is only beginning to be understood by sociologists, anthropologists, and to some extent by computer scientists (Gasser, 1986; Curtis et al. 1988). To a large extent it is unrecorded, often hidden, and often privatized.

Thus, in terms of actual practice, formalisms without accompanying analysis of work organization can introduce elements of bias, confusion, or lack of generality, where their actual use is not recorded, or where they cannot be easily revised. Rogers Hall (1989, 1990), for example, has shown that children solving algebra problems invent ingenious, situation-specific methods for solving traditional mathematical problems. They are, however, often ashamed of the results, and those who "test well" are those who can translate what they actually do for problem solving into the demands of the formally correct method.

Craft Work and Formal Representations: Quis formabit formatores?[5]

Simply put, what is lost when people work around formal representations is a clear understanding of what is left out when formalisms are created. Because formalisms tacitly encode biases (what gets left in and what gets put out as unimportant), and because ad hoc work-arounds are not documented or well understood, we know that they reinforce, in invisible ways, the prejudices of their designers.

In contemporary science and engineering, people who have a wide repertoire of work-around routines and who are good at dealing with the gaps left by formal representations are extremely valuable. There is, in fact, a whole scientific mythology of wizards, "gurus," laboratory technicians with "golden hands," as well as a plethora of myths about scientific and technical creativity that are closely linked with the ability to "restore the work" (Star 1992a, 1991b; Shapin, 1989).

In large-scale technology projects, especially those that are highly automated, much of this craft skill has moved from the ability to make a thing to managing information about a thing. Let us consider for a moment the case of computer chip design, or VLSI CAD.

Chip design is a constantly changing, and still relatively new, field. Many aspects of design have recently been automated for the first time. During the 1970s, design automation was primarily limited to fairly low-level technical support. Much chip design at the level of layout was in hand-done drawing as late as the early 1980s. Chip design shops had the equivalent of "master craftspeople" who were highly skilled in drafting. Computer-aided design obsoleted hand-done drawing, and vested the craft skill in things like choice of tools, ability to master varieties of tools and software, and inventive ways of managing the extremely complex flow of information in the design process itself, including the use of expert systems and artificial intelligence (Horstmann 1983; Kowalski 1986; Freeman 1984; Foulk 1984).

Making the tools VLSI CAD designers use "smarter" means changes in work organization and communication patterns. As with all automation processes, many of the gaps between documentation, communication, technical requirements, and workplace practices become visible as attempts are made to automate such procedures. As Arnold (1984: 9) describes the situation, "CAD users found they had to reorganise to cope with a computer making their procedures more explicit and often thinking for the first time in many years about working methods."

This explicit analysis of work would require that managers and engineers think reflectively about their work practices. Cooley (1980) notes that research has been ill defined about exactly what can and should be automated. The assumption has been that the creative part of the work will be "left over" after automation; however, explicit analysis of differences between creative and routine work is often rudimentary (see Gasser 1986). Furthermore, the analysis presented here shows that the design situation as a whole is equally informed by trade-offs among parts of the organization. The regimes of formal representation extend beyond and within rigid boundaries.

Spokespeople from the CAD industry see changes in work organization in terms of customized solutions—more customization available at the workstation level. Simoudis and Fickas (1985) foresaw that the advent of knowledge-based engineering for CAD meant that the users of CAD workstations became of necessity more involved in the design process; as well, multiple local solutions to design problems are attempted in an iterative fashion as users tailor solutions to local problems. Rezac (1984) states that the key problem in design debugging similarly involves multiple local solutions. How to decompose designs in order to create flexible, user-involved solutions is central, he notes, to changes in the CAD workplace. The input and analysis of designers also means integrating the concerns of workers from different spheres (Myers 1985). The push for integration and automation of different spheres means that integrating different approaches from various lines of work will be central to this effort (Kochan 1984).

In theory, the changes in workstations, networks, and automation as discussed above mean more distributed work patterns and a freeing up of bottlenecked computing resources. Cheaper CAD software and hardware amplifies this effect. It would seem that the past scenario of "master designers" (who were "wizzes" at technical design and coordination) heading a team should be changing. Everyone would in theory become an automated wizard.

Yet, in the empirical studies described below, such "master designers" had not disappeared. The ongoing process of interaction and coordination across layers and among disparate portions of the work team requires no less complex regimentation and articulation than that given to a chip drawing. For instance, several years ago I visited a satellite design team, another kind of high-tech work using CAD. On the wall of the room near the ceiling the team had pinned up a single chart; it covered the entire length and width of the room, approximately one-hundred square meters, with boxes about fifty centimeters square. Each box represented one part. One person was responsible for each part, with no duplication. There were two people ultimately responsible to top management for the articulation of the whole thing, called "responsible engineering authorities." They are managing a division of labor as complex (or more so) as a general's management of an army, and all studies show that their functions are far from automation as we now know it.

From this example it is clear that problems of the tension between formal representations and local contingencies is recursive. Human creativity and ability to find ad hoc solutions to ill-structured problems simply go to a different part of the work process when domains become formalized. This is not to say that automation does not restrict creativity or impose unsavory work conditions on the less powerful; of course it often does. Rather, *the question of who will formalize the formalizers is infinitely recursive.*

Organizational Responses to the Problems Posed by Formalism

In answer to the original question posed at the beginning of this section, the mess, as we have seen, goes into a number of organizational and personal strategies. In the following section, I discuss each of the strategies I have observed. This is a preliminary list and I invite additions:

1. *Freezing,* or holding constant, parts of layered representations, then making trade-offs with associates to counter the rigidity imposed by the constant.
2. *Relying on wizards,* or what are sometimes called gurus, to work around difficult design problems.
3. *Attempting to model "tacit knowledge."* As part of the press to formalize results, many researchers and scientists are trying to codify, formalize, and document the work-arounds and ad hoc solutions discussed above.
4. *Denigrating the social as impure.* Another response to gaps discussed above is to attempt to reach pure technological rationalism in design. This includes generating what John Law (1988b) calls "purity" stories and what Sal Restivo (1983b) calls the generation of parallelism, strategies that analytically separate the social and the technical.

Freezing Parts of Representations. One strategy employed by users and managers of layered representations is to attempt to freeze some component of the representation. This may involve combining a simultaneous probe of layers with a cross section of several parts of the representation. In chip design, this leads first to a version of the "make or buy" decision: Off-the-shelf commodities "well-structure" at least some part of the representation, rendering it more combinable with other commodities. In this example, a group of CAD and computer designers at a meeting try to come to an agreement about a common frame for one of the components of the chip and system they are designing:

> *First designer*: The trade-offs are this. One, we might waste some area, or have the wrong spec ratio. Then, there's simplicity in probing and bonding. There's a lot of work that you

> people don't see. It's hard to get chips bonded. If we could just simplify and standardize it would be cheaper. But the trade-off is that somebody might end up with a fifty-millimeter square chip inside a one-hundred-millimeter cavity! (7/14/87)

However, this strategy is not without its problematic consequences. It often simultaneously generates a need for customization, since many of the skills and demands that need to be met in the design of the chip are local (either client demands for customization or unique blends of skills). This is a central tension in the management of information in the development process, very much like the representation/work tensions in the re-representation process. The group continues its discussion, trying to figure out a way to agree about which thing they will hold constant:

> *First designer*: In my solution, I've been running alone for a long time. When it comes to pad design we're still running on superstition. Nobody knows how it works; it's still statistical, and it's still zapping us. It's fuzzy science. (They argue about whether to go with an outside manufacturer with widely adapted standards.)
>
> *First designer*: We have to trust somebody.
>
> *Second designer*: XYZ company stands by their work.
>
> *Third designer*: No, they fabbed (another component) wrong, it was way too wide. . . .
>
> *Second designer:* We made the last one too small. There wasn't enough space for them to see. We're just losing time. In the end, god is responsible, and he doesn't care.
>
> *First designer*: Mine has the advantage of being simple, but it may be dangerous.

They go on to discuss the pros and cons of adopting a slightly bigger, therefore slower, size that will be easier for the fabrication plant to "read" and easier for them to design to. They conclude the meeting by agreeing to one pad size, and one of them makes the comment while leaving: "If I only have to worry about it in one place I've won big." By dedicating one part of the location, they "freeze" a part of the design.

The advantages of freezing are clear. The cost of doing this is not always as immediately apparent as in this case, where they are sacrificing speed. Down the line, a new set of trade-offs appears as they attempt to export design. The reasons for freezing one component only hold for one local situation. As one respondent states: "The technical structure of the project is such that the only link is really the bus which is well specified. So we have an abstract, consistent definition there. It acts as a big anchor for the project. It is complex, and we pay for that, but it makes it easier to coordinate" (JM, 8/6/87).

In brain mapping, a similar connection between "freezing" and abstracting is noticed. Often a single area is identified as an *x* area: a famous case is a region thought responsible for speech, Wernicke's Area. In their classic article, "Wernicke's Region: Where Is It?," Bogen and Bogen (1976) show that the area shifts around depending on the representation. The same holds true for the earlier researchers I studied, who were able to reconcile many anomalies and come up with consistent maps, but only at the price of specificity. The more that common anchors are developed across disparate worlds to produce a layered representation, the more abstract the whole representation.

The danger that arises from freezing representations is that organizational commitments are made on the basis of original decisions, which may have been meant to be heuristic. This results in what philosopher William C. Wimsatt calls "generative entrenchment" (1986). That is, decisions made early in the life of a project ramify throughout development, in much the same way that small changes early in the life of an embryo have crucial downstream effects. And as with embryos, small mistakes early on may produce the organizational equivalent of teratagens or "hopeful monsters."

Relying on Wizards. One organizational result of the tension between formalism and empiricism is the generation of work-site wizards who can perform the probes and manage the difficulties of coordinating layers organizationally. These are not simply the "master designers" of the past; rather, they are managers and technicians who can cross layers. One respondent in the current study stated this metaphorically:

> Mead/Conway [a classic text on VLSI design] leads you to the idea that layout and design can be done by tall thin people, that is, people who can span the production process all the way from architecture all the way to silicon compilation. [By this he meant that the texts implied that the coordi-

> nation between layers was taken for granted; the ability to probe between layers, to understand the necessary technical skills at different layers of organization, was taken for granted but not realistic—none of us are that tall or that thin, he went on to say]
>
> Mead and Conway recommended spanning levels—the practice of it works out quite differently. Human beings can only deal with a certain amount of detail. They tend to black-box technology at all different levels. But sometimes they have to go inside the black boxes . . . so they invent magic with delays. That is, they attribute magic to the things they don't know about, but then things start slowing down, and it becomes magic with delays! (7/10/87)

In both the design of chips and in mapping the brain, we see the generation of "tall thin scientists" who can span two or more layers. These are what I refer to as "wizards": that is, they are both repositories of local knowledge about the social and technical situations, and simultaneously, they know enough of more than one layer to perform rare cross-layering coordination. By definition, this work is "interdisciplinary," and it is also necessary to the smooth production of the chip with all its layers.

There is always a trade-off between the division of expertise between layers and the representation of the whole, whatever that might be. This tension has been noted in the creation of geological maps. Rudwick (1976) and Bowker (1988) note that the visual representations created by geologists attempting to map layers of the earth are fraught with interdisciplinary controversy, conflict about degree of abstraction (which Rudwick calls a tension between "theory" and performance), and the eventual creation of abstract maps which must be locally interpreted by skilled "old hands." In the case of Schlumberger, analyzed by Bowker (1987, 1994), this process of negotiation actually took place in court as the courts attempted to define the nature of the intellectual property or knowledge embodied in the layered and abstracted representations.

Despite all attempts to capture and rationalize the knowledge of wizards and the wisdom of the group, much of it still remains elusive—a noted problem for designers of expert systems. Another respondent in the VLSI CAD engineers study says: "The more automated something is, the more sophisticated it is when something breaks. It's just hubris to think that you can design something not to break. The best case is for something to leave a broad enough audit trail so you can trace it easily" (7/10/87:5). And later:

> There are so many contradictory pressures and design constraints on a design that a novice couldn't possibly tell what was going to happen . . . and in fact they'd probably never attempt a design if they considered everything at the beginning, it would just be overwhelming.
>
> One example is design rules. Eventually, you develop some sense of which are critical, and which can be broken when. Design rules are made to be broken. If you have been, for example, through a fab facility, and you know you can collar somebody you've met before and say, hey, is this really going to work. . . . Several things cause design rules: the most important I can think of is mask shift. You develop tricks, as printers—you put lighter color on first, and then put the darker color second and define the edge with the darker color. A lot of life is learning about tolerancing (EM,7/10:13/87.)

The "causes" of design rules in this case are embedded in both human memory and the constraints of the technology. Wizardry means, as one person who makes design tools said, "walking the corridors to find a solution."

Curtis, Krasner, and Iscoe (1988:1271) talk about the relationship between what they call "gurus" and the tension between knowledge of the unique application domain and the formal specification of the software system being designed:

> On about one-third of the projects we studied, one of these individuals had remarkable control over project direction and outcome, and in some cases was described by others as the person who "saved" the system. Since their superior application domain knowledge contrasted with that of their development colleagues, truly exceptional designers stood out in this study, as they have elsewhere, as a scarce project resource. Thus, the unevenness with which application-specific knowledge was spread across project personnel was a major contributor to the phenomena of project gurus.

In large, high-tech development projects, information about changing contingencies, work-arounds, and other technical information is literally brought from site to site by people on roller skates. This is a very efficient way to get around a production site that may be as large as several hundred acres! There is also a metaphorical roller-skating, practiced especially by managers and secretaries in organizations, who are the ones responsible for "skating" between informational or sentimental gaps in large organizations.

The danger with wizards and gurus is that the information and skills they acquire become inaccessible or privatized, including information about the safety of critical technical systems.

Attempting to Model "Tacit Knowledge." Because wizardry and its attendant activities, roller-skating and articulation work, are clearly what makes projects go, they are also important for those who want to formalize and automate work processes. Harry Collins (1990) has talked about the designers of expert systems who metaphorically try to "strain the soup of expertise" and capture the matzo balls without getting the broth, or local/tacit knowledge.

Cambrosio and Keating (1988) have provided a good critique of the reification and misuses by sociologists of the concept of tacit knowledge, and note that most of this knowledge isn't really tacit, but rather codified in local practices and communication. As such, it is subject to negotiation, revision, and argument. If we think of it as out of the control or conscious attention of scientists, we mystify it and black box it out of our own reach. Kären Wieckert (1988) has documented numerous attempts to capture the knowledge of wizards in the design of expert systems. These attempts seem to reify certain parts of the expert's knowledge, but because they are devoid of organizational context, miss the essence of the expertise: the ability to improvise, to develop ad hoc solutions contextually. Similarly, Diana Forsythe (1992, 1993) has spoken of the reductionist epistemologies employed by expert knowledge engineers in imaging a knowledge devoid of situation.

The danger with attempting to capture tacit knowledge, especially in attempts to automate critical systems, is that the flexibility and smartness that comes from situated action (Lave 1988a; Suchman 1988) may be lost, in exactly the position where it is most needed. As systems become faster and design more automated, we also may have come to depend inappropriately on the roller-skaters, literal and metaphorical—only to find out that no one can skate as fast as technology is demanding. Weick, for example, argues that our current safety relies on the heterogeneity of organizations and the redundancy they provide—precisely through the ingenuity of work-arounds. He calls for "a better match between system complexity and human complexity" (1987: 112) and goes on to say:

> As team members become more alike, their pooled observations cannot be distinguished from their individual observations, which means collectively they know little more about a problem than they know individually . . . a homogeneous team does little to offset these limits. This line of argument, which suggests that collective diversity increases requisite variety which in turn improves reliability, may conflict with the common prescription that redundancy and parallel systems are an important source of reliability. That prescription is certainly true. But a redundant system is also a homogeneous system and homogeneous systems often have less variety than the environments they are trying to manage and less variety than heterogeneous systems that try to manage these same environments (p. 116).

Again, this is the issue of due process.

Denigrating the Social As Impure. Peculiar though it might seem, another reaction to the formal-empirical tension is the denigration of the social as impure. In my first day of a field study some years ago in a robotics laboratory, my respondent said to me: "A sociologist! It's too bad you weren't here Sunday!" "Why not ?" I asked. "It was the company picnic. There was so much sociology going on there!" he replied. This was not meant humorously.

In many engineering and scientific workplaces, political and organizational issues are put in a category separate from technology. They are seen as tainted. John Law (1988b: 1) writes of this in a discussion of the *Challenger* space shuttle disaster:

> And when the outline of a technical answer started to appear, then the questioning shifted. . . . The questions proliferated, and rapidly moved beyond the technical into the social, the economic, the political and the organisational. And fingers were pointed at greedy subcontractors, penny-pinching politicians, pressures from rival agencies. The technology had, it was suggested, been subverted from the earliest stages by such pressures.

Diane Vaughan (1989:32) writes of the *Challenger* disaster in similar terms, pointing to organizational constraints on individuals that pull them in unexpected ways. She concludes:

> The dynamics of organizational behavior often are subtle, and although it remains to be proved, we may be better at identifying, understanding, and fixing technical problems than organizational problems. Both preventive strategies and after-the-fact attempts to fix things are handicapped by the rudimentary state of knowledge on the causes of organizational deviance as well as by unsophisticated capabilities of converting theory and research into diagnostic skills in specific cases.

Here, a possible reason for the impurity stories is revealed: the lack of basic knowledge of such complex organizations, and the need for good communication across disciplinary lines.

The danger that comes from denigration of the "social" is that acquiring systematic knowledge of organizational behavior becomes nearly impossible in a situation where "the social" is at once reified and accorded lower status. Furthermore, the very people who have expertise at studying organizations will be alienated from working within engineering culture, and will find it impossible to obtain the necessary resources to do research, thus forming a self-exemplifying catch-22.

Summary of Part 2

I have argued here that organizations attempt strategies that will try to create organizational consistency in the face of strong tensions between formal representations and empirical experience. The tensions arise from the fact that the ad hoc strategies, work-arounds, and local knowledge that keep organizations going are first deleted from formal representations. When the formalizations become recipes for action, then further ad hoc work-arounds are necessary to make the prescription fit the local circumstances. This can be an infinitely recursive process, since at the top level, no one can formalize the formalizers. Where does the mess go? It doesn't go anywhere; rather, it is the formal representations that attempt to leave it behind.

PART 3: ETHICS, DISASTERS, AND FORMALISMS

As Perrow (1984) has cogently argued, we live now in a precarious world populated with risky technologies and a way of organizing work that amplifies their danger. The role of formalisms in that danger is central, for all the reasons outlined above, and for some others discussed below. Formal representations, including especially those used in creating and using information technologies, are changing the space of our world in ways we are only beginning to intuit. While priorities have always been arranged and rearranged politically, certain aspects of information technologies change the politics. These aspects include the fact that highly formalized representations seem neutral or objective to many, if not most, people—and the political decisions made in creating them invisible; that there is an illusion of completeness of information because of information technologies, whereas in fact information is highly decentralized and always incomplete; that we may delegate moral decisions or responsibilities to technology in ways that are blind. In the following sections, I discuss the general notions of artifacts with politics, then apply it to formalisms. I conclude with a discussion of the moral and ethical implications of delegation of morality to technology.

Do Formal Artifacts Have Politics?

At one level it is easy to think of artifacts with politics: weapons systems, surveillance technology, sex selection or other reproductive technologies. The more formal the representations in the technical system, however, the more difficult it is to understand their politics. Langdon Winner (1985) has written persuasively of the ways in which technical artifacts may embed political decisions within them. He uses the example of the design of highway systems between New York City and Long Island, deliberately constructed to exclude buses. The ramifications were that buses could not go out to Long Island—and neither could people who rode them, presumably of a lower social class than the Long Island residents. This is one degree of inference more than that required by the immediacy of, say, the atomic bomb. These are artifacts that are political in a second-order way. They are not overtly harming poor people. Rather, to see the politics, you have to know that buses are a certain height; that Robert Moses had certain political positions; that there is a certain socioeconomic status on Long Island.

Taking Winner's argument into the realm of formal artifacts requires a yet longer chain of inference. For example, to understand that statistics have values embedded in them you must know something of turn-of-the-century eugenics; of the rise of quantitative measures in human sciences and public policy; you must be able to understand their original purpose as the equivalent of the bridge meant to snag buses (MacKenzie 1981; Kevles 1985; Hornstein 1988). They represent a mathematics of difference, of separation, of distinction. But until you unravel their origins and their uses, they just look like a bunch of numbers—as the Long Island bridges just look like bridges.

A politics of formalism rests on several things. First, a formal representation is an abstraction: It takes away properties from a particular situation. Second, it is a simplification: It reduces the complexity of real life situations in order to make them formally (usually, but not exclusively, mathematically) tractable. Third, and most important, every formal representation contains choices about what to keep in (what is important) and what to throw out. All such choices are political (Bowers, in press).

The question becomes more complicated, however, when formalisms are transported from one place to another. Jean Lave (1988a, 1988b) and Rogers Hall (1989, 1990) have noted, for example, that students faced with a math testing situation do two things. First, they solve the problem in a situated fashion: drawing on familiar objects, places, and histories, they manufacture creative units of analysis, quantities, metaphors, and solutions. When they report the results of their work, however, they must contort those solutions to the canons of traditional algebraic language—often at a great cost. So, for example, teenagers at a bowling alley can keep perfect scores, but fail miserably on standardized math exams. Consider the teenagers at the bowling alley as the subjects and objects of science and technology. What is the relationship between scientific formalisms and their users?

Ethics, Formal Representations, and Organizational Complexity

Why is it important to understand the relationship between work organization and the technology that is produced? If Latour, Law, and Callon are correct in their analyses of the relationship between representations and long-distance control, then we cannot develop a robust theory of ethics and values to deal with the distribution of power until we understand how representations are controlled and distributed.

Organizationally consistent, layered representations mean that decisions are broken up into pieces; they occur at arm's length across components or sites. No matter what our personal system of ethics, there is no such thing really as a personal decision when it comes to large artifacts. Who will be in charge of the functioning of the satellite I just described? For legal purposes, it will be the "responsible engineering authority"—but he or she is shielded in responsibility by the corporate face. Individual engineers or doctors or scientists have some kinds of choices that are personal: Yes, I will work on this or No, I won't. But as the electronic world becomes ever more densely interconnected, even the sense of "to work on" shifts rapidly. Most of what happens in very large projects is a result of an ecological shift between the poles and tradeoffs I have listed above. It is not the case that wizards, by holding the "glue" information in a situation, are more responsible. It is not the case that a technician working on a single component or layer is less so. Bowers (in press) argues that such questions are particularly important for the very large systems now being designed to coordinate work across space and time CSCW (Computer-Supported Cooperative Work, or systems); I would add that all large technological systems of representation have these properties.

The image of the "banality of evil" described by Hannah Arendt (1963) is omnipresent in engineering and computer science. When the division of labor is so complex, and its management so obscured by our conceptions of bureaucracy, who can be responsible? Is responsibility fungible—in the case of the space shuttle, for example, was each software programmer and engineer one-millionth responsible for seven deaths? Such statements make no sense, but they do define one pole of engineering ethics.

Philosopher John Ladd (1970; 1986) has approached the problem of the ethics of information technologies and bureaucracy with a set of lucid observations about the nature of corporations and of morality. When we anthropomorphize corporations in order to imbue them with morality, he claims, we are in fact mystifying the actions of corporations and bureaucracies. They are not individuals, and they are not moral collectivities in the sense of, say, families, political action groups, or churches. Attempting to socialize them or apply moral suasion as toward a person is destructive; at best it devolves into moral chaos.

Furthermore, Ladd states, ethical responsibility and decision making always imply that we have full knowledge of what we are deciding. The distributed and complex nature of modern information systems makes completeness an obsolete notion. Adding the analysis presented in this paper, the invisible work jettisoned in the process of making and using formal representations is substrate for such decisions—and we have little information about it.

The work represented in this chapter, and in the related work of Ladd, Vaughan, and Law, is trying to find a different way to approach the question of ethics in designing very large-scale systems or in other complex organizational endeavors. By focusing not on individual decisions, nor on reified corporate pseudo-actors, we can hopefully find new units of analysis for ethics. Having systems of trade-offs, embedded in interlocking representational regimes, mean that it is the networks of people and things that are the proper locus for analysis.

We cannot find out how engineering values operate at different layers by "reverse engineering" the chip back into the workplace. To do so, there would have to be a direct mapping from the geometry of a piece of technology back to the division of labor of the design process. Given the shifting, changing, and partially obscured nature of layered, formal representations, this would be fatuous. But we can begin by refusing one final type of "great divide," which is between science and technology on the one hand, and the social on the other. Perhaps by focusing on the continuity of representations with design, we can develop new ways of thinking about risk, responsibility, ethics, and trade-offs.

SUMMARY: SPACE AND THE DELEGATION OF VALUES

Things that are far away as I have defined space in the first paragraph of Part 1 include those things that are invisible. When we accept formalisms without being able to trace their genesis or impacts, we effectively distance ourselves from knowledge. We also, by letting go of information that may be relevant to human safety and lives, distance ourselves morally. Wizards, tacit knowledge, layered and precise representations, and creating great divides in which the social is derided are very dangerous practices.

Everett Hughes provided the classic discussion of the delegation of values in his essay, "Good People and Dirty Work" (1971b: 87–97). The essay was written after a post-World War II visit to Germany. It asks the question that motivated so much of social science in the 1950s: How could a whole nation "go mad"? Were Germans worse than other people? Were they evil? Hughes poses the answer as one of delegation and enforced silence, as a matter of the functioning of in-groups and out-groups. To the extent that we delegate dirty work—such as running prisons, conducting warfare and dangerous technical experiments—and dissociate ourselves from the direct process, to that extent are we participants in creating a cult for whose powers we declare ourselves not accountable. In Hughes's words:

> We are dealing with a phenomenon common in all societies. Almost every group which has a specialized social function to perform is in some measure a secret society, with a body of rules developed and enforced by the members and with some power to save its members from outside punishment. And here is one of the paradoxes of social order. A society without smaller, rule-making and disciplining powers would be no society at all. There would be nothing but law and police (p. 97).

To the extent that we are delegating moral and ethical decisions to computer systems (especially distributed systems), we enhance the creation of cults which can operate without checks and balances. What exactly are we delegating, and to whom or what or where? Surely there are few more pressing questions about the place of knowledge.

Embedded in the question of delegation is the question of due process mentioned above. The simultaneous existence of multiple viewpoints and the need for solutions that are coherent across divergent viewpoints are driving considerations here. The term is borrowed from a legal phrase that refers to collecting evidence and following fair trial procedures. The due process problem in either a computer or human organization is this: In combining evidence from different viewpoints (or heterogeneous "nodes"), or in collecting evidence from different viewpoints, how do you decide that sufficient, reliable, and fair amounts of evidence have been collected? Who, or what, does the reconciling, and according to what set of rules? In order to model the delegation process, we must understand the different viewpoints, the insiders and outsiders, in a decision-making process.

To the extent that intelligence is seen as individual, those creating artificial intelligence systems will be vulnerable to the kind of secret cult "dirty work" Hughes is warning against. Given the tight coupling of artificial intelligence research with the military research budget, this is indeed a serious consideration. Kären Wieckert (1988) has noted that although expert systems designers have not successfully modelled human intelligence, they have in fact been able to create tacit islands of power in the organizations using the systems.

If we begin to rely on "smart," distributed systems without attending to questions of due process, we will tacitly instantiate and delegate the values of scientists and engineers—whatever they may be. Furthermore, we will be doing this blindly (in the same way that Ladd talks about moral anarchy). We have no understanding of ethical or organizational tolerances in the way that we have understanding of engineering tolerances for making things. Nor do we understand the interaction of moral and material tolerances. It seems to me that investigating how we are already delegating morality, plus responsible planning for the future, is crucial.

Implications of a Changed Relationship with Organizations: Redefining Space

What positive steps can we take to redefine the moral space described here? The following list is meant as a partial programmatic list, and I invite interested researchers to add to it:

1. Investigate the nature of distributed memory across organizations and computers.
2. Analyze the nature of objects or theories that inhabit multiple contexts of use or development.
3. Analyze the effects of drawing boundaries around information at different levels of organizational granularity.
4. Form the current tension and debate in science between formal and empirical methods into an object for analysis in both social science and other sciences.

NOTES

1. This work was supported by an award from the Fondation Fyssen and by a grant from the National Science Foundation, Division of Biological and Behavioral Sciences, Number EVS 83–62. This paper is a think piece about this set of relationships and priorities. The space it occupies is a journey with companions. I would like to thank Howard Becker, John Bowers, Geof Bowker, Michel Callon, Les Gasser, Joseph Goguen, Rogers Hall, Carl Hewitt, Gail Hornstein, Rob Kling, John Ladd, Bruno Latour, Jean Lave, John Law, Diane Vaughan, Sal Restivo, Anselm Strauss, Jeanne Pickering, Karen Ruhleder, and Kären Wieckert. May we continue to rush in where angels fear to tread. I am also grateful to my anonymous respondents for the generosity of their time.

2. Very Large Scale Integration using Computer Aided Design.

3. This concept was developed with E. M. Gerson.

4. Very complex systems of representations generate "shadow inscriptions," that is, inscription devices to track the ordering of representations themselves. Catalogues are different from lists in this way. Abstractions, or formalisms, order cascades that are too complex to track otherwise.

5. "Who will formalize the formalizers?" Thanks to Karen Ruhleder for Latin translation!

ROB KLING
C. SUZANNE IACONO

3

Computerization Movements and the Mobilization of Support for Computerization[1]

INTRODUCTION

There is a major mobilization for computerization in many institutional sectors in the United States. Computerization is a social process for providing access to and support for computer equipment to be used in activities such as teaching, accounting, writing, designing circuits, and so on. Computerization entails social choices about the levels of appropriate investment in and control over equipment and expertise, as well as choices of equipment. Many professionals and managers are adopting computing systems rather rapidly, while they often puzzle about ways to organize positive forms of social life around them. By the early 1990s, computing and telecommunications accounted for half of the capital investments made by private firms (Dunlop and Kling 1991, Kling, in press). However, the most fervent advocates of computerization have argued that the actual pace of computerization in schools, offices, factories, and homes is too slow (Papert 1980; Feigenbaum and McCorduck 1983; Yourdon 1986; also see Kaplan 1983).

Why is the United States rapidly computerizing? One common answer argues that computer-based technologies are adopted only because they are efficient economic substitutes for labor or other technologies (Simon 1977). Rapid computerization is simply a by-product of cost-effective computing technologies. A variant of this answer views computerization as an efficient tool through which monopoly capitalists control their suppliers and markets, and by which managers tighten their control over workers and the labor process (Braverman 1975; Mowshowitz 1976; Shaiken 1986).

A second answer focuses on major epochal social transformations and argues that the United States is shifting from a society where industrial activity dominates to one in which information processing dominates (Bell 1979). Computer-based technologies are simply "power tools" for "information workers" or "knowledge workers" as drill presses are power tools for machinists (Strassman 1985).

While each of these responses offers insight into computerization, we believe that they ignore some of the broadly noneconomic dimensions of computerization in industrialized countries. The market assumptions of these common answers have also shaped the majority of social studies of computerization (see Kling 1980, 1987 for a detailed review of the empirical studies of computerization). These studies focus on computerization in particular social settings that range in scale from small groups and workplaces (Shaiken 1986) through single organizations (Kling 1978a; Kling and Iacono 1984) to comparative multi-organizational studies (Laudon 1974, 1986). They usually ignore the ways that participants in the settings under study develop beliefs about what computing technologies are good for and how they should organize and use them (see, for example, Attewell and Rule 1984).

During the last fifteen years we have conducted systematic studies of computerization in diverse organizations: banks (Kling 1978b, 1983), engineering firms (Kling and Scacchi 1982), insurance companies (Kling and Scacchi 1982), manufacturing firms (Kling and Iacono 1984), public agencies (Kling 1978a), and schools (Kling 1983, Kling 1986). We have also been participant observers of four specific computerization movements: artificial intelligence (1966–1974), computer-based education (1974–1993), office automation (1975–1990), and personal computing (1983–1993). We have also learned about several computerization efforts at our home university as participants, first-hand observers, or coordinators of an assessment team.

Our research and participant experiences have taught us that the adoption, acquisition, installation and operation of computer-based systems are often much more socially charged than the adoption and operation of other equipment, like telephone systems, photocopiers, air conditioners, or elevators. Participants are often highly mobilized to adopt and adapt to particular computing arrangements through collective activities. These collective activities take place both outside and within computerizing organizations, and they share important similarities with various other social, professional, intellectual, and scientific movements. Strongly committed advocates often drive computerization projects. They develop and encourage ideologies that interpret what computing is good for and how people in these projects should manage and organize access to computing. They usually import these ideologies from discourses about computerization external to the computerizing organization (Kling and Iacono 1984).

In this chapter we examine how specialized "computerization movements" advance computerization in ways that go beyond the effect of promotion by the industries that produce and sell computer-based technologies and services. Our main thesis is that computerization movements communicate key ideological beliefs about the favorable links between computerization and a preferred social order, which helps legitimate relatively high levels of computing investment for many potential adopters. These ideologies also set adopters' expectations about what they should use computing for and how they should organize access to it. In this chapter we focus our attention primarily upon the character of the computerization movements and their organizing ideologies.[2]

We have selected five specific computing technologies to examine as the focus of computerization movements (CMs): urban information systems, artificial intelligence, computer-based education, office automation, and personal computing. Collectively, these specific CMs, along with movements organized around other computing technologies, form a general computerization movement. We identify a core ideology that supports the general CM and examine groups whose worldviews balance the pursuit of the most advanced computing technologies with alternative social values. We argue that computerization is a process deeply embedded in social worlds that extend beyond the confines of any particular organization or setting. Finally, we examine the character of CMs.[3]

COMPUTERIZATION MOVEMENTS

Sociologists have used the concept "movement" to study many different kinds of collective phenomena. The most common term found in this literature is "social movement," often used in a generic way to refer to movements in general. But sociologists also have written about professional movements (Bucher and Strauss 1961), artistic movements, and scientific movements (Aronson 1984; Star 1989a). From the diverse ways sociologists have used the concept, we found Blumer's (1969:8) general definition most helpful for our analysis: Social movements are "collective enterprises to establish a new order of life." This is an inclusive definition that allows us to consider elements relevant to computerization that other, narrower conceptions would rule out. We also found the work of Zald and his colleagues (McCarthy and Zald 1978; Zald and Berger 1978) helpful in that it allows us to consider social forms that are both highly and loosely organized as parts of computerization movements.

Computerization movements (CMs) are a kind of movement whose advocates focus on computer-based systems as instruments to bring about a new social order. The mobilizing ideologies of computerization countermovements (CCMs) oppose certain modes of computerization which their advocates view as bringing about an inappropriate social order. We will examine five CMs, some CCMs, and their mobilizing ideologies in the next sections. Various scholars who write about movements emphasize the potential importance of activist entrepreneurs who also help drive movements through books, speeches, and other actions. Such activists, who write for broad national audiences but who do not belong to a particular movement organization, play an important part in the computerization movements we describe here.

"Movements" serve as theoretical constructs. Our descriptions that "people join movements," "people speak for movements," "movements hold a particular ideology," or "one movement is a wing of a second, general movement" and similar attributions are conveniently concise ways of describing much more complicated and varied collective actions.

We also find two distinctions made by Blumer (1969) specially helpful for our analysis. First, Blumer distinguished between "reform" and "revolutionary" movements. The ideologies of revolutionary movements emphasize changing key social relations throughout a social order, while reform movements focus on change of a restricted set of social relations. Blumer also distinguished between "specific" and "general" movements. Specific movements are the various wings or submovements of a broader, general movement. Many movements, like those that advance feminism, Eastern religions, civil rights, quantification in the sciences, or computerization, are heterogeneous. The distinction between specific movements and general movements helps us characterize the relationships between distinct wings of a larger movement.

One theme in our discussion of computerization and the specific movements that help produce it is the importance of seeing how local practices and concerns, in workplaces, schools, or corporations, are linked to these external developments. By distinguishing between a "general" CM and several "specific" CMs (Blumer 1969), we want to draw attention to how similar conceptions about modes of computerization found across many organizations or social settings should be understood.

TECHNOLOGICAL MOVEMENTS AND COMPUTERIZATION MOVEMENTS

Some movements focus on the promotion of particular technologies as central to a vision of a preferred social order, such as the nuclear power movement (Useem and Zald 1982) or the "appropriate technology movement." These technological movements (TMs) stand out as a special class because their mobilizing ideologies promote an improved social order through the use of a particular family of technologies. Other movements, such as the antinuclear movement (Downey 1986; Walsh 1981), oppose the use of a particular technology. Following McCarthy and Zald (1977), we call these latter movements "technology countermovements" (TCMs). The kind of claims made by TM and TCM spokespeople differ considerably. TM advocates usually claim that the new technologies will improve society in the future. Unless the advocate has a reliable crystal ball, these claims are, of course, speculative. In contrast TCM advocates claim that some current use of a technology causes environmental or social problems today. The ideological content of TCMs brings them closer than TMs to social movement participants who claim that their grievances are anchored in real social conditions (Spector and Kitsuse 1977; Schneider 1985; Troyer in press).

While McCarthy and Zald's (1977) emphasis on movement organizations is a helpful insight for studying TMs or TCMs, we do not limit ourselves to this conceptualization. They characterize movement organizations very narrowly, as organizations that identify their goals with a movement or countermovement and that attempt to implement those goals. Some of the key advocacy for particular technologies comes from broad professional organizations which have subgroups acting as movement organization. But another important set of participants is activists who write for broad national audiences but who do not belong to a particular movement organization. The extent to which movement organizations play major or minor roles in a particular movement is then an empirical question.

Computerization movements (CMs) are one kind of TM that focus on computer-based systems as the core technologies which their advocates claim will be instruments to bring about a new social order. The mobilizing ideologies of computerization countermovements (CCMs) oppose certain modes of computerization which their advocates view as bringing about an inappropriate social order. The five CMs we examine below are at different stages of development. Each has different participants who meet and communicate in distinct, but sometimes overlapping, social worlds. The participants in one CM may have little or no concern for the outcomes of another. However, the participants in one CM are often sympathetic to some other CMs.

CMs are a major source of support for computerization in the United States today. How society changes during this period of computerization does not depend only upon an industry's ability to manufacture and market products and on individual consumers buying and using them. Moreover, other actors besides equipment vendors and consumers mobilize enthusiasm for increased computing. In the case of computer-based education, for example, the mass media, parents, and teachers play critical roles in mobilizing support for the spread of computers in classrooms. Instructional computing researchers also stimulate interest in the potentials of new technologies (Taylor 1980). Local organizations such as school boards and PTAs discuss policies and the practical implications of obtaining resources to implement such concepts as "computer literacy" in the curriculum. Moreover, the ways in which the promises of computer-based education are characterized are similar in many schools. We find movements as a useful way to explain the social mobilization for technologies where the same ideologies and debates recur across diverse social settings, such as in different schools.

Seymour Papert (1979:80), an influential advocate of a special approach to computer-based education in the early 1980s, wrote: "In 1973 Christopher Jencks . . . argued . . . that schools do little to redress the inequality of life chances. Certainly he could find no evidence that the introduction of TV, movies, language labs, and other educational hardware made a significant difference. Nor did the innovative curricula of the 1960s. My argument is that powerful computers could have done so." And, Papert continues, "Dewey, Montessori, and Neill all propose to educate children in a spirit that I see as fundamentally correct but that fails in practice for lack of a technological base. The computer now provides it" (85).

Papert's advocacy of the LOGO programming language in his popular book, *Mindstorms*, illustrates how a computing advocate can define the capabilities, options, and consequences of a particular mode of computerization that most of his readers are unlikely to fully understand. Since neither a writer nor a book makes a movement, we have to examine how each CM is a collective activity. But similar conceptions about mode of computerization found across many organizations or social settings serve as signs that some local participants are also adopting ideas about computerization from outside sources.

Danziger (1977) illustrated the starting point for such an analysis when he identified a "litany to EDP" (Electronic Data Processing)—a set of over-idealized promises about the virtues of computerization—that he observed in his studies of urban information systems. But Danziger didn't link the ideologies he characterized to a CM or CM organizations that disseminated them. This is the next key step for such an analysis. We (1984) examined the way that some managers mobilized support for a complex inventory control system. They had members of their employing organization attend indoctrination sessions sponsored by a professional association. Kaplan (1983) examined the ways that arguments about a "lag" in the rate of adoption and sophistication of medical computing technologies permeated the professional medical computing literature between 1950 and 1980. Laudon's (1974) analysis of computerization in police and social service agencies linked those efforts to professional reform movements of the Progressive Era, a link he himself did not pursue in subsequent research (Laudon 1986).

Few social analyses of computing have examined the mobilizing ideologies that local advocates employ or suggest that local advocates develop approaches to computerization through participating in CMs that extend outside of their home organizations (see literature reviews by Kling 1980 and by Attewell and Rule 1984). People become aware of modes of computerization that they may not have personally experienced through the activities and byproducts of CMs: advocates, public speeches and written works, popular stories, television shows, and magazine articles. For example, *Time* magazine became an agent for a CM when it named the computer the "Machine of the Year" for 1982 ("Machine of the Year" 1982). It helped stimulate interest in computing, rather than simply reporting on the promises, experiences, and problems of computing.

The collaboration of participants with diverse interests has mobilized CMs. Some CM participants represent themselves, while others represent their organizations. For example, computer scientists and researchers develop careers as "experts" in a certain computer-based technology and become associated with specific CMs. Other participants, such as consultants and computer users, may have temporary job assignments that attach them to a particular computer-based technology.

Some participants or groups may belong to only one or two CMs. For instance, parents and teachers may participate in the computer-based education CM; materials specialists who are interested in automated inventory control systems may participate in a special "Material Requirements Planning movement" (Kling and Iacono 1984); urban planners interested in quantitative simulations may participate in the urban information systems movement. Participants in each of these CMs may care little about other modes of computerization. Journalists and news reporters, on the other hand, have become central to the mobilization of computing in general as collaborators providing public exposure to

CM advocates. Public forums, trade shows, school board meetings, and similar events enhance interactions among participant groups.

Both specific CMs and the general CM have grown as a result of these interactions. The most active participants in a CM advocate computerization in addition to using computer-based systems. In this paper we pay most attention to a small but influential group of CM participants, those activists who lobby others to computerize. The most active CM participants try to persuade mass audiences and whole professions to computerize in a particular way. CM activists may not identify with the computer industry, as was true for the first personal computer hobbyists. By the mid-1980s, CM activists could be found in many institutional sectors of North American economies. Socially they were predominantly middle- and upper-class professionals.

The main alternative to CMs as explanations of sources of beliefs about computerization and support for adopting new technologies is an "industrial model" which focuses upon the actions of manufacturers and vendors in the computer industry to stimulate sales through advertising and other means. Consumers (including computer-using organizations) buy and use computer hardware and software. Computerization is seen as simply the by-product of collective actions in markets. The Marxist variant of this market explanation focuses upon the owners and managers of private firms as key agents who adopt systems to increase their profits and control over workers (Noble 1985). In the next section we illustrate how five CMs are organized such that they are not merely subsumed within a market. While vendors and various computer users clearly play important roles in stimulating interest in computerization and in shaping the technologies in use, we believe other forms of collective activity—what we call computerization movements—also shape expectations and stimulate demand that market analyses ignore.

FIVE SPECIFIC COMPUTERIZATION MOVEMENTS

Below we describe five specific computerization movements. These, along with other CMs, form a general CM.[4] We briefly indicate how the collective activities of each CM include the development of a mobilizing ideology and organized activity to promote it.[5] We examine key ideological elements of these CMs in the next section.

Urban Information Systems

Urban information systems process data about the activities of people in local jurisdictions, the government services they receive, and the internal operations of local governments. These include systems to support tax collection, police operations, municipal libraries, and urban planning. Support for urban information systems was continuous with progressive reform movements that sought to professionalize local government administration in the early part of the century (Laudon 1974). Local governments that have pursued these reforms are disproportionately likely to adopt computer-based applications (Danziger et al., 1982). The urban information systems CM is a relatively low-profile movement that forms one segment of a larger professional reform movement to "take politics out of government operations." Its adherents conceive of computerization as fostering government by skilled professionals instead of political appointees and bringing the purported rationality and efficiency of business to governmental operations.

Urban information systems activists have employed associations for local government officials and professionals as movement organizations. Professional organizations for tax assessors, finance officers, and other administrators have been strong supporters of computerization within their areas of expertise. Urban planners, social service professionals, and computer specialists have also found organized support for their computer interests within the Urban and Regional Information Systems Association, which has held annual national conferences since the 1960s. Other professional associations, such as the International City Managers Association, provide information and staff support to foster automation in local governments.

During the 1960s the federal government stimulated the growth of urban information systems through a wide variety of grants: the U.S. Department of Housing and Urban Development supported planning and social services (Kling 1978) and several massive demonstration projects (Kraemer and King 1978); the Department of Transportation supported transportation planning; the Law Enforcement Assistance Administration supported police systems. Though federal funding for specific urban projects has been substantially reduced since the late 1960s, urban governments now support their own information systems which have become embedded in the operations of many departments, especially tax collection, finance, police, welfare, and planning (Danziger et al., 1982). The urban information systems movement is highly institutionalized today.

Artificial Intelligence

The belief that computer-based technologies can be programmed to "think" about complex cognitive tasks is the central tenet held by enthusiasts of artificial intelligence (Turkle 1984). Early conceptions of artificial intelligence (AI) were framed within an abstract scientific discourse, emphasizing the formal modeling of different domains of knowledge or the study and simulation of human cognitive processes. In the early 1980s, some AI enthusiasts replaced abstract models of universal cognitive processes with the social construct of "expert systems" (Feigenbaum and McCorduck 1983).

In one of the most enthusiastic accounts, Feigenbaum and McCorduck (1983) assert that almost no meaningful intellectual work will be possible in "the world of our children" that does not depend upon knowledge-based (e.g., AI) systems. This idea, a controversial one within academic computer science, argues for a vision of the world its proponents hope others will help them construct. Public proselytizing for AI, so conceived, has transformed it from a technological-scientific movement into a CM.

AI spread as a scientific movement within the computer science profession itself. In the 1960s and 1970s interest in AI was focused around the *Artificial Intelligence Journal* and the triennial International Joint AI Conference, originally organized by the secret Artificial Intelligence Council. By the mid-1980s, other organizations were established, including a "special interest group" on Artificial Intelligence (SIGART) within the Association for Computing Machinery and the American Association of Artificial Intelligence (with its monthly publication, *The AI Magazine*).

Large-scale AI research began in the early 1960s under sponsorship of the Advanced Research Projects Agency of the Department of Defense, although other military agencies and the National Science Foundation also supported some of this initial research. This funding was used to create major laboratories at a handful of universities and research institutes. In the late 1970s AI became commercialized as established industrial firms such as Schlumberger and General Motors made substantial investments (Winston and Prendergast 1984). AI specialty firms emerged along with expert systems and practical natural language processing. In the 1980s AI was transformed into a CM as the mainstream business magazines (e.g., *Fortune, Business Week*) began to aggressively promote a fantasy definition of powerful and accessible AI being "here today."

Media accounts in the United States usually describe the value of current technologies in terms of their potential capabilities. They often exaggerate their level of development, the quality of product engineering, and the extent to which they have been adopted. They have also exaggerated the attributes of the few commercial applications that do exist by suggesting that these systems "think" like people rather than simply performing the symbolic manipulations on formal representations of the tasks usually performed by people. The American press often promotes AI in sensationalist terms (cf. "Artificial Intelligence Is Here," 1984; Applegate and Day 1984). A representative example appeared on the cover of the February 15, 1988 issue of *Insight on the News*. It heralded its cover story about AI with: "See machines think. Think, machine, think. The world of artificial intelligence." More accurate stories sometimes appear in magazines and newspapers, but they are less common than the sensationalized stories. Despite sensational claims in the press, none of the higher performance AI systems, with the exception of a particular checkers-playing program, has "learned" to modify its behavior, let alone learn "the way that humans do."

Computer-Based Education

Computer-based education includes both computer-assisted instruction programs that interact with students in a dialogue and a broader array of educational computer applications such as simulations or instruction in computer programming (see, for example, Taylor 1980). There is a major national push for extended application of computer-based education at all educational levels. For example, in the mid-1980s several private colleges and universities required all of their freshmen students to buy a specific kind of microcomputer, and others invested heavily in visions of a "wired campus." There was also a major push to establish computer literacy and computer science as required topics in the nation's elementary and secondary schools.

Computer-based education has been promoted with two different underlying ideologies in primary and secondary schools (Kling 1983). Some educators argue that computer-based instructional approaches can help fulfill the traditional values of progressive education: the stimulation of intellectual curiosity, initiative, and democratic experiences. For example, Cyert (1984) has argued that computerized universities are qualitatively different than traditional universities: college students with microcomputers in their dorm rooms will be more stimulated to learn because they will have easy access to instructional materials and more interesting problems to solve. Papert (1979) argues that in a new computer-based school culture, children will no longer simply be taught mathematics; rather, they will learn to be mathematicians. These visions portray an enchanted social order transformed by advanced computing technologies. Other advocates are a bit less romantic, but not less enthusiastic. For example, Cole (1972:143) argues:

> Because of . . . the insatiable desire of students for more and more information at a higher level of complexity and more sophisticated level of utilization . . . more effective means of communication must be used . . . computers can provide a unique vehicle for this transmission."

Others emphasize a labor market pragmatism that we label "vocational matching"(Kling 1983). In this view, people will need computer skills, such as programming, to compete in future labor markets and to participate in a highly automated society; a responsible school will teach some of these skills today. Advocates of computer-based education promote a utopian image of computer-using schools as places where students learn in a cheerful, cooperative setting and where all teachers can be supportive, enthusiastic mentors (Kling 1986).

The computer-based education movement is not a well-organized national movement. It is far more diffuse and localized than other CMs. In some regions, such as the San Francisco Bay area, consortia of teachers who are interested in computer-based instruction have formed local movement organizations. At the university level, the Apple Computer Inc. has formed a consortium of faculty from schools that have adopted Apple computers on a major scale. While this movement organization is linked to a particular vendor, its participants advance a more general vision of computer-based education. Some magazines, such as *The Computer Teacher*, also promote the movement. Regional conferences, usually hosted by schools of education in state universities, are held for school teachers and administrators interested in computer-based education.

Academic researchers have been developing instructional courseware for primary and secondary schools since the 1960s. However, such products only became viable when computing equipment costs declined with the advent of microcomputers. By the 1980s, the promotion of computer-based education surged substantially. On one hand, there was growing public belief that public schools were having chronic problems in educating children. On the other, the personal computer industry was reducing the prices of basic equipment and promoting educational applications. Steve Jobs, cofounder of Apple, lobbied hard and visibly to receive tax advantages by giving a free Apple IIe microcomputer to each public school district.

One microcomputer for a school of five-hundred students has little educational value, but the industry's marketing efforts and press promotion stimulated public interest in educational computing. Popular literature stresses the capabilities of equipment, ignoring the lack of high-quality courseware and inadequate teacher training in computer use (Kling 1983). Many parents are concerned that their children be exposed to computers in school, though they know little about the details of computerized education.

The computer-based education movement received a symbolic boost in the spring of 1983 when the President's Commission on Excellence in Education released an urgent report which recommended that one semester of "computer science" be added to high school graduation requirements. This one report is simply a high-profile example of many local activities throughout the nation, much as the mobilization for the Equal Rights Amendment was one national-level activity that captured the sentiments of many local variants of the woman's movement. In the 1980s there was continuous ferment at state and local levels, with coalitions of administrators, teachers, and parents banding together to lobby for various computer-based education programs in the public schools.

Office Automation

In the 1950s and early 1960s, "office automation" was synonymous with the introduction of computer-based technologies in offices—batch information systems, in that period. By the mid-1980s, office automation (OA) became a diffuse term that usually connoted the use of text-oriented computer-based technologies in offices (Uhlig et al., 1979).

OA technologies have two different kinds of roots. Standalone word processors evolved from magnetic card and magnetic tape typewriters. The organizational side of these office technologies evolved from kinds of secretarial work and the administrative services departments that commonly controlled organization-wide secretarial pools. The second root is computer systems that tied together text processing and electronic mail with general purpose computing capabilities (e.g., workstations). These differences in technology become less significant since specialized word processors were marketed with a wider array of information-handling functions. However, the social worlds that support these technologies still differ, even though they overlap.

Visionaries of automated offices used the term "office of the future" (Uhlig et al., 1979) as one of their major rallying points. An "office of the future" could never be built, since there is no fixed future as a reference point for these technologies. However, a more prosaic conception lies beneath this vision: a terminal on every desk that provides text processing, mail, calendar and file handling, communications, and other computing capabilities with a flexible interface. The scenarios emphasize the deployment of equipment, while OA advocates portray social relations as cheerful, cooperative, relaxed, and efficient—better jobs in better environments (Giuliano 1982; Strassman 1985).

These scenarios gloss the realities of work life in a highly automated office. Some office automation has led to deskilling and highly-pressured jobs while work has been upskilled in other offices (Iacono and Kling 1987). More seriously, the clerical workforce is likely to retain jobs near the bottom of the American occupational structure in terms of pay, prestige, and control of working conditions (Kling 1990). This is true even if the content of clerical jobs requires vastly more complex computer-related skills because most clerical jobs specialize in less discretionary, delegated work (Iacono and Kling 1987). The changes in professional work in automated offices are less clear; but it is clear that professionals cannot count on working with ample resources in a cooperative, cheerful environment (Kling 1987). Computerization is a complex social and technical intervention into the operations of an organization. Conventional CM ideology emphasizes the power of new equipment and downplays the kinds of social choices that can allow powerful equipment to facilitate better jobs. When journalists criticize OA, it is often because the equipment does not deliver the miracles promised by the more enthusiastic advocates (Salerno 1985).

Several national professional organizations promote OA. Associations of professional administrators and computer specialists have expanded their activities to include OA within their domains while the more specialized Association of Information Systems Professionals has a direct interest in the diffusion of OA. A large number of trade magazines and several academic journals, such as *Office* and *ACM Transactions on Office Information Systems* support the OA movement. A strong office products industry also develops and promotes OA equipment. This industry includes major vendors of mainframes, microcomputers, and specialized office equipment. Since the late 1970s the American Federation of Information Processing Societies has sponsored an annual trade show and conference of academic and professional OA activists.

Personal Computing

Personal computers (PCs), like video recorders, became middle-class luxury appliances of the early 1980s and commonplace in the 1990s. The PC movement began in the early 1970s as groups of hobbyists built their own primitive computers (Levy 1984). Apple Computer and Tandy Corporation grew quickly in the late 1970s by providing two of the first microcomputers for which the purchaser did not have to solder parts and continually fiddle. The PC movement is one of the few CMs that has a distinctly "mass public" audience, although an elite audience of wealthy organizations that have also adopted and institutionalized PCs in some of their operations has developed as well.

Some of the early writings about PCs emphasized personal and social transformations that would accompany widespread PC use (Kay 1977; Osborne 1979). By the last 1980s, the mobilizing ideologies of the PC movement shifted to an instrumental pragmatism based on the capabilities of PCs for professionals. However, many PC enthusiasts believe that "almost everybody" should have a PC. For example, Jim Warren, the founder of a popular computer magazine and a series of immensely popular regional PC fairs, commented:

> I continue to feel that computers in the hands of the general public are crucial tools for positive social change. The only hope we have of regaining control over our society and our

> future is by extracting the information we need to make informed, competent decisions. And that's what computers do (quoted in Goodwin 1988:114).

While PC applications such as word processing, financial analysis, and project scheduling are useful for professionals, "home applications" such as checkbook balancing, recipe storage, and home inventory are perhaps too marginal to justify a PC on instrumental grounds. The assumption that "almost everybody" should have a PC reveals the ideology of the PC movement. The PC movement is national in scope and is popularized by high-circulation, national publications. By the mid-1980s, about twelve PC magazines were widely sold on newsstands in the United States. Many were specialized for audiences who own microcomputers made by particular vendors, such as Apple, Tandy, and IBM. Others were specialized for business applications or some particular kind of technology (e.g., programming languages, UNIX). One very popular multivendor magazine, *Byte*, circulated approximately 500,000 copies per monthly issue.

The main national-level activities are the trade fairs that travel from region to region. The software and hardware vendors in the microcomputer industry now play a major role in these ventures. The PC movement has numerous local movement organizations. The computer clubs in metropolitan areas are usually segmented along vendor lines (e.g., Apple, IBM, Atari) and are still dominated by hobbyists. In addition, computerized "bulletin boards" are operated by commercial firms and hobbyist amateurs. The operators of large commercial computer bulletin boards, such as Compuserve and Prodigy, serve businesses, professionals, and hobbyists nationwide. The operators of amateur-run bulletin boards are relatively transient and aim their services at computer hobbyists in their metropolitan areas. In the mid-1980s there were over two-hundred amateur-run bulletin boards in the United States, and the number mushroomed in the 1990s. The PC movement overlaps with the office automation movement and the instructional computing movement, since microcomputers have come to play a significant role as focal technologies for each. Some specialized organizations link these movements. For example, in the Los Angeles area, the PC Professional Association holds monthly meetings for PC coordinators who work in business organizations.

The mass media has become a major promoter of the PC movement. Newspapers such as the *Los Angeles Times* run a weekly PC column. It is difficult to separate the effects of journalistic promotion from industry advertising. As the PC industry has grown, firms like IBM, Apple, and Commodore have placed advertisements on prime time television. The PC industry's advertising is significantly enhanced by CM promotion. In the case of the PC movement, the mass media have played a major role. It is relatively rare to find the mass media investigating PC use, yet, dissatisfaction with the service and support provided by major computer sales chains is commonplace among PC enthusiasts. Journalists are much more apt to write about poor service in auto sales than in home computer sales.

IDEOLOGICAL ELEMENTS OF CMS

We have found ideologies of computerization especially developed in two kinds of writings. Some accounts focus on the coming of an "information society" and treat computerization as one key element in that transformation. In this view, the computerization of America will be an apolitical, bloodless "revolution." All will gain, with the possible exception of a few million workers who will be temporarily displaced from "old-technology" jobs such as telephone operators and assembly line welders. Prophets of a new "information age," like Bell (1979), Toffler (1980), Dizard (1982), Naisbitt (1984) and Strassman (1985), argue that this transformation is an inevitable and straightforward social process. Human intention, pluralities of interest, and large-scale conflict play a minor role in these predictions of substantial social transformation.

A second kind of writing in which we found computerization ideologies focusing on specific CMs is of greater interest to us. Few CM activists, including those who publish their arguments, assert their key ideological themes directly. They can be located in books and articles through a "symptomatic reading" through which the reader identifies relevant themes that are absent from a document as well as those it, however implicitly, contains. For example, if an author argues that every household should have PCs for all family members, she is arguing in effect that families should restructure their budgets such that computing investments are not compromised.

We have identified five key and related ideas from such symptomatic reading. In addition, as students of and participants in several CMs, we have had continuing discussions with movement activists in which some of these ideas recur. As a system of beliefs, these five themes help advance computerization on many fronts.

1. Computer-based technologies are central for a reformed world.
2. Improved computer-based technologies can further reform society.
3. More computing is better than less, and there are no conceptual limits to the scope of appropriate computerization,
4. No one loses from computerization.
5. Uncooperative people are the main barriers to social reform through computing.

Computer-Based Technologies Are Central for a Reformed World

Many CM activists assert (or imply) that computer technology provides a historically unique opportunity for important social changes. For example, Papert (1979:74) advocates a special mode of computer-based education (e.g., LOGO programming) and strongly criticizes other modes of computer-based education as reinforcing "traditional educational structures and thus play[ing] a reactionary role." Papert clearly articulates the belief that computing is special and different from all other educational innovations of the 1960s in its potential to redress serious widespread social inequities. Papert's analysis is largely individualistic and focuses upon children's cognitive abilities. He makes no attempt to ask whether social inequities are tied to a structured economic system rather than simply to the distribution of skills in society.

Similarly simple assertions about the special role of computing can be found in literature for each of the CMs. In the case of urban information systems, Evans and Knisley (1970) argue that a wide array of pro-social services can be specially supported by urban information systems. Giuliano (1982) suggests that office automation provides a unique historical opportunity to organize offices without tedious clerical work. Hayes-Roth (1984) suggests that AI-based expert systems will radically transform the professions by providing a unique capability in taking over cognitive expertise previously monopolized by people. In each case, these analysts imply that the computer-based technologies they advocate have a historically powerful and unique role.

CM activists often argue that computers are a central medium for creating the world they prefer. This belief gives proposals for computerization a peculiarly technocentric character: Computer technologies are central to all socially valuable behavior. This belief is exemplified in characterizations of computer-based education in which any meaningful learning is computer mediated (e.g., Papert 1980).

One variant of this argument holds that computing is essential for modern organizations to compete effectively through increased productivity. Productivity, an economic conception, is linked to social progress through economic advance, and computerization, like prayer, will always have this desired result if it is "done properly." In addition, advocates attribute "productivity gains" primarily to new computing technologies, even though other elements, such as work organization and reward systems, play a critical role.

Concerns for productivity are closely linked to concerns for reducing costs. These concerns, and the relatively high cost of "state-of-the-art" computing which helps set expectations, give the practice of computerization a relatively conservative political character. New systems of socially significant scale require the approval of higher level managers who are unlikely to approve arrangements that threaten their own interests (Danziger et al., 1982).

Improved Computer-Based Technologies Can Further Reform Society

Advocates often portray the routine use of ordinary computer equipment as insufficient to reap the hoped-for benefits. Rather, the most advanced equipment is essential. This ideological theme is most clear when CM activists emphasize the continual acquisition of advanced equipment. They give minor attention to how to organize access to it or how to use it for social change. Nationally, the belief is reflected in funding priorities for computing research and development: the overwhelming support is for the development of new equipment. In the organizations we have studied, we have found relatively little money and attention spent on learning how to humanely integrate new computer-based technologies into routine social life.

CM activists often define computing capabilities as those of future technologies, not the limits of presently available technologies. Many accounts promoting computing reflect this future orientation. For example, Kay (1977) advocates book-sized personal computers that handle graphics, play music, and store vast amounts of data. Hiltz and Turoff (1978) advocate nationwide computer conferencing systems that would connect every household and office, much like telephones. During the last twenty years, computer-based technologies have become substantially cheaper, faster, smaller, and more flexible. Doubtless, computer-based systems will improve technologically during the next two decades similar to the ways in which cars, airplanes, typewriters, and telephones improved between 1910 and 1980. However, CM activists usually dismiss contemporary technologies, except to the extent they foreshadow more interesting future technologies.

In any year, only a few organizations can purchase the state-of-the-art equipment that CM activists recommend. With rapid changes in technological capabilities, today's technological leaders are surpassed by tomorrow's, unless they recycle their equipment so rapidly that they never "fall behind." Like people who purchase a new car or stereo with every model change, the heroes of this vision invest heavily and endlessly.

Computer hardware has become faster, cheaper, larger in scale, and more reliable in the last thirty years. Software support systems (like programming languages) have become more powerful and flexible, albeit at a much slower pace. Finally, computer applications have also improved technologically, at a still slower rate. But a focus on future technologies helps deflect attention from the problems of using today's technologies effectively, while offering the hope of salvation soon.

More Computing Is Better Than Less, and There Are No Conceptual Limits to the Scope of Appropriate Computerization

This theme goes beyond the previous theme that "more computing can help reform society." CM activists usually push hard on two fronts: people and organizations ought to use state-of-the-art computing equipment, and state-of-the-art computing should become universal. In their writing and talks, CM activists usually emphasize the claim that certain groups use the forms of computing they prefer, rather than explaining carefully how people alter their social lives to use or accommodate to new technologies. For example, it is common for enthusiasts of computer-based education in schools to report that there are now several hundred thousand microcomputers in use, but to spend relatively little time examining how children actually partake in computer-oriented classes.

CM activists imply that there are no limits to meaningful computing by downplaying the limits of the relevant technologies and failing to balance computing activity against competing social values. They portray computing technologies as mediating the most meaningful activities: the only real learning, the most important communications, or the most meaningful work, whether now or in the future they prefer. Other media for learning, communicating, or working are treated as less important. Real life is life online (e.g., Feigenbaum and McCorduck 1983; Papert 1980; Hiltz and Turoff 1978).

No One Loses from Computerization

Computer-based technologies are portrayed as inherently apolitical. While they are said to be consistent with any social order, CM advocates usually portray their use in a cheerful, cooperative, flexible, individualistic, and efficient world. This allows computer-based technologies to be shown as consistent with the most cherished social values. Computerization can enable long-term societal goals such as a stronger economy and military for the nation. Any short-term sacrifices that might accompany these goals, such as displaced workers, are portrayed as minor unavoidable consequences.

CM activists rarely acknowledge that systematic conflicts might follow from computerizing major social institutions. Those that are acknowledged are defined as solvable by "rational discourse" and appropriate communication technologies (see, for example, Hiltz and Turoff 1978). Many activists ignore social conflicts in their discussions of computerization and thus imply that computerization will reduce them. Some authors explicitly claim that computerized organizations will be less authoritarian and more cooperative than their less automated counterparts (Simon 1977). In most of the accounts of office automation, staff are cheerfully efficient and conflicts are minor (see, for example, Strassman 1985; Guiliano 1982).

Similarly, in the literature on computer-based education, cheerful students and teachers who are invariably helpful and understanding populate computerized classrooms (see, for example, Taylor 1980). Spitballs, paper planes, and secret paper messages that pass under the desks do not appear in this literature (Kling 1983). Teachers who are puzzled by new technologies, concerned with maintaining order in their classrooms, or faced with broken equipment, competitive students, and condescending consultants are also ignored in the published discourse. Occasionally, an advocate of one mode of computer-based education may criticize advocates of others (e.g., Papert 1979) but these are only sectarian battles within a CM.

Most seriously, the theme that computing fosters cooperation and rationality allows CM activists to gloss deep social and value conflicts that social change can precipitate (Kling 1983; Kraemer et al., 1987). In practice, organizational participants can have major battles about what kind of computing equipment to acquire, how to organize access to it, and the standards to regulate its use (Kling and Scacchi 1982; Kling 1987).

Uncooperative People Are the Main Barriers to Social Reform through Computing

In many social settings we have found CM advocates arguing that poorly trained or undisciplined users undermine good technologies (Kling and Iacono 1984). Even when they are making their procedures even more complex or automate more exceptions and special contingencies, these advocates argue that the limitations of their coworkers, rather than problems in their strategy of automation, are the major impediments to nearly perfect computer-based systems. Computerization is difficult when compared with the imagery of easy use advanced by CM advocates. In short, people place "unnecessary" limits on the complexity of desirable computer-based technologies and must be properly trained and taught to reorganize their activities and institutions around the new technology.

These central themes of computerization movement ideology emphasize technological progress and deflect competing social values. They are a foundation for social visions that include the extensive use of relatively advanced computer systems. In this vision, computer users should actively seek and acquire "the best" computer technologies and adapt to new technologies that become available, regardless of their cost. In this moral order, the users of the most advanced technologies are the most virtuous. And as in melodramas where the good triumphs in the end (Cawelti 1976:262), only developers and users of advanced computing technologies find the good life.

These five ideological themes shape public images of computers and computerization. We do not claim that computer systems are "useless" or merely a fad. But these ideological positions help activists to build commitment and mobilize resources for extensive computerization in organizations that adopt computer technologies *beyond the value that mere utility justifies.* It is ironic that activists often employ the imagery of science and objectivity (e.g., "knowledge") to advance their CMs. They deflect attention from what other analysts claim are the social problems raised by computerization—problems of consumer control, protection of personal privacy, quality of jobs, employment, and social equity, among others.

COUNTER-COMPUTERIZATION MOVEMENTS

CMs generally advance the interests of richer groups in the society because of the relatively high costs of developing, using, and maintaining computer-based technologies. This pattern leads us to ask whether CMs could advance the interests of poorer groups or whether there are countermovements that oppose the general CM. Many CM activists bridle at these questions. They do not always value "helping the rich." In our fieldwork we have found that CM advocates sometimes see themselves as fighting existing institutional arrangements and working with inadequate resources. We do not suggest that "relative deprivation" triggers CMs. We are suggesting that CM participants can argue that they have nonelite positions even though they work with elite organizations. CM activists often develop coalitions with elite groups that can provide the necessary financial and social resources. The elite orientation of the general CM is sufficiently strong that one might expect some systematic "progressive" alternative to the major CMs.

There is no well-organized opposition or substantial alternative to the general CM. A general counter-computerization movement (CCM) might well be stigmatized as "Luddite." CM activists portray computing simply as a means to reform a limited set of social settings. Therefore, a specific CCM would have to oppose all or most computer-based education developments or most office automation or most PC applications, and so on. A successful ideological base for such opposition probably would have to be anchored in an alternate conception of society and the place of technology in it. In practice there is no movement to counter computerization in general, though some writers are clearly hostile to whole modalities of computerization (Braverman 1975; Mowshowitz 1976; Reinecke 1984; Weizenbaum 1975). These writers differ substantially in their bases of criticism, from Frankfurt School critical theory (Weizenbaum) to Marxism (Mowshowitz, Braverman).

The major alternatives to CMs come from social movements that are organized to criticize problems—some specialized aspect of computerization. For example, civil libertarians are critical of those computing applications that most threaten personal privacy but are mute about other kinds of computing applications (Burnham 1983). Consumer advocates may be highly critical of strategies of computerization that place consumers at a disadvantage in dealing with supermarkets on pricing or that increase consumers' liability in case of bank card problems. But neither civil libertarians nor consumer advocates typically focus on problems of computerization in employment. Union spokespeople are especially concerned about how computerization affects the number and quality of jobs (Shaiken 1986) but mute about consumer issues. Antiwar activists criticize computer technologies that they view as making war more likely but are relatively mute on all other computerization issues. Consequently, some analysts try to envision computer use that is shaped by with other values—such as consumer control over electronic payments or improvements in working life. They view appropriate computerization as something other than the most technologically sophisticated computer use at any price.

Major policy initiatives in these directions have come from other social movements. For example, consumer groups, rather than CMs, have been the main advocates of allowing users of debit cards protections such as ceilings on liability when cards are stolen, clear procedures for correcting errors, reverse payments, and stop payments (Kling 1983). Civil liberties groups, such as the ACLU, have played a stronger role than CMs in pressing for the protection of privacy in automated personal record systems. Labor unions, rather than CMs, have been insistent in exploring the conditions under which work with VDTs has possible adverse health consequences. Each of these reform movements is relatively weak and specialized. Moreover, each initiative by CCMs to place "humane" constraints on laissez-faire computerization may be met by well-funded opposition within parts of the computer industry or computer-using industries. Consequently, the general drift is toward increased and intensive computerization with equipment costs, technological capabilities, and local organizational politics playing major enabling/limiting roles.

DISCUSSION

We have argued that the computerization of many facets of life in the United States has been stimulated by a set of loosely linked CMs guided by mobilizing ideologies offered by CM activists who are not directly employed in the computer and related industries (Kling 1983). We have characterized CMs by their ideological content, shown how five computer-based technologies are the focus of specific CMs, characterized core beliefs in their ideologies, and examined the fragmentary character of CMs. Our analysis differs from most organizational analyses of computerization by considering CMs that cut across the society as important sources of mobilizing ideologies for computing advocates. Their publications and meetings provide channels of communication for computing enthusiasts outside of the organizations that employ them, and that they try to aggressively computerize.

But much more should be done in examining particular CMs. We need to learn in more detail about their participants and social organization and to better understand their relations with computerizing organizations, interest groups, the media, other CMs, and different segments of the computer industry. We hope that this analysis will encourage scholars to examine the CMs in other specific social settings and the activists who push them. These activists play a critical role in setting expectations about what a particular mode of computing is good for, how it can be organized, and how costly or difficult it will be to implement. These expectations can shape participants' attempts to computerize in a specific social setting, such as a school, public agency, hospital, or business.

CMs play a role in trying to persuade their audiences to accept an ideology that favors everybody adopting "state-of-the-art" computer equipment in specific social sectors. There are many ways to computerize, and each emphasizes different social values (Kling 1983). While computerization is rife with value conflicts, CM activists rarely explain the value and resource commitments that accompany their dreams. And they encourage people and organizations to invest in computer-based equipment rather than paying equal or greater attention to the ways that social life can and should be organized around whatever equipment is acquired. CM activists provide few useful guiding ideas about ways to computerize humanely.

During the last twenty years, CMs have helped set the stage on which the computer industry expanded. As this industry expands, vendor organizations (like IBM) also become powerful participants in persuading people to automate. Some computer vendors and their trade associations can be powerful participants in specific decisions about equipment purchased by a particular company or a powerful force behind weakening legislation which could protect consumers from trade abuses related to computing (Kling 1983). But their actions alone cannot account for the widespread mobilization of computing in the United States. They feed and participate in it; they have not driven it. Part of the drive is economic and part is ideological. The ideological flames have been fanned as much by CM advocates as by marketing specialists from the computer industry. Popular writers like Alvin Toffler and John Naisbitt and academics like Daniel Bell have stimulated enthusiasm for the general computerization movement and provided organizing rationales (e.g., transition to a new "information society") for unbounded computerization. Much of the enthusiasm to computerize is a by-product of this writing and other ideological visions.

Most computer-based technologies are purchased by organizations. The advocates of computerization within specific organizations often form coalitions with higher level managers to help gain support and resources for their innovations, which they present as professional reform movements with limited scope. Office automation "only" influences general office practices. Advanced computerized accounting systems "only" influence the financial department and those who manage revenues and expenses. Computer-based inventory control systems "only" influence materials handling in manufacturing firms. In each administrative sector that is professionalized, some related group has taken on the mantle of computerization as a subject of reform. But these professionals do not identify with the computer industry; they identify themselves as accountants, doctors, teachers, or urban planners with an interest in certain computer applications.

By attempting to alter the character of social life across the society, the general computerization movement is basically revolutionary. A few computing promoters exploit a slick "revolutionary" image. As Langdon Winner (1984) aptly observes, Americans have been marketed "revolutionary" toothpastes, home entertainment centers, and plastics. None of these consumer items has "revolutionized" the social order. According to Winner, the sensible observer would treat claims about a "computer revolution" as marketing hype. We agree with Winner that shallow promotional claims dominate the public "revolutionary" discourse. Moreover, the social changes that one can attribute to computerization are often socially "conservative." But we suspect that the pervasiveness of computerization is having quiet cumulative effects in American life.

However, certain technologies are relatively plastic and substantially extend people's range of action. Basic technologies for transportation, energy, and communications, such as automobiles, electricity, and telephones, have become central elements of social life in advanced industrial societies. We suspect that computer-based technologies will be as socially important as these other technologies, rather than peripheral, like plastics, processed foods, and hair spray.

There is unlikely to be a general CCM. More seriously, it is unlikely that humanistic elements will be central in the mobilization of computing in the United States. Central humanistic beliefs are "laid onto" computerization schemes by advocates of other social movements: the labor movement (Shaiken 1985), the peace movement, the consumer rights movement (Kling 1983), the civil liberties movement (Burnham 1983). Advocates of the other movements primarily care about the way some schemes for computerization intersect their special social interest. They advocate limited alternatives to particular CMs but no comprehensive, humanistic alternative to the general computerization movement. In its most likely form, our "computer revolution" will be a conservative revolution that will reinforce the patterns of an elite-dominated, stratified society.

EPILOGUE 1994

Why do organizations adopt new computing technologies? The conventional answer to this question focuses upon two kinds of central social actors: computer vendors who devise and manufacture products for sale, and consumers (often managers in organizations) who purchase computer systems and services because they meet some "need" which can be understood by examining their task structure and business. Organizations generally adopt the computer systems that work for them, and they are generally effective at implementing and using the systems that they acquire. In this view, the organizations that adopt computerized systems act as closed rational (task) systems (Scott 1992). Other actors, such as trade associations for the computing industry, professional societies, regulatory agencies, and the numerous technical journalists who write about innovations in computing are assumed to play minor roles. This viewpoint has a strong grounding in both the traditional Weberian view of organizations in American sociology, and in conventional economic analysis.

We based our alternative conception of the social processes that drive computerization on an open natural systems conception of organizations and the kinds of meanings that many managers and professionals impute to specific forms of computerization. We suggested that certain professional associations were serving as arenas in which people developed beliefs about what computing was good for and how specific forms of computerization could fit their own organizations. Professional associations also enable their members contact with other people in similar organizational niches with whom to "network"—to discuss common problems, to gain emotional support, and to help locate other jobs in their industries.

We developed our open natural systems conception of computerization when we studied information systems in high-tech manufacturing firms. We found that a particular professional society (American Production and Inventory Control Society) played a strong but subtle and indirect role in the ways manufacturing firms adopted complex computerized inventory control systems. It also inculcated members with a similar set of beliefs about what these new inventory systems could do and gave them a shared language with which to communicate their beliefs. In other computer worlds, such as those related to PCs or instructional computing, there are numerous active professional associations and hobbyist groups and a level of social activity that goes well beyond any one professional association. We coined the term *computerization movement* (CM) to connote the active mobilizing role of the groups, including professional associations and allied journalists, who promote specific forms of computerization. In our empirical studies, we have found that we can learn about the direct influence of CMs upon the ways that specific organizations computerize if we probe directly in interviews (Jewett and Kling 1990).

We did not organize our study to empirically examine CMs' relative influence in different organizations. We were concerned with identifying CMs as social forces, examining their mobilizing ideologies, and examining the possibility of counter-computerization movements. If we were writing the chapter in 1994, we would make a few changes that are worth identifying.

We examined five key ideas that form a core of the ideologies of CMs:

1. Computer-based technologies are central for a reformed world.
2. Improved computer-based technologies can further reform society.
3. More computing is better than less, and there are no conceptual limits to the scope of appropriate computerization.
4. No one loses from computerization.
5. Uncooperative people are the main barriers to social reform through computing.

It is simpler to characterize the ideologies of CMs as forms of "technological utopianism" (Kling 1994; Kling and Lamb in press; Kling in press). *Technological utopianism* is a rhetorical form that places the use of some specific technology, such as computers, nuclear energy, or low-energy low-impact technologies, as key *enabling elements* of a utopian vision. Sometimes people will casually refer to exotic technologies—like pocket computers that understand spoken language—as "utopian gadgets." Technological utopianism does not refer to these amazing technologies. It refers to *analyses* in which the use of specific technologies plays a key role in shaping a benign social vision. In contrast, *technological anti-utopianism* examines how certain broad families of technology facilitates a social order that is relentlessly harsh, destructive, and miserable.

We would alter the set of CMs that would be central to our analysis. Social movements rise and fall in their influence, and some CMs also seem to be much livelier at particular times. It's worth discussing some alterations in the office automation CM and the rise of some new CMs: virtual reality and computer networking.

The office automation CM has lost its force, partly due to the predominance of PCs and related movements. The OA CM also focused upon relatively individualistic conceptions of the value to be gained from office computing systems. The themes of many popular management consultants shifted toward more socially rich conceptions, like teams, in the early 1990s, and the OA CM's conceptions seemed more marginal. However, parts of the OA CM have been transformed into a CM for "computer-supported cooperative work" (CSCW) (Kling 1991). Like other important computing terms, such as artificial intelligence, CSCW was coined as a galvanizing catchphrase and later given more substance through a lively stream of research. A community of interest formed around the research programs and conferences identified with the term and advanced prototype systems, studies of their use, key theories and debates about them, and university courses. CSCW denotes at least two kinds of things: special products such as computer conferencing systems (groupware), and a movement by computer scientists who want to provide better computer support for people, primarily professionals, to enhance the ease of collaborating. In practice, many working relationships can be multivalent and mix elements of cooperation, conflict, conviviality, competition, collaboration, commitment, caution, control, coercion, coordination, and combat. Privileging collaboration is one of the key ideological elements of CSCW.

According to Doug McAdam (1988), three sets of factors shape the ebb and flow of social movements: the level of organization of the participants (indicating indigenous organizational strength); the assessment of the prospects of success; and the structure of opportunity (opportunity structures may be more or less flexible or vulnerable to change). It would be interesting to conduct a careful empirical study of the decline of the OA CM with special attention to the opportunity structures for its participants.

The virtual reality CM is relatively young and more fragile. Like some other technology-based CMs, like AI and CSCW, people begin to participate in the CM while the focal technology is still ill defined and subject to several alternative conceptions. And, like many other CMs, its participants are mobilized, in part, through enthusiastic stories in the popular and technical press, as well as by specialized conferences and publications. The virtual reality CM's strongest influence today appears to be in the shaping of specific youth and arts subcultures, rather than in business firms.

The computer networking CM has greater national significance than the CSCW CM and the virtual reality CM. The computer networking CM has been active for well over ten years, but it was relatively esoteric and of interest primarily to computer specialists. The Internet Society, for example, promoted the development of the Internet. However, in the last few years two related events helped strengthen its appeal: the expansion and growing commercialization of the Internet and the promotion of technological support for a National Information Infrastructure by the Clinton Administration. These events are, in part, a by-product of lobbying by groups such as members and officers of the Internet Society.

Several organizations are the focus of the computer networking CM in the United States. The Coalition for Networked Information (CNI) was founded in 1990 to "promote the creation of and access to information resources in networked environments in order to enrich scholarship and to enhance intellectual productivity." It is a joint project of the Association for Research Libraries, CAUSE, and EDUCOM, and is one of their most active. CNI's members are organizations, and it currently includes over 160 colleges, universities, publishers, network service providers, computer hardware and system companies, library networks and organizations, and public and state libraries. CNI sponsors several working groups. For example, one group called for descriptions to identify "projects that use networking and networked information resources and services in the broadest possible ways to support and enhance teaching and learning" (Coalition Working Group on Teaching and Learning 1993). This public call for descriptions also noted:

> The Coalition will use the project descriptions it receives in response to this call: (1) to build a database that can be used to share information and experience in this area; (2) to promote awareness of individuals, institutions, and organizations making important contributions to the state-of-the-art in this area; (3) to attract attention to and mobilize resources for this area; (4) to plan a program in this area for the EDUCOM 93 . . . conference in Cincinnati . . . and (5) to otherwise encourage individuals, institutions, and organizations to use networks and networked information resources and services to support and enhance teaching and learning.

Clearly, the organizers of this call are not timid in making their interests in mobilizing support for network applications very explicit. The nature and magnitude of the influence, and the influence of the participants of any CM, is a separate issue.

The set of social activities that we identify as a CM includes people in diverse social locations who are promoting the widespread use of some computing technology with a technologically utopian vision. Not every CM thrives, and the members of CMs are selectively influential. The virtual reality CM, for example, may have little influence on organizational practices in the United States. Our argument focused on the influential CMs rather than upon CMs that did not mobilize much support. We believe that when one studies the sites where new computing applications are being adopted, it is common to find the influences of CMs. Members of the adopting organizations or people who consult to them are likely to belong (or have belonged) to CMs that promote that form of computerization. These ideas open up rich lines of empirical research.

In 1988 we observed that there was no significant CCM to the general CM. Such a movement would have to rest on a technologically anti-utopian vision of computerization in social life. There are a few technologically anti-utopian books or articles (Mander 1991), but no visible general CCM.

However, some groups have emerged to articulate "socially responsible" alternatives to the majority of computerization movements. The Computer Professionals for Social Responsibility (CPSR) was founded as an antiwar group and it opposed certain kinds of computer-based systems (i.e., StarWars). In the last five years it has shifted its focus to examine a wider array of public issues of computerization: workplace democracy, civil liberties issues in networking, and broad public access to national computer networks. CPSR has become an active participant in national policy negotiations about information technology policy. The Electronic Frontier Foundation supports "litigation in the public interest to preserve, protect and extend First Amendment rights within the realm of computing and telecommunications technology." It sponsors influential conferences, and some of its officers write op-ed articles for influential international computing publications. However, it is not clear how much of a movement is represented by these groups and a few other similar groups.

Despite these three changes, the concept of CMs has proven rather durable. Sociologists who are interested in social movements might also find CMs to be an interesting subject of study, since they work somewhat differently from many of the movements that are traditionally studied. Leaders of mass social movements, like the environmental, civil rights, and woman's movements, are often seeking changes in social practices legislation and public policy by mobilizing strong and explicit public sympathy. Thus these groups within these movements often organize activities to attract the news media and bring other forms of public attention. In contrast, the participants in CMs are usually trying to create social changes in settings where strong and explicit public sympathy is of little help. Consequently, they are more apt to mobilize support through professional and organizational social networks of interested parties. These activities leave less of a public trace than do the activities of groups that rally, picket, and organize mass activities.

CMs, like the traditional social movements, organize activities for their participants to be agents of purposive social change. Viewing the mobilization of computing in terms of CMs allows analysts to more readily enter the social worlds of the managers and professionals who are actively mobilizing support for computerization and who are trying to make a computer revolution (Kling and Iacono 1991).

NOTES

1. Mark Poster helped clarify some of our ideas about ideologies, and Kenneth Kraemer helped explore the relationships between utopian computing ideologies and ideologies of utopian urban developments. Leigh Star and Kären Wieckert helped us clarify the overall analysis. This research was supported under NSF Grants DCS 81–17719 and IRI8709613.

2. We are paying much less attention to the second part of our thesis—the ways that movements serve as a source of ideological inspiration and as a social world for computing advocates who participate in diverse organizations which they try to get to computerize in the ways they prefer.

3. We do not provide fine-grained data about the links between CMs and computerizing organizations. We hope that our analysis will stimulate scholars who are interested in the social dimensions of computerization to develop new lines of inquiry and more complete analyses.

4. Not all applications of computer technologies are the focus of CMs; for example, payroll systems are not promoted through CMs. Other computer technologies, such as electronic funds transfer (Kling 1978,

1983), robotics, and supercomputing are also advanced by CMs. But we don't have room to examine them here.

5. Data that show how key players mobilize support for computerization in specific organizations and how they draw their arguments from the relevant CMs would make our argument more conclusive. In our study of the development of an inventory control system in a particular manufacturing firm, we reported that the local ideologies of computerization were imported from a movement organization (Kling and Iacono 1984).

STEVE WOOLGAR

4

Representation, Cognition, and Self: What Hope for an Integration of Psychology and Sociology?[1]

What are the prospects for the integration of the psychology and the sociology of science? Leaving aside (for now) the question of whether or not such integration is desirable, this chapter asks whether or not these disciplines are compatible. I suggest the answer depends critically on what we take to be the strategic value of the sociology of science and, in particular, on our attitude to the concepts of representation, cognition, and self.

Although this chapter makes a preliminary attempt to address the question of the compatibility of sociology and psychology with specific reference to the compatibility of the sociology of science and the psychology of science, I suspect the implications are wider. When considering whether or not there is compatibility with respect to science, one is in a sense treating what could be called the hardest possible case (Collins 1981). If it turns out there is no compatibility when treating scientists in particular, then the prospects for general compatibility between the two disciplines are further diminished.[2]

I address the compatibility question in the following way. First, I review a key perspective of recent work in social science: the notion of inversion and its anthropological correlate, ethnographic distance. I press the argument that inversion is not just an advantageous theoretical gambit, but a necessary attitude in the attempt to understand issues and themes germane to science. Second, I argue that the application of inversion (ethnographic distance) in the social study of science provides important insights into the interrelated notions of representation, cognition, and self. Third, I (briefly) survey the origins and genesis of varieties of sociology of science in order to show that the evolution of perspectives is based on well-entrenched antipathies, in particular towards the notion of cognition. Fourth, with specific reference to the AI debate, I examine the idea of cognition and discuss the nature of the interrelationship of representation, cognition, and self. Finally, I conclude on what we stand to gain or lose by integrating the sociology and psychology of science, and offer some speculations about relationships among disciplines in general.

ON INVERSION

In the spirit of anthropological distance that informs his novels (notably *Erewhon*), Samuel Butler (1970 [1872]) poses an intriguing inversion of central, basic, taken-for-granted features of our perceptual world. In *Erewhon*—(which is almost "nowhere" backwards)—Butler satirizes the Victorian society in which he lived by portraying an imaginary society whose central beliefs and value systems are the inverse of our own. In particular, rights and responsibilities pertaining to illness and crime are inverted; what we call sickness (misfortune, hard luck, loss of any kind) is punishable because such states of affairs are reckoned to be the responsibility of the individual; whereas what we call crime (or deliberate acts against others or their property) is viewed as beyond the control of the individual; it is explicable in terms of theories of contagion and is (thus?) treatable by highly trained practitioners and their prescriptions.

Significantly, the (inverted) views of the Erewhonians are much more than the complacence to dictates issued by those with authority. Butler's society is no simple mechanism wherein the general populace match their actions and behavior to some central value system. Instead, the central values and beliefs of *Erewhon* are highly institutionalized in practice, in education, the law, religion, and language more generally. Statements of (and philosophical, political discussions about) these values and beliefs neither determine nor force individuals' actions, but merely reaffirm them; declarations of belief constitute the post hoc justification for practices already deeply entrenched in social and cultural structure.

Stories like *Erewhon* thus suggest the possibility of a radically inverted society: that a society can function (or, at least, can be *described* as functioning) with central values directly contrary to those we take for granted in our own existence. Butler's society seems to work. In other words, the actions, values, and beliefs of members of the society could be seen to cohere, however bizarre they appear to our own (inverted) way of looking at things. To be sure, certain features of Erewhonian behavior appear to give rise to what might be construed as inconsistencies. But for each such example, it is possible to find the counterpart "inconsistency" in our own ("working") society. The major achievement of *Erewhon* is to highlight the essential arbitrariness of our own value systems and of the hierarchy of perceptual order that informs our lives. Of particular interest is the suggestion that our own institutionalized notions of the relationship between mind and body are essentially arbitrary. Butler once remarked that our minds need clothes as much as our bodies. Could a society possibly exist in which perceptions were organized such that everyone could see directly (visibly, obviously) what we are thinking (knowing, remembering, etc.), but where some measure of detection (reflection, consideration, expertise, inquiry, scientific procedure) was required to determine what we look like? In such a society, the usual mundane skills of deducing thoughts and motives on the basis of facial and bodily appearance, language and action, would be stood on their head (as we say).

INVERSION IN THE SOCIOLOGY OF SCIENCE: THE EXAMPLE OF REPRESENTATION

One of the most important aspects of the recent sociology of science is its stance of ethnographic inversion with respect to certain "key" features of scientific practice. It is important to stress that these are features *perceived* as crucial to scientific practice. A prime example is the notion of representation: the idea that there are ways (often sanctioned methods) for producing a connection between a representation (or image) and a transcendent reality which supposedly resides beyond the outside of the representing agent.[3]

Examples of the relationship between representation and object (between document and underlying reality), construed in both general and more specific terms, are given in Table 4.1.[4]

Table 4.1
The Relationship between Representation and Object

Representation	*Object*
document	underlying reality
signifier	signified
appearance r	eality
trace on the graph	voltage
swing of galvanometer needle	current
interview responses	respondent's attitude
diary entry	diarist's feelings
what was said	what was meant
words	meaning
action	motive(s)
behavior	interests
gestures	significance
oscilloscope trace	pulsar
photograph	the photographed scene

The ethnographic stance insists that representation is not something to be taken for granted when studying science. It is instead an aspect of the culture of science which requires examination. The corollary, of course, is that it is inappropriate to adopt the notion of representation in one's own study of science. It is topic rather than resource or more exactly representation is both topic and resource rather than just resource. To adopt or take for granted a notion so central to scientific practice would be a direct disavowal of ethnographic inversion; it would be to go native before the study had begun.

The principle of ethnographic distance is especially important when studying science because science is a phenomenon which in certain respects is not at all exotic. We may very likely experience the esoterica of a particular area of science as strange and unfamiliar. But the general principles of operation may be very familiar; scientific practice involves attempts to use, attribute, and assign meanings to representations. We observe scientists trying to create order in their own very specialized fields, but we observe them deploying principles of order making essentially no different from those of everyday life. Whereas with, say, the Navaho Indians we readily find everything about their practices exotic, at least in the sense that they seem to defy basic dictates of Western logic and explanation, our scientists are themselves exemplars of Western representational practice. In crude terms, the anthropologist makes sense of the Navaho Indians by recourse to what she takes for granted about her own culture. The Navaho "automatically" seem exotic. With science, by contrast, one has to work at exoticism. One's own beliefs and fundamental preconceptions about explanation, argument, evidence, and so on both support and derive from science itself. In this sense, science is an exemplar for rationalist culture as a whole. Hence the anthropologist of science is not studying an alien tribe so much as one (privileged) sector of his own tribe: the high priests of representation whose work is supported by, the exemplificatory of, the root epistemological assumptions of the anthropologist's own culture.

Later we shall return to the use of ethnographic distance/inversion. For now the point is that, as sociologists concerned with the basis for scientific practice, we can not afford uncritically to adopt any of the baggage that comes with accounts of how scientists typically proceed. This is a wide-ranging prescription. It means, for example, that the basic discourse associated with scientific practice must be treated skeptically. Thus I mean to include perlocutions such as "he had an idea . . .," "he reasoned that . . ." and so on. It is precisely the obvious character of these kinds of accounts of scientific and other interpretive work that alerts the sociologist of science, in her anthropological mood, to exercise caution. Where such accounts seem obvious, unproblematic, not deserving a second thought, and so on—that is precisely the point at which the anthropologist must raise an analytic eyebrow. Given that such apparently brutish features of scientific discourse are *essentially* arbitrary, what makes members of this culture treat them otherwise?[5]

EVOLVING PERSPECTIVES IN THE SOCIOLOGY OF SCIENCE

A rough and ready version of the evolution of the sociology of science is given in Figure 4.1.[6]

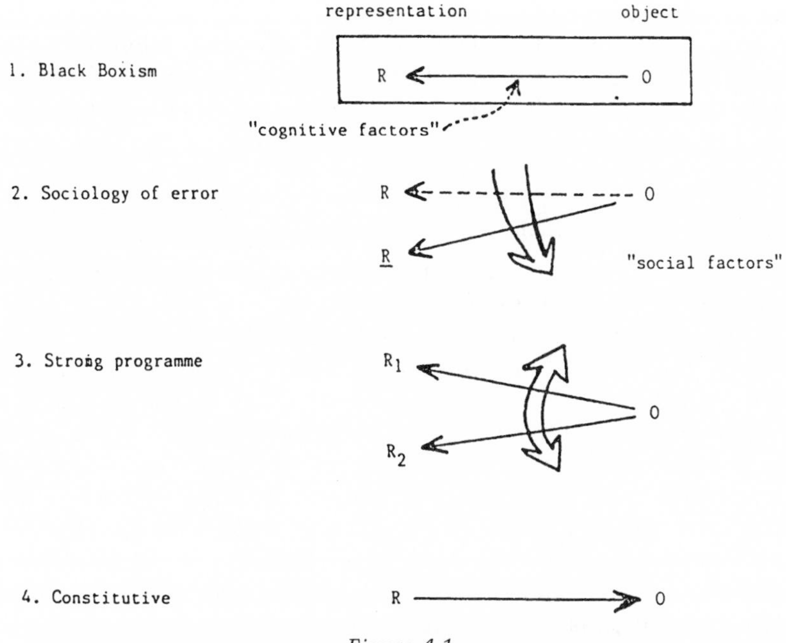

Figure 4.1
Evolving Perspectives in the Sociology of Science

The significance of this perspective is that the sociology of science has now developed a program which can interrogate representations across a variety of disciplines. More importantly for our purposes, this perspective has developed by way of a series of evolving positions. In order to demonstrate this evolution and its importance for conceptions of cognition, it is necessary to present a (slow and dirty) picture of the recent evolution of social studies of science.

Prior to the modern age of the sociology of science, science was generally construed as a product untouched by the social and cultural forces often accepted as bearing upon all other (lesser) forms of knowledge. There was no sociology of scientific knowledge, precisely because the scientific production of representations (out of their corresponding objects) was reckoned to involve no social factors. The relationship between object and representation was treated as a black box by sociologists (and other theorists of knowledge). In particular, the processes reckoned to affect the connection between object and representation were those of the individual scientist and, yet more particularly, those relating to the "mental operations" of such scientists. At the risk of oversimplifying, we shall refer to these as "cognitive factors": cognitive factors were essentially the main way in which individual scientific operations could be understood. They were very poorly studied and badly understood, to be sure. But the main point here is that they stood in opposition to sociological investigations of the generation of scientific knowledge. As long as scientific investigation could be explained in terms of cognition, it was neither necessary nor desirable to use a sociological explanans: R (representation) followed from O (object) as long as the appropriate thought processes (cognitive activities) were in operation.

A second related perspective did admit the relevance of social factors, but only in circumstances where erroneous scientific knowledge was seen to result. This perspective is tantamount to a partial opening of the black box. The question posed—What led to the generation of this or that incorrect representation?—allowed for sociological speculation, but only in instances where something went wrong (i.e., the generation of *R* rather than R). The otherwise smooth operation of cognitive processes in the process of connecting objects and their representations was thought to have been disrupted, distorted, deflected by the intervention of things social. Classically, then, fraudulent behavior or pressures of competition for instance, are invoked with respect to instances of perceivedly erroneous scientific knowledge. Sociology gets a look in here, but only in relation to acts that deviate from the normal cognitive operations associated with scientific thinking.

A third perspective tried to redress the asymmetry of the second scheme by suggesting that both true and false scientific knowledge be equally amenable to sociological analysis. Under this rubric no black box need remain unopened. The central rationale for this was that historical evidence shows overwhelmingly that what counts as true knowledge varies, both over time and between different groups, societies, and cultures. This means that it is not possible to distinguish (analytically) between true and false knowledge. And this in turn means that we can not distinguish between items of knowledge production (knowledge claims) affected by social factors and those only affected by cognitive factors. The distinction between cognitive and social, indeed, the very existence of these as separate categories, mirrors and reproduces the distinction between true and false knowledge.[7]

The strong program (Bloor 1976) in particular argued that we need to be impartial and symmetric in our analytic treatment of knowledge claims. Unfortunately, one drawback to this position is that it still implies the antecedent existence of objects in the world, prior to the work done to produce representations "of them." Given this fundamental (ontological) commitment, there remains the danger of giving way to the contemplative mood which informs the very conception of object as prior to representation; the implication is that it is possible to conceive of a nonsocial activity, one in which the antecedent object makes its presence known in a passive, neutral (cognitive?) way.

A fourth, more recently developed variant of the sociology of scientific knowledge addresses this problem by inverting the supposed connection between object and representation. Social practices are construed as actively constituting objects in the world; in other words, representations create the world rather than reflect it. In a sense, this emphasizes the practical, situated character of representative practices whether scientific or otherwise, and hence tends further to deny the cognitivist stance: the idea that mental or other inner processes enable the rightful perception of an already preexistent world. In this model O (object) comes into existence in virtue of the furnishing and juxtaposition of R (representation). Whether or not this is a correct representation, and whether or not cognitive and/or social factors have interceded, is a separate issue.

The point of this brief, crude caricature of the recent history of the sociology of science is to emphasize that the increasing skepticism about representation has been accompanied by a decreasing emphasis on the importance of cognition. The suppression of the cognitive is a deliberate and hard-won outcome of a long series of evolutions of the perspective we now know as the sociology of scientific knowledge. The denigration of the cognitive has been a costly victory, fought against (what used to be) a well-established orthodoxy (Mertonian sociology of science in its various forms) and—perhaps more significantly—against enduring preconceptions about the fundamental nature of scientific methods. In the face of this background, we can not expect sociologists to react favorably to the prospects of its resurrection in calls for the integration of psychology and sociology of science.

THE ROLE OF THE AGENT

The sociology of science culminates in the view that a critique of science is empty, or at least severely restricted, if it fails to address the ideology of representation. One of the most important aspects of the ideology of representation concerns the role of agency. The key relationship to be negotiated is that between the objects of the world and their representation in terms of signs, records, and so on. Agents of representation are those entities (actors, actants) that mediate between the world and its representation. Their role is the relatively passive one of enabling or facilitating representation. However, there is an interesting asymmetry with respect to an agent's relationship to the world and to its representation. Agents are considered passive in the sense that they are not thought capable of affecting the character of the world. According to the ideology, the mediator does not intrude to the extent that she/he is in any way responsible for the character of the de-sign-ated object. However, the agent *is* held responsible for the character of representations. While correct mediation amounts to author-itative speech about the objective world, incorrect mediation can be said to be the source of distorted representations of the (unchanging) world.

The alleged passivity of the agent vis-à-vis the facts of the world is well captured in the idea that facts are neutral, that they are there to be discovered by anybody. But the alleged irrelevance of the agent provides an interesting awkwardness when it comes to acknowledging and rewarding individual scientists for their contributions to science. For these occasions provide a celebration both of the ideology of representation and the part of the (honored) individual. The dilemma is that the honored individual is held to be especially capable of obtaining representations of the world, but that such representations do not arise solely in virtue of the individual agent. This accounts for the rather coy, "lucky bystander" tone of Nobel Prize acceptance speeches: Thank you very much, I couldn't have done it without the help of numerous others and (most significantly for our purposes) I *just happened to be in the right place at the right time.* The appeal of the ideology of representation is the notion that, given the right circumstances, any other agent could equally have produced the same results, facts, insights, and so on. This is the corollary of the view that the same facts were already there, enjoying a timeless preexistence, merely awaiting the arrival of a transitory agent.

But it is crucial to note that the scientist is only one kind of agent held responsible for mediating between the world and its representations. The scientific laboratory is populated by a wide variety of inanimate agents: experimental apparatus, oscilloscopes, measuring instruments, chart recorders, and other inscription devices. Not all agents of representation share equal responsibility in the business of furnishing representations of the world. Some are reckoned more capable than others, some particularly good at certain kinds of interpretative work, others as having outlived their usefulness, and so on. At any time, the culture of the laboratory comprises a highly stratified moral universel of rights and entitlements, obligations and capabilities differentially assigned to the various agents. The *moral order* can change, for instance, with the introduction of a new agent into the community. For example, in a study of solid-state physics investigations into the properties of amorphous metal alloys, it was found that several person-days' effort was devoted to assessing the capabilities and performance of a device for measuring changes in electrical resistance during isothermal annealing (Woolgar 1988b). The deliberations included negotiations among various agents (the company representatives, the head of the laboratory, putative users of the device) over the capabilities of the device. Even after its eventual purchase, the machine was put through several further tests before being granted the trust necessary for its participation as an adequately socialized member of the community.

The hierarchy of rights and responsibilities is characterized by a particular relationship between human and inanimate agents of representation. Neophyte scientists similarly undergo socialization into the community, imbibing the ethos of representation, but learn to regard inanimate agents as "machines," that is, as technologies of representation of a different order from themselves. While these machines are often credited with the capability of producing direct representations of the world—their rhetorical power is that their representations appear "automatic," apparently untainted by human intervention—they are presumed to remain under the control of the human agents. The power ascribed to any particular technology of representation can be understood in terms of the amount of work that would be required of an (unaided) human operating in its place.

The capacity (or ability) attributed to various agents and technologies of representation is a reflection of the solidification of past results and knowledge claims. In Knorr-Cetina's (1981) terms, the scientific laboratory comprises materializations of earlier scientific decisions and selections. The knowledge products which stabilize and solidify are embodied in technologies that set the scene for the next set of interpretive decisions (representations). This is of significance in three ways. First, established facts become further solidified in the sense that they are no longer just facts by virtue of their utterance; they are embodied in mechanisms that (are said to) enable further work (experiments, inferences, measurement, data collection). Facticity is thus enshrined in terms of instrumental value. Second, the rhetorical importance of materialization is that the previous results are turned into a technology that in the course of further representational practice, can be apprehended as a set of merely passive, neutral instruments. An inscription device (Latour and Woolgar 1986: chap. 2; Bachelard 1953) with the capability of making an apparently direct connection between, say, the radiation received by a radio telescope and the shape of a pulse on the oscilloscope, in fact encapsulates years of work on the theory of electronics, the wave form of radiation, and so on. Third, the instrumentalism and rhetorical neutrality of these technologies conspire to define the next upshot of representational activity. Our apprehension, description, and classification of the physical world depend on the technologies that make these activities possible. In other words, the world is constituted in virtue of the technologies of representation available.

THE MORAL ORDER OF REPRESENTATION

Let us look more closely at the nature of the moral universe of representation. Exactly in what ways does the differential assignation of rights and responsibilities to entities affect/sustain the ideology of representation? We can gain a glimpse of the nature of this moral order by invoking that staple character of inversion maneuvers: the metaphorical alien anthropologist figure (Maaf) who comes upon Western knowledge and its proponents for the first time.[8] Consider, then, the newly arrived alien anthropologist confronted with the intricacies and machinations of the peoples of the Western world. As is evident from his preliminary field notes, our alien anthropologist is most immediately struck by an attitude that seems to pervade almost every aspect of Western thought and behavior that he encountered. Without exception, every act of thinking, language, inscription, and so on seems to be associated with individual and bodily bounded organisms. For the particular class of organisms called "human," actions of thought and behavior are specifically attributed to (reckoned to originate from) individuals, and to specific regions within the bodies of those units. Thinking comes from the brain and this is reckoned to manifest itself in actions "performed by" the mouth, face, arms and so on.[9]

Maaf's first attempt to make sense of the perceptions and conceptions of his tribe leads him (her/it) to produce a sketch of conceptual structure (Figure 4.2). He realizes, of course, that this is an extremely preliminary effort to make sense of what is going on. He is likely to substantially modify these jottings at a later stage, if not discard them altogether in the face of skeptical reactions from his fellow alien anthropologists. They are nonetheless good enough for our introductory purposes.

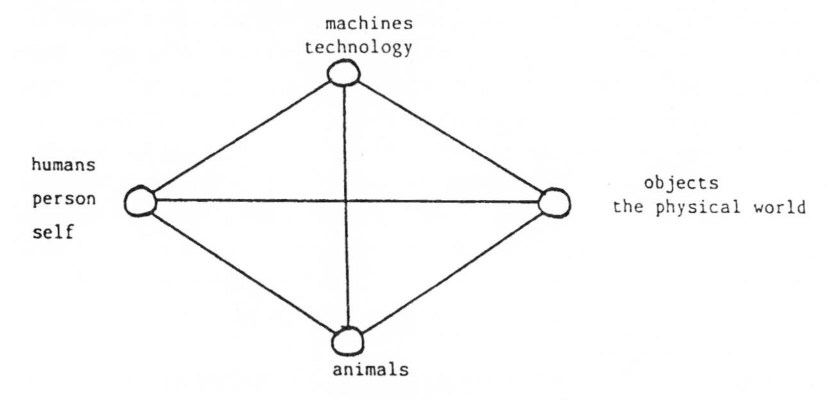

Figure 4.2
The Moral Universe of Western Thought (by Maaf)

The moral universe of Western thought is populated by several discrete classes of entity. Figure 4.2 construes this moral universe in terms of the conceptual structure of relationships between the various entities. Natural order comprises perceptions of the nature and characteristics of members of each class, the rights and obligations that accrue to each class, the similarity and distinction between classes, and the relationship between them. The spatial structure is a (crude) attempt to represent the nature and characteristics of members of various different classes of entity (humans, persons, machines, animals, physical objects). In particular, their mapping on this two-dimensional space is an attempt to indicate the relative similarity and difference between them; the nature of the relationships between them. For example, the relative exoticism of animals, from the point of view of humans, results in a distance between these two categories on the map. The relative distances also bear upon the ease with which, whether, and how a member of one class can represent (know) a member of another. This in turn raises questions about the rights and obligations that accrue to each class of entity.

On the basis of his further observations, our alien anthropologist surmises that there are rules and prescriptions about the correct way of representing other entities. He infers from the constant disputes and arguments about knowledge-generating practices that there exist heavily sanctioned procedures for describing and explaining the entities of another class and their actions. Maaf is starting to fill his notebook, but he is not entirely happy. . . .

A first source of dissatisfaction with this scheme is the rather monolithic picture that this little diagram supplies. In fact, Maaf sees a wide range of gradations of entity within each of the classes conveniently denoted as such on the diagram. Thus, for example, he discovers that certain kinds of animal are reckoned more similar to humans than others; that certain machines are reckoned to more nearly mimic the capabilities of human action than others; and so on. He is especially amused to discover that the class of entities called humans themselves include a body of researchers—the exact human equivalent of the alien anthropologist—devoted to documenting the differences among different kinds of humans. Unfortunately, all such sources of variations were conveniently expunged by our alien anthropologist in the interests of retaining the neatness of the diagram.

More importantly for our purposes, the alien anthropologist quickly notices that the whole system is in a state of flux. It is not just that different kinds of animals (or machines) are reckoned to have varying attributes, more or less like those of humans. But that the fate of these entities, in particular their relative standing vis-à-vis the supposed attributes of humans, changes over time. For example, our anthropologist discovers that an intense debate had taken place a few years previously over whether or not chimpanzees could talk. Concern focused over the extent to which language could be considered the exclusive province of human behavior, in the light of claims that certain well-trained chimps exhibited capacity for language. The implications of this debate for the structure of the moral universe are highly consequential. If the outcome of the debate is some consensus that certain animals do have the capacity for language, this places animals less distant from humans. It changes the rules for how we should think of animals, what procedures we can or should undertake for the study of animals. More profoundly, it affects the moral order of rights and entitlements as distributed between humans and animals. In other words, the discovery of language in animals asks that we revise what we supposed special and distinct about nonanimals (i.e., humans).

THE CHANGING MORAL ORDER: THE EXAMPLE OF THE AI DEBATE

Of course, the animal language debate is just one example from a wide range of disputes and discussions involving the characterization of the various entities that populate the moral order (What can humans do; What is the essence of personhood; and so on). Many other debates had focused on the character of animals and the similarity/difference between animals and humans (for example, Darwinism, the creationist controversy, the "clever Hans" effect, and so on). Maaf also discovered that discussions about new technologies also involved considerations of humanness. Discussions about technology—its capacity, what it can and can not do, what it should and should not do—are the reverse side of the coin to debates about the capacity, ability, and moral entitlements of humans. Attempts to determine the characteristics of machines are simultaneously claims about the characteristics of nonmachines. Differences in interpretations of technology are both expressive of, and give rise to, competing preconceptions about the essential qualities of humans. In discussing and debating new technology, protagonists are reconstructing and redefining the concepts of "man" and "machine" and the similarities and differences between them. As well as providing a tangible focus for continuing debates about the uniqueness of humans, technology can also act as a catalyst for changing conceptions of the nature of humans. Assessments of the character and success of new technology can both reify existing assumptions and provide powerful images for further attempts to establish (construct) "the character of man."[10]

Throughout the history of our fascination with artificial devices, the substance of these debates has varied according to the perceived capacity of the technologies under discussion. For example, when technology predominantly comprised prosthetic devices (functional additions to the mechanical abilities of the human body), it could always be argued that humans were unique by virtue of their intellectual faculties; no prosthetic device would emulate a human's ability to reason, know, and understand. However, the work of AI attempts to develop a technology that emulates action and performance previously credited to uniquely human intellectual abilities. Consequently, the advent of computers, and of AI in particular, has raised questions about the uniqueness of the human in a slightly different form. For example, in some discussions, "emotions" is now invoked as the category of attributes which testify to human uniqueness, just as "intellect" was invoked when the debate focused on prosthetic technologies.[11]

Maaf discovered that the controversy over AI is the latest chapter in the long-running debate about human uniqueness. This dispute is especially important because it constitutes an argument about the difference/similarity between humans and machines in terms of cognition. Very roughly, the dispute is between proponents and opponents of cognitivism—the general doctrine that behavior can be explained by reference to cognitive or mental states. On the side of the proponents are researchers in the field of cognitive science (theory), a specific form of cognitivism that seeks to develop explanations of conduct in terms borrowed from computer science. Cognitive science thus deploys a model of human action that, while long entrenched in the Western human sciences, is now being reified in the form of new computational artifacts (Suchman 1987). It claims both to support and be reinforced by the effort of AI to design and build artificially intelligent machines.[12]

LANGUAGE AND THE INSTITUTIONALIZATION OF MORAL ORDER

One of the most insightful aspects of *Erewhon* is Butler's description of the way values and beliefs are institutionalized in language use.[13] Even at the level of everyday interaction, Erewhonians display, teach, and retell (and thereby reaffirm) the cultural basis for their social practice. A typical greeting is "Hope you are well-behaved today"; a typical and conventionally acceptable excuse is "Sorry, I had the socks" (being a colloquial contraction of "Sorry, but I had been stealing some socks"); the typical admonition from parents to their children: "I wish you wouldn't play with him, he's so unlucky." This inverted organization of morality, Erewhonian views on responsibility and nonresponsibility, is institutionalized in language use.

Similarity, we find that language use figures prominently in the AI debate. The various diverse research efforts subsumed under the AI rubric make use of a wide variety of mental predicates.

These include terms such as reasoning, thinking, knowing, believing, deciding, seeing, learning, understanding, and problem solving. To the extent that the claims of (and on behalf of) AI articulate the "cognitive" character of machine performance, they depend on the basic proposition that these kinds of behaviors can be explained in terms of cognition. In claiming the "cognitive" character of behavior, in describing certain behavior as, for example, "intelligent," one effectively construes a task of requiring intelligence for its performance. The task's performance is thus presumed to derive from some cognitive state or ability. This is the basis for the computational theory of mind that lies at the heart of cognitive science. Changes in cognitive states are said to be effected by various computational procedures and the behaviors (actions, the performance of tasks) are to be understood as the outcome of these computational procedures. Of course, the very designation of actions as "behaviors" that can be explained in this way is itself contentious. The relationship between terms like *performance, action,* and *behavior* is at the heart of the dispute between cognitivism and its critics. In other words, the struggle over language use is a reflection of the struggle to renegotiate institutionalized conceptions of behavior.

What counts as behavior, and what distinguishes human from other kinds of (machine, animal) behavior, is crucial to the dispute over cognitivism. As already mentioned, "behavior" can refer to, and is implicated by, a wide range of other descriptive terms: action, conduct, performance, and so on. Not surprisingly, parties to the debate each use these terms in quite different ways. For example, action for the neo-Wittgensteinian sociologist is not what the cognitive scientist conceives of as action. Clearly, this is more than just an argument about the causes of certain well-observed phenomena. Parties to the dispute are offering quite different definitions of the phenomena to be explained.

Descriptive terms pertaining to behavior are "loaded" in virtue of expectations about their typical usage. For example, it seems straightforward to refer to a machine's "performance" and only slightly less natural to speak of its "activity." But some awkwardness sometimes arises when we refer to a machine's "action" or "behavior." The oddity of applying mental predicates to machine performance is equally revealing of the prejudices built into our use of these terms: "the machine thinks . . .," "the machine reasoned that . . .," and so on. This kind of awkwardness is sometimes exacerbated by certain pronominal usage. For example, "machines who think" (McCorduck 1979).

In general terms, what we experience as "awkwardness" can be understood as violations of our expectations about the "correct" order of performance, the moral order of abilities and entitlements. Our everyday linguistic usage reflects the institutionalization of these expectations. By the same token, debates occasioned by the advent of AI can be understood as attempts to renegotiate this accepted moral order. The successful institutionalization of the concept of, say, "thinking machines" might be reflected in the future "normal" usage of many phrases that now appear strange.

However, the situation is more complicated than this general account might lead us to believe. For it does not need much reflection to see that there is no simple one-to-one correspondence between categories of actor and particular mental predicates. There are many occasions of use when potential awkwardness fails to materialize. Thus, we are all familiar with cars that "refuse" to start, with "temperamental" television sets, and with space shuttles that (at least until *Challenger*) are "behaving perfectly." On these occasions, predicates often connoting features of human intentionality are unproblematically applied to nonhuman objects. Conversely, there are many occasions when humans unproblematically appropriate descriptions of their behavior sometimes used to connote the absence of "humanness." For example, Turkle (1984) relates how depictions of a person's "machine-like" behavior can confer great esteem within the community of AI "hackers."[14]

THE RAMIFICATIONS OF THE DEBATE

The AI debate shows how fundamental conceptions of the qualities of humans are institutionalized in language. The corollary is that changes in language use mirror changes in institutionalized conceptions of humanness. Changes in language use index-changing conceptions of the relationship between the different entities on Maaf's grid. Perhaps most important for our purposes is the fact that changes in conceptions of humanness have implications for many interrelated features of the moral universe. As we suggested, both with the example of *Erewhon* and with our discussion of the language of behavior, a whole series of institutionalized features of the social and cultural world support and reaffirm embedded cultural values, including "what humans are like."

Consequently, a change in this particular value will be attendant upon changes in other associated institutions. In other words, any single change in the moral order has all kinds of spin-off/knock-on effects.

We can illustrate the point by returning to (Maaf's) diagram (Figure 4.3). We see that the system in a stable state is sustained by the interlinking of the various struts. Each strut represents the conceptual distance between the entities. Now, to borrow the language of structural engineering, any change in the length of one particular strut will entail stress in at least one other strut, and will likely set up stresses throughout the entire network. By analogy, any change in the relative distance between entities in our moral universe will have implications for (perhaps all) other relationships. More concretely, the perception that machines are more like humans than previously supposed may necessitate (as a way of reestablishing stability) the idea that machines are more unlike animals. Less simplistically, perhaps, conceptions about the moral rights and obligations of various entities may change, and this will have consequences for what are perceived to be correct behaviors. For example, if you are persuaded that animals are more like machines than you previously thought, this may lead you to revise your conceptions about the treatment of animals, their feelings, and so on.

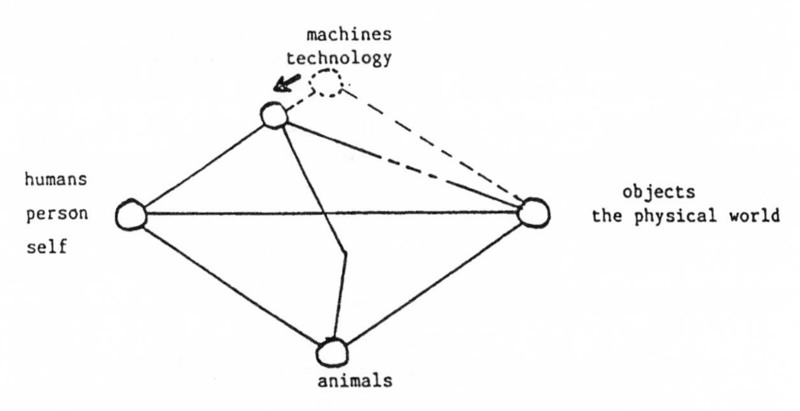

Figure 4.3
The Changing Moral Order (by Maaf)

In the case of the AI debate, the ramifications involve much more than changes in ideas about whether or not machines can be intelligent. It provides the possibility of a critical test of fundamental assumptions about the nature of human conduct. In particular, basic assumptions about the nature of knowledge and reason are at stake. These assumptions have gripped Western thought for two thousand years and they underpin ingrained ideas about the character of human beings, their abilities and potential. If AI turns out to be a feasible project, this would vindicate those philosophies that hold human behavior can be codified and reduced to formal, programmable, and describable sequences. If the project to devise artificial reason is adjudged a success, it will vindicate those objectivist and rationalist philosophies that are thought to have guided the rise of science and technology. On the other hand, the perceived failure of AI would amount to a victory in the eyes of the humanist antireductionists. Its failure will buttress those (phenomenological and interpretivist) philosophies that hold that calculation and measurement are largely irrelevant to human capacity for intelligent behavior.

To return to Maaf's thumbnail cosmology, the idea of a critical test promises the possibility of (once and for all) cementing the nature of distinctions, similarities, and relationships between at least two main constituent entities. The moral order will receive an injection of stability, at least in respect of one of its dimensions. Positions adopted in the longstanding and fundamental debate about the nature of human action—whether or not humans are essentially rather complicated machines, whether social study should proceed on the basis that the human being is really just an animal, and so on—all stand to be revised or modified in light of the outcome of the current massive research effort in AI. If this is right, we might anticipate that in the efforts of AI several significant philosophical pigeons are coming home to roost (cf. Woolgar 1985, 1986).[15]

THE QUESTION OF SELF

But the ramification of changes in the moral order has yet a further dimension. Maaf suddenly realized that in all his efforts to make sense of the Western moral universe, he had unthinkingly adopted one of its central and deepest assumptions. All the attributions of humanness, animality, mechanism; the conception of distance between these entities; their relative rights and obligations; all these hinged on the notion that the human entity was doing the attributing and describing. The grid he had drawn was in fact a map of the way humans see things. Perhaps if he had originally concocted a starlike configuration, instead of a trapezoid, with the human entity placed at the center of a hub of radial connections to other entities, the point would have become obvious more quickly.

It became clear to Maaf that despite all the apparent changes in the moral order, the one element that seemed most entrenched was the privileged position of the human in representing the characteristics of all other entities. It was, in other words, a Self-centered view of things. True, much was said and argued over in relation to the nature of the human, self, and personhood. But this seemed not, in general, to impinge upon the ability of humans to go on doing representation. In discussions of human nature and how to account for it, the self as a concept seemed to be disengaged from the Self in the act of representation.

It gradually dawned on Maaf that the sociology of scientific knowledge (SSK) occupied a position of cosmological significance in the moral order. Not only had SSK problematized the notion of representation, and in so doing had overcome the previous reliance upon ideas about "cognition" it was also beginning to take issue with just this concept of Self. Some of the more recent contributions in the field took the line that a thoroughgoing critique of representation should also attempt the deconstruction of the Self in the text. In exploring "reflexivity," a small but growing body of writers were in fact exploring ways of revising conceptions about humans as privileged representation-generating entities. More than any other attempt to renegotiate the moral order, the AI debate, theories of evolution, and so on, this intrepid band of reflexivists was embarked on the deconstruction of the very idea of a moral order.

Maaf was excited. This seemed an attractive and intriguing route to pursue. But then it occurred to him that perhaps he only liked this way of thinking because it gelled with his own view of the relative status of aliens and humans. But who was representing the character of this relationship? Maaf thought he would leave that one alone for the time being. Instead he would look forward to meeting and talking with the reflexivists. He had no idea, of course, of the consternation he would experience upon discovering that one of the reflexivists was already putting together a story about him (cf. Woolgar, this volume)!

CONCLUSION

I have attempted to broach the question of the integration of sociology and psychology by suggesting that we need to appreciate the broader context in which fundamental assumptions about action and behavior have currency. I have done this in three ways. First, we saw the importance of inversion as a way of reminding ourselves of the essentially arbitrary nature of some of our more fundamental assumptions, especially about the nature of scientific reasoning. From the perspective of the anthropologist, it is simply inappropriate to adopt and/or use language that embodies these assumptions. Inversion has been crucial to the development of sociological studies of scientific knowledge. By problematizing the concept of representation in science, SSK deliberately moved away from reliance upon assumptions about cognition. Moreover, by taking issue with the concept of representation, recent work in the field is challenging conceptions of Self—the idea that we can continue usefully to conceive of the individual (scientist) as the origin of (scientific) representations. Second, we showed, with particular reference to the AI debate, that ideas about cognition and representation are currently the focus of heated controversy. This debate once again reinforces the arbitrariness of what we tend to take for granted about ideas like cognition. Third, we used the fable about Maaf to demonstrate the interconnectedness of the ideas of representation, cognition, and self, and to show how all three support a particular set of ideas about the character of the human agent.

Given this perspective on cognition, it becomes less than satisfactory to pursue a program aimed at articulating the operation of "cognitive processes." We need to recognize, instead, that "cognitive" is an essentially odd notion. What is so remarkable is that it is so popular, deeply ingrained in culture and practice, even at the level of gesture and touch. Members of our culture signal their agreement by moving the (presumed) center of cognition up and down ("nodding the head," as we say). When wishing to indicate that "thought processes" are taking place, they place the hand or fingers at or near the same receptacle on top of the head, or they screw up their facial features to suggest that work is going on inside.

So why is explanation in terms of cognition necessary? At best, such an approach seems *premature.* To adopt the line taken by writers like Coulter (1979, 1983), we have as yet little understanding of the "grammar of situated use" of terms like understanding, knowing, deciding, doubting, remembering, and so on. To appreciate how thinking happens, we need to see what kinds of actual social practice give rise to attributions of "thinking." Note here the emphasis on "cognitive" activities construed as interactional accomplishments. Each such activity can be formulated for a variety of social interactions. For example, "deciding" need not be construed as solely the result of conscious (internal) deliberation.

Explanation in terms of cognition may also be *superfluous*. Certainly, if it's the case that all such "cognitive" activity can be adequately accounted for using sociological accounts, then a psychology of cognition (or a psychology that utilizes the concept of cognition) is unnecessary. Why not let there be a ten-year moratorium on cognitive explanations? If there is anything left to explain at the end of this period, then we sociologists promise we will resort to explanations in terms of cognitive capacities (Latour and Woolgar 1986)!

Explanation in terms of cognition may be *needlessly mystifying*. To adopt cognition as a central explanatory concept is to reinforce the exoticism of the tribe being studied. Instead of taking some anthropological distance, we are using the key concepts of the tribe to account for their behaviors. This is precisely to fly in the face of the earliest moves to develop a strong program in the sociology of scientific knowledge. To embrace the "cognitive" is to miss the skepticism about arguments that exclude the sociologist.

We begin to see the basis for suspicion about calls for a pluralistic attitude toward the relation between sociology and psychology. Such pluralism appeals to the idea that we can all work together, that it's not a question of having to decide between approaches: everyone has something to contribute. My first objection to the call for pluralism is that it trades on the vicious imperialism of the triangulation myth. The notion of pluralism implies that all contributors are contributing toward the study of a common object. Triangulation insinuates a common object antecedent to the various work and perspectives of different researchers. But of course this directly contradicts a basic tenet of relativist sociology of scientific knowledge: Objects do not preexist their study; rather, the work of studying them is how they get constituted. It follows that different approaches are not unproblematically different approaches to the same object. Rather, they each constitute their own objects of study. The likelihood of thesc turning out to be (that is, being rendered as) the same object is small.

My second objection is that the apparently benign call for pluralism in fact belies the implicit status of "cognition" as a prior concept. The call for pluralism seems so reasonable that it is positively coercive: anyone who does not want to divide up the researched world seems unreasonable, imperialistic, difficult. But it turns out that these calls for pluralism conceive of cognition as the precondition, the prior concept to the social. By analogy with the division between the contexts of discovery and justification, the call for psychology and sociology to work together again leaves sociology with the residue phenomena.

My third objection is that the call for pluralism implies a division of labor with regard to the units of analysis. Psychology deals with individuals; sociology deals with groups. But this is a transparently obvious maneuver for making sociology peripheral if you conceive of cognition happening in individuals. The position developed here is not the kind of sociology that "deals with groups" in the usual sense. Rather, the units of analysis for sociological inquiry are the moves being made in the language game. Whether or not such moves originate from individuals or groups is part of the phenomenon to be investigated, not something to be assumed before we start.

So is there hope? Yes and no. Yes, if we conceive the sociology of science as a way of studying *what counts as* "cognitive" activities. This is what might be called the "situated practice" solution alluded to earlier (Coulter 1979, 1983; Suchman 1987). The program is to produce descriptions of the circumstances and ways in which "cognition" is attributed and has currency in the course of social action. The dangers of this program are twofold. First, the tendency is to presume the codifiability of phenomena. Hence, the danger is that this leads to a program that aims to substitute a social for a cognitive mechanism as the basis for explaining human action. A social mechanism, but a mechanism nonetheless. This unfortunately overlooks the kind of point made by Turner (1989): The revealed realm of the tacit is itself a contingent object. The second danger is that this approach privileges the notion of the individual as the originator of attributions of description. By concentrating on the way members do cognitive work, this tends to ignore the more radical thrust of the critique of representation that challenges the concept of Self as responsible for representational action.

The answer is no (there's no hope for integration) if we take the direction of SSK as a serious and fundamental challenge to the idea of representation. This radical (postconstructivist) project is not compatible with the psychology of science precisely because it takes as topic what cognitive science takes as resource.

In sum, my answer to the compatibility question is yes (the pluralist response) and no (the epistemic response). The psychology and sociology of science are compatible if we view the sociology of science as providing illuminations about a series of factors (social factors) that bear upon scientists' knowledge-generating activities; this is the less radical thrust of the recent sociology of scientific knowledge. But they are incompatible if we recognize SSK as urging a radical reconceptualization of conventional modes of explanation. The basic assumptions of both psychology of science and (the less radical) sociology of science engage in and take the ideology of representation as unproblematic. In the latter view of SSK, the ideology of representation and conceptions of cognition and Self are themselves under attack.

Finally, we should note that this latter response does not mean there should be no dialogue between sociology and psychology. Even if we take the view that attempts of integration may blunt the cutting edge of the sociology of scientific knowledge, much is still to be gained by attempting further to articulate the differences between the two disciplines.

NOTES

1. An earlier version of this paper was presented at the University of Colorado at Boulder, 24 November 1987. My thanks for participants' comments then and since.

2. Elsewhere I have examined the effects of the observations that the "hardness" of Collins's (1981) hardest possible case is itself a local accomplishment/construction (Woolgar 1988d).

3. This notion of representation is "the original sin of language, that separation of speech and world we know as the disjunction of words and things" (Tyler 1987).

4. Cf. Woolgar (1988a: chap. 2).

5. Notably, including those representations that sustain its own discipline. See the work on reflexivity by Ashmore (1985), Mulkay (1985), and Woolgar (1988a).

6. For other authoritative statements on the matter, some of which include their own self-deconstruction, see Woolgar and Ashmore (1988), Mulkay (1985), and Woolgar (1988c).

7. Indeed, there is a good argument that as soon as you recognize that everything is social, then the term becomes redundant as a discriminator. To say something is social may have a useful antagonistic function, but it also implies that it is possible to have something which is nonsocial. This leads to the suggestion that the use of the terms *social* and *cognitive* should simply be banned (Latour and Woolgar 1986).

8. The Maaf is a device for injecting ethnographic distance, a stranger's perspective designed to highlight what we otherwise take for granted about the world we both live in and try to investigate. It also permits one to put one's own ignorance to good use by beginning at the beginning: my argument is very much that of a distanced newcomer to the phenomenon of cognition.

9. Maaf's (initial) concentration on Western cosmology prevented him (as it did most Westerners) from discovering major divergences from this orthodoxy on the part of certain obscure and "primitive" non-Western tribes. For example, some Western anthropologists knew that the Orikuya construe mind as located in their left elbows, that the Preenash think of their essential selves (samos/as) as residing in the nearest akhran tree, and so on.

10. Feuerbach has been credited with noting that man creates technology in his own image. However, the point here is not simply that the nature of technology is driven by conceptions of the human state; rather, that in the development of (and discussion about) technology, prevailing conceptions of the human state and technological capacity are being renegotiated.

This approach has some affinities with Hughes (1987) when he refers to "institutional structures that nurture and mirror the characteristics of the technical core of the [technological] system." If we allow that these "institutional structures" embody competing notions about human uniqueness, intellectual capacities and so on, it is perhaps not surprising that AI is so controversial. The exact technical characteristics of this particular technology are much disputed precisely because institutionalized notions about the character of the human are at stake. The present approach also has some resonances with Callon's (1987b) argument that technology is a "tool for sociological analysis." A close examination of discussions of the character of technology—what it is, what it can and can not do—shows that discussants themselves perform "societal analyses." In the present case, "analyses" performed by discussants of AI constitute claims about basic human character, abilities, and capacities. By following protagonists in the AI controversy, we follow the construction and reconstruction of different models of human's basic attributes.

11. The strategy of involving a "reserve" category of attributes has in general been relatively unsuccessful in averting fears about the erosion of human uniqueness by the creation of machines. Attempts to establish alternative criteria of uniqueness have not easily overcome deeply held preconceptions about the special character of the human. Certainly, this kind of argument did not prevent considerable alarm over the ever increasing capacity of devices to mimic human mechanical abilities—consider the climate in which Butler published his marvelous parody—nor did it prevent speculation and considerable controversy about the fundamentally mechanical character of all human action—for example, the discussion of La Mettrie's (1784) work by Needham (1928), Rignano (1926), Rosenfeld (1941), and Vartanian (1960).

12. A main aim of AI is the design and construction of machines to perform tasks assumed to be associated with some cognitive (sometimes "intellectual") ability. This is not, however, a uniform position within AI. Some AI researchers flatly declare their disinterest in "cognitive abilities." For them, the machine's performance of tasks is the sole technical goal of their research, independent of whether or not such task performance would require intelligence in a human. Nonetheless, others, both practitioners within AI and spokespersons on its behalf (philosophers, marketing entrepreneurs), explicitly see the *raison d'etre* of AI as the attempt to design machine activity that mimics what they construe to be "cognitive" behavior. This latter position is referred to as the "strong" version of AI (Searle 1980).

13. This is, of course, to reformulate Butler's language in terms appropriate to our current (sociological) interests.

14. The fact that "awkwardness fails to materialize" suggests an interesting reflexive tie between categories of "machine" and nonmachine predicate." For example, in saying that a space shuttle is behaving perfectly, we provided for a hearing of the something-more-than-a-mere-machine quality of the space shuttle; in this usage, its "behavior" can be heard as connoting its sophisticatedly technological (intelligent?) qualities: much more than a lump of metal.

15. More generally, this suggests the intriguing idea that we might try to turn any philosophical assumptions into a technology as a way of revealing their limits. If we need to assess the value of a philosophical argument, the acid test will be whether or not a technology founded on its basic assumptions can be demonstrably successful (cf. Dreyfus 1979). Unfortunately, the flaw in this suggestion is that "acid" or "critical" tests are rarely so straightforward: the sociology of scientific knowledge shows us the sense in which the outcome of tests and demonstrations are always contingent or temporary (cf. Woolgar 1986).

PART 2: MATERIALS AND SPACES

ADELE E. CLARKE

5

Research Materials and Reproductive Science in the United States, 1910–1940[1]

In order to observe or produce the phenomena they study, all working scientists must obtain and manage research materials. In the United States at the turn of this century, the shift of emphasis from descriptive morphological approaches to experimental physiological approaches in the life sciences radically altered scientists' needs for research materials and had significant consequences for the organization of such materials.[2] This chapter focuses on the consequences of that shift for the acquisition and organization of research materials in the case of reproductive physiology during the first half of this century.

By 1900, physiology had emerged as an independent professional discipline in the United States, as it had a generation or two earlier in Europe. Its central problem structure had been defined and delimited, and American physiologists could claim their own society, journal, and university departments. The terms *physiological* and *physiologist* were now used as identity claims for a variety of reasons under a variety of conditions. Physiological approaches, combining a set of processual questions about nature with experimental methods, were spread throughout the life sciences, far beyond the discipline of physiology itself and its official university departments.[3] As the research agenda in American biology, medicine, and (soon after) agriculture began to extend beyond structure and form—the traditional preserve of anatomists and morphologists—entirely new lines of scientific work emerged, several of which were developed into new disciplines or specialties.[4]

The new approaches were found in a variety of institutions, old and new. Several of the new departments, research institutes, sponsoring organizations, and marine biological stations and museums have already been subjected to scholarly examination (Maienschein 1985a, 1985b; Reingold 1979; Werdinger 1980). However, historians and sociologists of science have almost entirely ignored the emergence of new infrastructural organizations to support the altered agenda for research in the life sciences—among them, biological supply firms, technical equipment industries, animal houses, and research colonies.[5] In this chapter special attention is given to new infrastructural arrangements for supplying a variety of live and fresh materials for the experimental investigation of mammalian reproductive physiology. Interestingly, many of the reproductive scientists who now adopted experimental approaches in their research had been trained not as physiologists, but as zoologists or anatomists (Long 1987). They, and "reproductive biochemists" soon after, quickly integrated and absorbed physiological experimental approaches into their routine working procedures (Kohler 1982; Geison 1979).[6]

In 1979, Charles Rosenberg (1979b) called for a historical "ecology of knowledge" that would include disciplines and institutional contexts. The larger aim of this chapter is to suggest that a still richer ecology of knowledge would result from giving closer attention to such issues as the organization of research materials and the development of instrumentation and techniques. This approach would enlarge our historical appreciation for the day-to-day practices of scientists and bring them within the purview of recent literature on the sociology of work, organizations, and professions.[7] We would thereby expand our understanding of the social and material conditions of scientific research and their consequences for the production of scientific knowledge.[8] Just as local institutional, disciplinary, and departmental conditions may shape scientific work, infrastructural organization may also be consequential. Research is, after all, constrained or enhanced by the accessibility, cost, organization, and pacing associated with specific materials, instruments, and techniques.[9]

MATERIAL CONSEQUENCES OF PHYSIOLOGICAL APPROACHES

In terms of research materials, the shift of emphasis from descriptive morphological work to experimental physiological approaches had three major consequences: (1) increased demand for live and fresh materials; (2) increased demand for large quantities of the same species rather than limited numbers of specimens of several different species; and (3) development of colonies for on-site access to desired research materials.[10]

Life scientists who adopted the new experimental approaches in their research had to explain and justify the changing nature of their work to a variety of audiences in order to secure support for their research and related materials.[11] In 1914, for example, Frank R. Lillie, chairman of the Department of Zoology at the University of Chicago, wrote to the president of the university in a budget request for a vivarium—a facility for housing live animals for research: "The study of living animals has replaced to a considerable extent the study of their dead bodies, and a vivarium is as much a need of the modern zoological laboratory as a museum . . . zoology is becoming an experimental science. A department that is well equipped for the older descriptive zoology and embryology, as ours is, may yet be deficient in facilities for pursuing the newer lines of work which must necessarily claim most of our interest. The development of experimental methods of studying zoological problems has been gradual; it is no sudden fashion but a normal development. And it is as certain that it is a permanent direction for research for zoology as for chemistry. A science cannot retreat from the vantage ground of experimental methods."[12] Lillie's request illustrates in striking fashion some of the new and very different requirements for materials, space, and laboratory organization that scientists and their institutions faced because of the new experimental approaches.[13] These new requirements and their implications for the daily work of the scientist deserve some specification and elaboration here.

Demand for Large Quantities of Same-Species Materials

Nineteenth-century morphologists and comparative anatomists generally sought to acquire representative samples of a wide variety of species, mainly for descriptive interspecies comparative work. They might then look for homologues (parallel structures) across species, seek interspecies characteristics for finer classificatory distinctions, and so on. The goal in acquiring materials for morphological work was, therefore, to secure one or at most a few specimens of each species. These specimens were generally (although not invariably) dead and complete organisms. A single specimen could theoretically be utilized for hundreds of studies, given adequate organization of access to it.[14]

The new experimental approaches entailed very different requirements for research materials. Both live and fresh cadaver materials became essential. Moreover, large quantities of such materials were typically sought, and often from the same species. Two general considerations lay behind these new requirements for research materials. First, live and fresh materials were needed because biological processes and their histological and cytological manifestations could rarely be adequately observed in dead or preserved specimens; preservatives and stains often altered materials in damaging ways (Corner 1981; Amoroso and Corner 1975). Second, large numbers were needed because the new standards of research included an attempt to confirm results through adequate sample size. In particular, it was hoped that one might be able to control "accidental" variations between individual specimens by using large quantities of homogeneous materials, preferably from the same species.[15] The crucial if unexamined assumption here was that basic biological processes were parallel in different species.[16] Scientists who made this assumption naturally focused on the physiological processes rather than the species or material per se. The goal was to work out a particular process in one species, developing a model or analogue that would subsequently be refined across species (Holden 1985; Lane-Petter 1963).

Gaining Access to Materials

The means of gaining access to mammalian materials changed dramatically with the shift in goals from traditional descriptive work to the new experimental approaches. In the older tradition, the primary means of access consisted of collecting single or limited numbers of each specimen, preserving them, and establishing collections of a wide variety of specimens.[17] Collecting expeditions were the major means of acquiring morphological research materials.[18] The fundamental social processes were (1) the development of and reliance upon both amateur and professional collectors in desired locations to obtain specimens[19] and (2) the creation of professional scientific networks to share and/or pass along desired specimens. Such networks enhanced access to research materials both geographically and in terms of the range of collected species.

The new experimental approaches entailed drastically different means of gaining access to research materials. Life scientists, and especially reproductive biologists, now sought reliable supplies of fresh and live materials, a task that was in fact a large and time-consuming part of their daily work.[20] They relied on five major means of access to appropriate mammalian materials for their reproductive research: (1) collecting exotic animals and specimens (materials native to foreign habitats, and/or rare to the environment in which the research is conducted, and/or difficult to maintain alive); (2) using mundane (local, easily accessible, and easily maintained) materials such as domestic animals (e.g., swine and cattle) or other routinely available materials; (3) using medically supplied materials (e.g., aborted and stillborn embryos, surgically removed organs of the reproductive system, etc.); (4) using animals and specimens obtained from biological supply companies; (5) establishing and maintaining on-site research colonies.

At first, the new breed of reproductive biologists and medical researchers primarily collected and used exotic and mundane materials from the field. Acquiring such materials was generally a do-it-yourself project for the research scientists themselves. Technical assistance was usually very limited (consisting mainly of students) and at times nonexistent. The selection of materials was therefore often drastically limited to what came easily and cheaply to hand. As Robert M. Yerkes commented in 1916, "Most investigators are either impelled or compelled by circumstances to work on easily available and readily manageable organisms" (Yerkes 1916).

There was thus a catch-as-catch-can ethos about the selection of and access to experimental materials at the turn of the century. Indeed, the new band of experimental biologists imitated the practice of their more traditional predecessor by developing quite elaborate amateur and professional networks to enhance their access to materials, even though they sought live and fresh materials as opposed to dead specimens. At times, such "networking" led to joint acquisitions and research. Such networks remained important even as life scientists came increasingly to rely on biological materials suppliers, medically generated materials, and specifically designed biological colonies. Patterns of access to and use of particular materials were often passed down from teacher to student or from one scientist to another at a given research site. In such cases, and indeed more generally, there was a tendency to continue to use a single animal material once a stable means of access to it had been achieved and as the intricacies of its maintenance had been mastered. As their organizational commitments became entrenched,[21] scientists often explored a variety of research problems by using an already available animal rather than seeking new materials for new problems. This is no small wonder, since the quest for access to appropriate materials represented such a major investment of resources in terms of time and energy as well as money. The availability of and familiarity with particular biological materials surely did much to shape the pace, direction, and even the content of research in reproductive physiology and no doubt in other fields as well.

Colonies As Means of Access to Research Materials

The task of acquiring large quantities of live materials was challenging. Here agricultural scientists involved in reproductive and other physiological research had distinct advantages. Most agricultural institutions maintained herds of various animals as a matter of course in their efforts to find and teach improved methods of animal production. Scientists had fairly easy access to these herds, which were supported by state and federal funds. Moreover, well-established agricultural scientists could also often make use of the private herds of farmers and ranchers through coordination with extension agents of the United States Department of Agriculture.[22] Inspired in part by such agricultural models, biomedical scientists gradually articulated their need for large numbers of a given species immediately at hand for ongoing research.

In 1892, Charles Otis Whitman, first chairman of the Department of Zoology at the University of Chicago, put forward one of the earliest proposals in the United States for a major biological colony for research purposes (although his proposal was not published in *Science* until 1902) (Whitman 1902a, 1902b). Whitman was a transitional figure; raised on descriptive morphology and experimental zoology, he nonetheless encouraged the adoption of physiological approaches at Chicago and at the Marine Biological Laboratory at Woods Hole (Maienschein 1987; Pauly 1984). Toward this end, he proposed the establishment of animal and botanical colonies of both experimental and field types for multiple purposes at one site.

Whitman's proposal contains this striking expression of the shifting priorities and approaches of American biologists at the turn of the century:

> The biological laboratories of to-day, in design, equipment and staff, are almost exclusively limited to the study of *dead* material. Living organisms may find a place in small aquaria or vivaria, but they are reserved, as a rule, not for study, but for fresh supplies of dead material. It is no disparagement of the laboratory to point out a broad limitation in its ordinary functions and the pressing need of new facilities for observation and experiment on *living* organisms (1902a:4, emphasis in original.)

Whitman further insisted that:

> The functions to be fulfilled by a farm are . . . [now] prescribed by the . . . deeper and broader needs of pure research on *living* organisms. . . . The biology of today . . . has not too much laboratory, but too little of living nature. The farm will certainly do much to mend this great deficiency (1902a: 305, emphasis in original).

Like later advocates of biological colonies, Whitman argued for the creation of such farms partly on practical grounds, predicting the development of useful animal and plant hybrids. He also assumed that such colonies would encourage the wider adoption of one of the most significant features of the new scientific agriculture—advanced record keeping and monitoring (Rosenberg 1976, 1979a; Rossiter 1979). Such modern farming methods, Whitman insisted, could provide scientists with the "history of the material to be investigated." These "exact records . . . will thus render a most important service to laboratory records" (Whitman 1902b:507).[23]

The leading models for Whitman's project were agricultural experimental stations, the Experimental Stations of the Carnegie Institution of Washington, DC, and Luther Burbank's establishment in Santa Rosa, California.[24] Saying that "there is no satisfactory name for the new plant [i.e., institution] required for such work," Whitman settled on the term *Biological Farm,* capturing the notion of plants and animals under cultivation.[25] He proposed that several different university departments should contribute to and use the large colony simultaneously. His proposal thus stands out in the history of materials acquisition as an early call for a multicontributor, multipurpose, multi-user biological colony—a new institutional form.

During the first quarter of this century, American reproductive biologists established ambitious colonies of a wide variety of research materials. They developed two major types of colonies: research-only colonies (in which additional materials were purchased or obtained as needed) and research-and-production colonies (in which some animals were reserved and maintained for breeding, while experimental supplies were also made available). Research-and-production colonies were usually more expensive to operate, since breeding could be difficult to achieve and larger numbers of animals were needed to meet both goals. The duration of gestation also affected production colony costs: rapidly reproducing animals such as guinea pigs and rats were less expensive to colonize than primates, for example.

Figure 5.1 provides a preliminary conceptual flow chart of the development of biological colonies. Further research is needed to clarify the nature of such relations in specific contexts and circumstances.

HISTORICAL ILLUSTRATIONS

The rest of this chapter illustrates the analytic themes previously identified by discussing patterns of access and use of specific mammalian materials in reproductive science ca. 1910–1940. Table 5.1 presents a classification of the major types of materials and lists specific examples from reproductive science that will be discussed in the next section.

Mundane Fresh Specimens

A variety of both domesticated and wild animals are mundane materials in the sense that they are native and/or local and can be easily obtained. In this section, I discuss access to three domesticated animals commonly used in reproductive science: swine, cat-

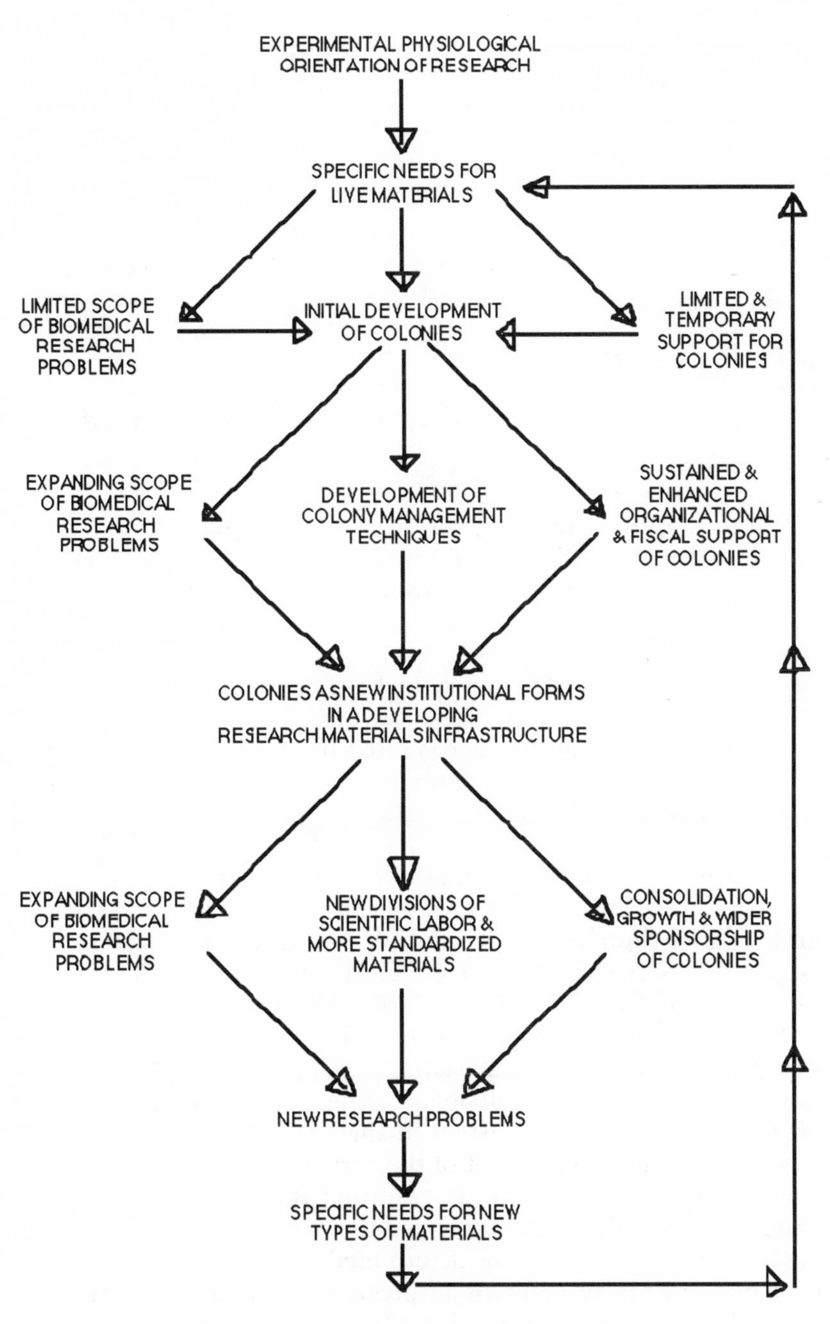

Figure 5.1
Preliminary Chronology of Colony Development

cated animals commonly used in reproductive science: swine, cattle, and horses.

Table 5.1
Access to Major Types of Physiological Materials for Reproductive Research

	Major Types of Materials	
	Mundane Materials	*Exotic Materials*
Fresh specimens	Pig and cow ovaries and uteri	Human embryos
	Cow urine	Human ova
	Mare blood	
Live materials	Cats and dogs	Opossums
	"Laboratory animals" (rats, mice, guinea pigs)	Nonhuman primates

Swine. Starting in the 1890s, pigs or swine were used in mammalian reproductive biological research in Franklin Mall's laboratory at Johns Hopkins. As Florence Sabin, Mall's biographer, noted in 1934:

> The anatomical laboratory in Baltimore was situated only a block away from an abattoir, where a large number of pigs were killed each day. It has often been said that the embryo pigs which were so abundantly obtained from this source made Mall's department (p. 213).

Mall's students later filled about twenty-five chairs in anatomy in the United States, and many of those students utilized pig embryos in their research, notably including Herbert Evans, George L. Streeter, Florence Sabin, and George W. Corner (Amoroso and Corner 1975; Sabin 1934:200–201).

Corner thus inherited access to and a pattern of use of a particular mundane material from Mall. By 1910, Corner was investigating the embryological development of the pancreas in pig embryos. He then became Mall's assistant in anatomy at Hopkins, where he first studied the development of the corpora lutea in the ovaries of sows, partly as a means of determining the age of early human embryos. Specifically, Corner hoped that the stage of development of the corpora lutea could be determined in relation to the state of development of the embryos in pregnant pigs and that this data could then be applied to humans who could not be studied directly, since appropriate human ovarian and embryonic materials were "very scarce" (Corner 1981:21, 43, 86, 102–106). The study failed to solve the problem of dating early human embryos, but Corner was nonetheless delighted in the long run:

> I had chanced to begin my study of the mammalian reproductive cycle with a species whose cycle turned out to be so comprehensible as to constitute a model against which to compare the more complex cycles of other species on which I was later to work (1981:106).

Corner again used mundane materials from sows in his ovarian cycle research from 1915–1921. According to his sometime colleague Carl Hartman, "many's the day which he and his assistants spent in the local packing house laboriously searching for appropriate stages through the entrails of freshly slaughtered sows" (Hartman 1956: 8).[26] Corner's autobiography is full of reminiscences of making and remaking arrangements with local pork slaughterhouses or abattoirs to obtain their discarded reproductive organs in a fresh and timely fashion. As Corner moved among several universities during this early part of his career, he became familiar with slaughterhouses in Baltimore, Berkeley, and Rochester.

In 1919, Corner was doing research on sows to examine the developmental cycle of corpora lutea in the ovary over their twenty-one-day estrus cycle. This investigation required three sequential weeks of frequent observations, tagging the animals at various stages of the cycle and then slaughtering them. In his autobiography, Corner noted, "Paradoxically, I could do the work far more easily and less expensively on the outskirts of urban Baltimore than on the University of California, Davis farm" (Corner 1981:13). This was true at the time because, due to the high costs of proteins and fats during World War I, pig farms had been established near urban garbage dumps and citizens were encouraged to save kitchen scraps to feed the pigs. Corner easily gained access to pigs for his research at one such establishment. Then:

> The Johns Hopkins University . . . went into the pork business, with me as its agent, by purchasing the [twenty-two] sows for $800. I had them hauled to Hohman's slaughterhouse . . . near the medical school . . . Hohman's butchered the lot at the end of a morning's run, slowing the pace of the killing floor routine so that I could collect the ovaries and

> uteri systematically. The university lost a hundred dollars or so in the deal.[27]

Corner spent the summer of 1920 with Charles Davenport at the Carnegie Institution's Station for Experimental Evolution in Cold Spring Harbor, New York. A student there went daily by train to and from a New York City abattoir (a trip of several hours) to obtain an adequate number of fresh sow uteri (one thousand over the summer) at varying stages of pregnancy for another of Corner's research projects (Corner 1981: 136).

Corner was hardly alone in the quest for such research materials. His sometime student and colleague Karl M. Wilson, for example, also secured sow uteri from a "near-by abattoir, the genitalia being procured entire for examination."[28] Wilson described as follows some of the constraints and opportunities that were provided by such research materials:

> There are certain obvious disadvantages connected with the use of such material. The sexual history of the animal is, of course, unknown, and, furthermore, many of the animals are very young, and these doubtless have rather irregular cycles during the early period of their sexual activity. . . . On the other hand, these obvious disadvantages are at least partially compensated for in the abundance of material which one is able to obtain for study (1926:419).

Thus swine became a commonly used research material in the development of modern reproductive physiology due to both their ready availability and their appropriateness for the pursuit of embryological and reproductive problems. Research on sows was apparently curtailed by the removal of abattoirs from locations convenient to scientists as refrigerated trucks and trains became widely used.[29]

Cattle Fetuses and Reproductive Organs. Scientists also commonly obtained fresh cattle materials for reproductive research from the killing-room floors of local abattoirs. For scientists at the University of Chicago, for example, the stockyards of the meatpacking industry were a major materials resource. During World War I, Lillie, chairman of the Department of Zoology at Chicago, went so far as to seek draft exemption for a technical assistant, who "has done all of our collecting in the stockyard and in various localities for many years."[30] The technician made "at least two regular trips a week to the stockyards, and frequently additional trips" for which Lillie requested a department car in 1919.[31] The famous Chicago stockyards were relatively close, and the Department of Zoology had connections with nearby farms as well. Lillie identified both the stockyards and the farms as sources of freemartin cattle embryos for his classic paper on this topic (1917: 371–452; Newman 1948:215–39). The stockyard resources thus served as a striking example of industrial contributions to reproductive science that are often difficult to specify (Clarke in press a chap. 7).

Cow Urine and Mare's Blood. After ca. 1925, the core of the problem structure of reproductive science shifted from physiological to endocrinological research (Clarke in press a chap. 7). The biochemical nature of endocrinological research then generated a new set of needs for fresh materials. Once certain reproductive hormones were detected in urine, stable sources of urine were sought for endocrinological research (e.g., Zondek and Finkelstein 1966).[32] In agricultural institutions, cattle herds became a key source.[33] Graduate and undergraduate student/technicians collected the urine by placing pails under the cows and then tugging on the cow's labia to induce urination.[34] Distilling the urine to obtain the desired hormones remained a highly scented project until the development of radioimmunoassay techniques.

Pregnant mares came into demand once gonadotropins were detected in their blood (Cole and Hart 1930; Zondek and Finklestein 1966). Adequate supplies of blood could be obtained routinely without harm to the mares. Agricultural scientists obviously had the initial advantage in terms of access to pregnant mare serum gonadotropins. Researchers at the University of California, Davis, for example, later sold their patents on the process to pharmaceutical companies through the university, and royalties were fed back into local research budgets.[35]

These examples illustrate that even in the case of mundane and relatively accessible materials, reproductive scientists spent considerable time and energy in the quest for appropriate research materials.

Exotic Fresh Specimens

More exotic materials were also sought for reproductive research. In particular, highly prized human embryos and ova were obtained largely from cadavers and surgically removed organs and tissue. Access to such materials was uneven during the period 1910–1940. This result was due not only to the contingent nature of death and surgery but also to inadequate preservation methods and a lack of any systematic means of distributing specimens and cadavers. The following examples illustrate how access to such materials was achieved, how scientists organized their acquisition, and how surgical events were restructured to provide desired materials for research.[36]

Human Embryos. As late as 1938, E. Scipiades (1938:97) stated that "The science of human embryology is still incomplete, for the reason that research in this field has been hampered by two obstacles." The first obstacle was poor microscopy and inadequate histologic techniques. The second was the scarcity of young and adequately preserved human embryos.

The most ambitious early effort to collect and study human embryos was that initiated by Mall at Johns Hopkins in the 1890s. Although he began by collecting research materials independently, Mall's collection was subsequently sponsored by the medical school at Hopkins, where he was chairman of the Department of Anatomy. When he became director of the Department of Embryology of the Carnegie Institution in 1913, Mall transferred his collection there.[37] Shortly after he arrived at Hopkins, Mall made an appeal to local physicians through the *Bulletin of the Johns Hopkins Hospital* to send him embryos derived from miscarriages. He noted that "Poor specimens . . . are better than none at all, therefore in all cases the ova [sic] should be preserved, even if there be but little hope to obtain a good specimen" (1893:115). He then provided instructions for the preservation of the embryos in alcohol. Ten years later Mall (1903:29) reported that "The collection consists now of 208 specimens . . . over one-half of which were sent to me by physicians residing in Baltimore." In return for such donations of research materials, Mall and his assistants often sent reports on the embryos to physicians who sought information on the cause of miscarriage. Fairly elaborate forms were filled out on each embryo collected.[38]

In 1903, Mall again called for contributions to his collection of human embryos:

> Although embryologists have emphasized again and again the importance of preserving carefully early human embryos for study, it is necessary from time to time to remind physicians

> not to throw away the valuable material which is constantly coming into their possession. There are numerous questions which may be answered were there an abundance of specimens at hand, and they can be procured only through the cooperation of physicians in active practice (pp. 29–30).

In the same article, Mall also published a list of the collected embryo specimens with the names of the physicians who had contributed each one.

A decade later, in 1913, Mall made a formal "Request to the Graduates of the Johns Hopkins Medical School for Embryological Specimens."[39] Instructions for preservation were included and messenger service for embryo pick-up was offered for the Baltimore area. Two similar letters were sent out from the Hopkins Department of Anatomy by Evans at about this time. In one such letter, Evans wrote: "Our success depends absolutely on you. . . . And so we are asking you to remember us; to give us anything you can collect from an abortion case."[40] Evans's letter reflects some refinements in preservation techniques:

> In must be preserved fresh. If the case happens in town and a telephone is at hand [number to call provided] let us send for it immediately at any hour of the day or night for we want to use special fixatives. A few hours' delay is serious and shows in the appearance and staining of all the tissues afterwards.

After Mall's death in 1917, efforts to obtain embryos were continued by Streeter, who succeeded him as director of embryology at the Carnegie Institution. Today this embryo collection of the Carnegie Institution is housed at the University of California, Davis, where it is used in conjunction with the Primate Research Center.

Comparative embryology did not, however, become a focus of the Baltimore group until about 1920.[41] A. J. Schultz noted of his early years as research associate of the Carnegie Institution's Department of Embryology at the Johns Hopkins Medical School:

> I obtained an abundance of fetuses of different human races but I found very little interest in and no material of nonhuman primates to satisfy my main desire for the sort of comparative studies which I had learned to appreciate in Zurich (1970:4).

Hospital surgeries and local physicians were routine sources of human embryos and fetuses.

About 1924, while at Rochester Medical School, Corner became interested in obtaining human embryological materials from the operating room. He spoke of this goal to the chief surgeon and others there. Such informal requests were the regular means of acquisition. In his autobiography, Corner gives this firsthand account of the receipt of one such embryo:

> One evening about nine o'clock Dr. Jameson had called me to say that he was operating on a woman for a ruptured tubal pregnancy . . . hastening to the hospital I found the patient still on the table. The specimen he had saved for me, a small blood clot, was in a basin on the steam radiator, where he had ordered it put to keep warm. I had arrived just in time to save it from cooking. I took it out to my room in the animal house, to which the night watchman admitted me (1981:205).

This turned out to be a perfect embryo. G. W. D. Hamlett (1935:273) carefully reported a similar case of receiving a tubal embryo: "I received the unopened ovum, still in the injured tube, shortly after the operation and less than three hours after the tube's rupture."

Fetal materials of multiple conceptions were particularly prized because of their rarity, multiple placentation, and the potential interrelation of the fetuses.[42] In 1935 A. W. Diddle and T. H. Burford reported extensively on premature quadruplets (which they said occur only once in every 654,455 labors). They wrote as follows of two such pairs of twins and two placentas: "Unfortunately, the placentae were not weighed nor preserved with the fetuses" (Diddle and Burford 1935:282). Surgeons providing materials were, in this instance as in the Corner case, not necessarily aware of the particular needs of scientists or of appropriate means of preserving such precious specimens.

Human Ova. Scientists who sought unfertilized human ova could not easily rely on physicians to supply medical wastes. Fallopian tubes removed for pathology were not a promising source because of the pathology itself, and it was considered clinically inappropriate to remove healthy tubes (Allen et al. 1930). Stimulated by primate research, [43] and developing their technique initially on dogs, Edgar Allen, J. P. Pratt, Q. U. Newell, and L. J. Bland reported in 1930 on a joint program through which they had recovered unfertilized ova from women. Since it was "important to tap abundant sources of material," they initiated a cooperative investigation at two hospitals (one in Detroit and the other in St. Louis), "utilizing all available cases from both these services" (p. 48).[44]

The method that Allen and his associates had developed allowed ova to be flushed in situ from normal fallopian tubes during otherwise necessary abdominal surgery "without danger of injury, which might result later in obstruction" (Allen et al. 1930:48). The procedure involved clamping off the cervix, directly injecting a saline solution into the uterus, and permitting the fluid to escape from one fallopian tube at a time while collecting it in a series of watch glasses. Ultimately materials were obtained from over ninety operations, with sufficiently normal tubes on at least one side in thirty-five cases. These cases yielded five identifiable and successfully sectioned tubal ova, along with several dubious specimens. The corpora lutea that were believed to be associated with the ova were also excised from the ovaries for examination (Allen et al. 1930:55).

In another restructuring of surgical events, Arthur T. Hertig and John Rock reported in 1939 on the "feasibility of recovering human embryos by the simple device of scheduling unavoidable hysterectomies according to the fertile period, as worked out on the monkey" (pp. 26–27; Hartman 1939:676).[45] This line of work, which Hertig and Rock pursued for fifteen years, was dubbed the "egg hunt." During the course of the project, "some thousand eggs" were "harvested" from women surgical patients undergoing total hysterectomy (McLaughlin 1982: 59–60). In what "unquestionably was a delicate ethical maneuver," Hertig and Rock asked women awaiting hysterectomies to keep rhythm charts and also note times of intercourse, scheduling the surgeries shortly after ovulation (McLaughlin 1982:63).[46] Over the next fourteen years, Hertig and Rock collected thirty-four fertilized eggs, "gleaned from 211 unknowingly pregnant women," representing the first seventeen days of life (McLaughlin 1982:66).

Not only were surgeries restructured but also new surgical methods were developed and applied in the quest for research materials.[47] On the whole, human embryo and ovum specimens were neither routinely available nor well distributed. Scientists actively recruited such material for many years to form complete collections. Even so, as Yerkes pointed out in 1925, "Generally, human subjects are more readily commanded for medical needs [than primates] because of clinical hospital resources and facilities" (p. 269).

Mundane Live Materials

Mundane live materials, though often acquired and used on an occasional basis, were also organized into colonies for routine research access. Here I examine cats, dogs, and other mundane species that became "laboratory" animals.

Cats, Dogs, and Surgical Precedents. Cats and dogs were routinely obtained for the animal houses of medical schools, where they were used in the teaching of anatomy, physiology, and especially surgery.[48] Local suppliers and other sources such as ASPCA, dogcatchers, pounds, and nearby farms associated with the laboratories were the usual providers.[49] The establishment of reproducing colonies of cats or dogs for research does not appear to have been pursued until the late 1930s.[50] Fresh supplies were therefore routinely needed.[51] Reproductive scientists, especially those in medical school settings such as Allen and his associates, enjoyed established and very easy access to cats and dogs as mundane materials through the animal houses of their medical institutions (Allen et al. 1930; Dawson and Friedgood 1940:411–29). However, the fact that the cat ovulates after coitus and the relatively long gestation periods of both cats and dogs made them less desirable than guinea pigs, rats, or mice as laboratory animals during the early twentieth century.

On the Making of Laboratory Animals. Several species of mundane materials were transformed into laboratory animals during the early twentieth century. Colonies of both research-only and research-and-production types were established, and their use became routine in the life sciences. Chief among the species that became laboratory animals in reproductive science were guinea pigs (Loeb 1911; Stockard and Papanicolaous 1917; Wright 1931), rats (Griffith 1942; Long and Evans 1922; Smith 1927:158–61), and mice (Allen 1922; MacDowell and Lord 1926). All of these materials were relatively inexpensive to acquire (often from local pet shops and cottage industry suppliers) and to maintain (feed, house, and care for). Harry Steenbock provides an example from outside reproductive science; he established "the first rat colony in history for nutritional experiments" at the University of Wisconsin in 1908 "by purchasing at a cost of six dollars from his own pocket a dozen albino rats from a pet-stock dealer in Chicago" (Schneider 1973: 1238).[52] Reproductive scientists could easily establish small rat colonies in a single room and fund them through normal departmental research budgets.[53]

One such modified animal material, the hypophysectomized rat, became "[one of the] most widely used tools of investigation in endocrinology and reproductive physiology" (Fulton and Wilson 1966:399–400). Philip E. Smith, who developed and refined the technique of hypophysectomy, showed that removal of the anterior pituitary resulted in cessation of growth, loss of weight, and atrophy of the reproductive system, the thyroid gland, and the adrenal cortex. The injection of appropriately active hormones could "repair" these disabilities.[54] The hypophysectomized rat was, in many respects, a tabula rasa for the testing of a variety of hormone extracts. Smith's technique, along with the small size, rapid gestation, and relative inexpensiveness of rats for colonies, made them ideal "laboratory" animals.

Exotic Live Materials

Exotic live materials were also organized into research colonies. Here I describe early opossum and primate colonies.

Opossum Colonies. Opossums were relatively mundane materials in terms of acquisition since they were native to many parts of the United States and could be easily trapped. Yet they were exotic in terms of the difficulties they posed for the establishment and maintenance of living and reproducing colonies. In 1912, Professor J. T. Patterson of the University of Texas offered Hartman an instructorship in his Department of Zoology, along with the suggestion that he study the embryology of the opossum (Vollman 1965). Why the opossum? Hartman answered:

> For the same reason that Patterson selected the armadillo: The prairie and the creek bottoms were full of both species and nobody had yet bothered to study the development of these, the most bizarre of all the American mammals. . . . [For thirteen years] I hunted the animals in the field. I also took care of them (at first at my home), operated on them, secured, fixed, and stained their eggs and photographed the sections myself (Vollman 1965:v).

Hartman, building upon J. P. Hill's research on the kangaroo, ultimately studied nearly one-thousand female opossums, from which several thousand eggs were obtained. Hartman then described the main features of the anatomy of the sex organs, the estrus, ovulation, and fertilization, the early embryology, the birth, and postnatal development in the pouch of the opossum (Vollman 1965: 407, vi).

The marsupial (pouched) opossum offered a particular opportunity or advantage for reproductive biologists, described as follows by Carl R. Moore and David Bodian (1940:319) of the University of Chicago:

> The American opossum (*Didelphys virginiana*), the only native marsupial on the North American Continent, presents decidedly unique mammalian material for many biological problems, particularly for those concerned with late embryonic development in which operative treatments are desirable. This is due to the fact that intra-uterine development is extremely short (12–13 days post-copulation) and later development occurs in the marsupial pouch where young are directly available. At birth posterior appendages are useless paddle-shaped structures but over-development of anterior appendages provided with claws permits the young to crawl among ventral hairs and reach the pouch. . . . The young find and attach to the nipples of the mammary glands and once attachment is gained fixation to it is uninterrupted until approximately day 60 to 70.

Thus not only is the opossum relatively accessible, but so are its embryos for most of the gestation period—providing striking and unique research opportunities.

Hartman established and maintained an early colony of opossums while at the University of Texas from 1912 to 1924 (Vollman 1965:3). He caught the opossums himself, and essentially created a minor and temporary colony while generating extensive specimens for study. Other zoologists and embryologists also attempted to colonize the opossum, but all met with failure until a major effort was launched during the experimental period of the Wistar Institute farm in Philadelphia (Moore and Bodian 1940:320). Dr. M. J. Greenman, then director of the Institute, had an "unflagging interest in the opossum over a number of years." In fact, it took five years of effort (1930–1935) by Edward McCrady to develop the colony into a reliably [re]productive one (McCrady 1938).[55]

McCrady's main challenge in colonizing the opossum was the prevention of rickets. His colony was routinely devastated by the disease (despite a daily diet including milk, eggs, and even cod liver oil as sources of vitamin D) until he discovered, by opening the stomachs of fresh-caught animals, that they ate small vertebrates, bones and all. The addition of bone meal to their daily diet cured the rickets. Even then, however, the opossums failed to breed because of ovarian atrophy. A hint in Hartman's earlier work led the Wistar group to redesign the opossum cages to require exercise, which ultimately resolved this problem of ovarian atrophy (McCrady 1938:11–14).

Moore and Bodian of the Departments of Zoology and Anatomy at the University of Chicago also established an opossum colony in the late 1930s, but in the city. An article of 1940 indicates that they accomplished this goal—at very low cost in an area not immediately surrounded by opossums—through a network of individual collectors and a biological supply house (Moore and Bodian 1940). Their collectors, located in several southern and midwestern states, obtained special permits for trapping during the closed season to acquire females with pouch young. Moore and Bodian thus overcame the problems of acquiring this especially useful animal through careful organization of collectors and suppliers, as well as through their own breeding colony.[56] Opossum colonies thus represented a serious and often long-term investment of scientists' resources.

Primate Colonies. Primates became particularly attractive as research materials in the post-Darwin era, as their close relation to humans became increasingly clear (Schultz 1971:2; Schmidt 1972:1–22).[57] While many scientists had kept a few live primates for observation and postmortem study, no serious research colonies existed until the 1920s (Corner 1981; Zuckerman 1970).[58] At that time, Yerkes and Corner initiated two major efforts in this direction.[59] According to the eminent primatologist Schultz (1971:7), these efforts signaled the shift of primate study from "what had previously been regarded as merely the hobby of a few into a respected field of interest that began to receive adequate support and due academic recognition in some leading universities"—a new line of scientific work. Primates were required for the effective pursuit of certain problems in reproductive science. Corner (1923), for example, required primates in order to study menstruation, which occurs only in such species. Yerkes (1935, 1943) focused on problems of psychobiology, including reproductive function and behavioral studies. For both Corner and Yerkes, expendable primate materials were essential.

Sociologically, primate colonies are of special interest because of the difficulty and expense of establishing and maintaining them. The expense of such colonies far exceeded normal university research budgets. The early colonies of Corner and Yerkes, like most later ones, drew on the fiscal resources of two or more major institutions, usually a university and a major foundation. If in the physical sciences it was the need for elaborate and costly equipment that commonly pushed researchers to seek new sources of outside funding, it was the need for research colonies that commonly prompted such action in the life sciences.[60] The high costs of live materials drew biomedical scientists into new relations with external funding sources and into the arena we now call "Big Science."[61] Primate colonies were the extreme example of this process within American biomedical science between the two world wars.[62]

The Hopkins/Carnegie Colony. Of the two initial efforts to establish primate colonies, the more famous was Yerkes' Yale Laboratories of Primate Biology, Inc., established at Orange Park, Florida, in 1929 (Yerkes 1935:618). As early as 1921, Corner, from the anatomy department at the Johns Hopkins Medical School of Baltimore, had established a colony of monkeys for reproductive scientific research. In his autobiography, Corner (1981:163–64) provided the following account:

> Obviously the best prospect was to use the one species, the Indian monkey, Macaca rhesus, that could be purchased at moderate cost and was readily kept in American zoos. . . . Dr. Weed found the money to buy eleven female monkeys. . . . This was the first monkey colony in the United States for long-term physiological experimentation.

Zoo staff were often the only resource to which scientists could turn for information on maintaining live animals in captivity. Corner wrote as follows about his preparatory work for starting his colony in Baltimore:

> I learned that at the National Zoo in Washington and the Philadelphia Zoological Society's zoo, rhesus were being bred successfully. I went to Philadelphia to talk with the superintendent of the zoo, Mr. Ned Hollister, his head keeper, Mr. Blackburn, and the zoo's pathologist, Dr. Herbert Fox. They showed me a large colony of rhesus monkeys living to my surprise out of doors in a converted bandstand year around. With such animals I could work at my home laboratory with all the necessary facilities ready at hand (p. 163).

Significantly, Corner specified the zookeeper by name. If they lacked formal education, the keepers nonetheless had "tacit knowledge," extensive, lifelong, and even intergenerational experience of pivotal significance for scientists seeking to establish colonies.[63] Animal keeping had been a quasi-inherited position in many private collections developed by wealthy individuals whose servants' whole families often worked for them over several generations.[64]

Corner (1923), completed the first major study of the menstrual cycle with these rhesus monkeys, stimulating parallel research on humans (e.g., Bartelmez 1933, 1937; Hartman 1939; King 1926). When Corner left Hopkins for the University of Rochester, Streeter, who became director of the Carnegie Institution's Department of Embryology at Hopkins after Mall's death, wanted to begin breeding rhesus for embryological and other reproductive investigations as part of a new and more experimental research thrust of the Department of Embryology.[65]

In 1925, Streeter invited Hartman, who was then planning an expedition to the Philippines to gather monkey embryos, "to join his laboratory and study the monkey embryology the safe and sane way, bringing the monkey to the laboratory."[66] Hartman inherited Corner's monkey paddocks and "got acquainted with the monkey" (Vollman 1965:7). Drawing upon his prior experience with opossums, Hartman succeeded in maintaining a large breeding rhesus colony in Baltimore.[67]

> [Hartman] provided the dated embryos that Heuser skillfully sectioned and Street analyzed and described. The result was one of the greatest achievements of American biology in its day, being the first complete description of the development of a primate from the one-cell stage through the whole embryonic period (Corner 1981:284).

Biomedical scientists used this colony in a wide variety of work, especially on reproduction and embryological development (e.g., Bartelmez 1935; Hartman 1932, 1939; and Heuser and Streeter 1941).

Yerkes' Promotion of Primate Colonies. Although the Hopkins/Carnegie colony preceded the Yale colony, it was Yerkes who most actively promoted primate research and colonies in the United States. He conceived such work at the turn of the century, and in 1916 published a famous article in *Science* on "Provision for the Study of Monkeys and Apes" (Yerkes 1916, 1963). Here Yerkes made a strong argument for primate research colonies, asserting first that the study of primates had been neglected:

> We have but fragmentary and unsatisfactory knowledge. . . . [Primates] are relatively difficult and expensive to obtain by importation or breeding, and to keep in normal condition. It is clear from an examination of the literature on these organisms and a survey of the present biological situation that the neglect by scientists of systematic study of all the primates excepting man is due, not to lack of appreciation of their scientific value but . . . to technical difficulties and the costliness of research (1916:231).

Such technical difficulties, Yerkes argued, could be addressed by careful planning and programming of research.

Yerkes was sensitive to the high costs involved in establishing and maintaining a primate research colony. He estimated the costs of a colony in southern California to be $50,000 per year, requiring an endowment of $1 million (1916:231). Asserting that his program would be cost-effective through the conservative use of expensive resources, Yerkes (1916:233) sought to justify major multi-institutional investments in primate research colonies on the grounds that they would ultimately help in solving practical human problems:

> I wish to emphasize the important relation of the plan I have outlined to strictly human interests and problems. It is eminently desirable that all studies of infrahuman organisms, and especially those of the primates which are most similar, structurally and functionally, to man, should be made to contribute to the solution of our own intensely practical, medi-

> cal, social and psychological problems. . . . [Such problems] . . . may be solved, at least in large measure, most directly and economically through the use of monkeys and anthropoid apes.

Finally, Yerkes (1916:234) called upon other biologists to support his plan, arguing that the task of securing adequate provisions for systematic and long-term study of primates is "by far the most important task for our generation of biologists, and the one which we shall therefore be most shamed by neglecting."

Yerkes' promotion of primate research colonies was an ambitious undertaking. He initiated this effort in the scientific community in 1916, promoted it throughout his tenure at the National Research Council (ca. 1919–1924), garnered Carnegie funds for preliminary investigations (ca. 1924–1925), and in 1925 obtained a Rockefeller Foundation grant to establish a pilot laboratory and purchase several chimpanzees (Yerkes 1916, 1925, 1935, 1963). Significantly, Yerkes drew special attention to international scientific competition in making his case for primate colonies in the United States. In his book, *Almost Human* (1925), Yerkes went to great lengths to compare the primate colonies and related research of W. Koehler in Germany, Nadie Kohts in the new Union of Soviet Socialist Republics, and the Pasteur Institute in French Guiana with the relative absence of such facilities in the United States.[68] These and other arguments were successful. In 1929, the Rockefeller Foundation, in cooperation with Yale University, funded the Yale Laboratory of Primate Biology in Orange Park, Florida (Yerkes 1935).

CONCLUSION

The shift of emphasis to experimental physiological approaches in the life sciences at the turn of this century in the United States led to radically different needs for research materials. This chapter examined some of the constraints, opportunities, and resources involved in pursuit of such materials in the case of early twentieth-century reproductive science. Access to materials was initially achieved by scientists themselves for both exotic and mundane materials. Inexpensive and easily acquired materials predominated.

Physicians, biologists, and agricultural scientists often "inherited" particular research materials and established routes of access to them from their professors and/or at a given site. Staffs of zoos, museums, hospitals, and other personal and professional networks also helped scientists achieve access to desired materials. Gradually, these scientists pioneered in establishing colonies for routine on-site access to materials. The acquisition of adequate supplies of materials represented a considerable investment of scientists' resources.

Once means of access were arranged for a particular material on an ongoing basis, scientists seem to have developed working commitments to continued use of these arrangements and research materials—continued cultivation not only of the actual materials, but of access arrangements as well.[69] In other words, once successful access to given research materials was established, scientists seemed reluctant to abandon these materials unless and until their usefulness was exhausted through examination of a wide variety of reproductive biological problems. Certain materials, such as the rat (hypophysectomized or not), were conventionalized to such a degree that they became laboratory animals in reproductive and other life sciences.

Laboratory colonies were the major institutional solution to the problem of routine access to live and fresh materials. The importance of biological colonies lay not only in their immediate convenience in terms of access, but also in the broader scope of the research problems that could be addressed through standardized live materials maintained over time. The expense of such biological colonies drew life scientists into the arena of "Big Science" through sponsorship by major foundations.

The centrality of research materials to the production of biomedical knowledge may seem evident, but the topic has received almost no scholarly attention. Modern biomedical research requires a highly organized materials infrastructure, including animal colonies. Infrastructural organizations qua emergent social institutions deserve historical and sociological examination, especially in terms of their interaction with the social organization of scientific work and the development of scientific problem structures over time. To date, the historical sociology of research materials remains itself embryonic. The study of the relations among scientists, their research materials, their home institutions, their external funding sources, biomedical problems, emergent lines of scientific work, and even general political issues should prove to be a highly rewarding arena of investigation.

EPILOGUE: STUDIES OF RESEARCH MATERIALS (RE)VISITED

In the eight or so years since I wrote the above paper on research materials in the case of reproductive sciences, studies of scientific practice have flourished, not least the study of research materials as key elements in the situations where sciences and technologies are pursued. There have been several veins in this recent work: 1) studies examining materials as infrastructure for various kinds of scientific research; 2) studies examining pro- and antivivisection movements and related ethical/political issues; 3) studies centered on particular tools and their appropriateness and roles in specific lines of research; and more theoretically driven studies which take up 4) the (re)presentation, symbolism, display, and deployment of materials including research materials as tourist attractions, and 5) the historical and contemporary constructions of human, nonhuman, hybrid, and cyborg in relation to the designation *research material.*

Each of these veins foregrounds different questions in science studies broadly defined, and some studies combine several veins. All are requisite for a broader understanding of materials in scientific research. Here, in epilogue to my earlier paper, I offer a very brief overview of these veins, and only as pursued in very recent science studies. My goal is to provide sites of entree for researchers interested in these topics, especially bibliographic entree, however partial. I conclude with some suggestions for further research.

Research Materials As Science Infrastructure

Rats and mice have certainly been at the heart of twentieth-century life sciences research. However, they have not, until recently, often been examined as key and dramatic infrastructural elements—with some exceptions, such as Holstein's (1979) study of the first fifty years at the Jackson Laboratory. Begun in 1929, as other such materials production institutions were also starting up (e.g., Lederer 1985), the Jackson Labs today remain a pivotal infrastructural element. Karen Rader (1992, 1993b) has just begun an ambitious project on the standardization of *Mus musculus* at the Labs, especially through Clarence Little's promotion of cancer research in the 1930s.

Another major infrastructural organization for rats was the Wistar Institute in Philadelphia. Clause (1993) provides us with an excellent institutional history of its development (see also Brosco 1991) and the refinement of the Norway rat as a "laboratory workhorse." Clause takes up the key issues of standardization of animal materials, and describes the trademarking of the Wistar rat as a standardized mammal. Recombinant processes were not requisite for TM status!

Hendrickson (1988) offers a wider if more popular overview, and Lindsey (1979) a more technical history of the rat in the lab that, as Clause (1993:331) argues, deserves further attention. On the West Coast, Herbert McLean Evans's laboratory at the University of California at Berkeley was a key distributor of the "Long-Evans rat," a locally bred refinement. Evans's dense archival materials, held at the Bancroft Library, have not even been sorted much less used for historical research, which would also allow comparative attention to the rise of the American West as research site initially lacking adequate infrastructure. Like others, Evans and his colleagues used sales of rats to support their research shop; this is also an understudied phenomenon despite such labs having been "obligatory points of passage" (Latour 1987) for many other researchers.

The institutionalization of primate research materials has also begun to be examined. Haraway's *Primate Visions* (1989) considers the establishment of early (pre-World War II) primate colonies. She situates these colonies in their appropriate colonialist contexts, examining both laboratory and field research sites (see 1989: chap. 2, 4, 5; 1991: chap. 3). Hagen (1990) has examined problems in the institutionalization of tropical biology, focusing on the Barro Colorado Island Biological Laboratory. Backman (1983) focuses on more recent studies of reproduction using primates, attending to research networks in relation to the Regional Primate Research Centers via citation analysis. Nationalism, colonialism, and other competitive elements in the establishment of research colonies and these Centers deserve much more attention, especially given some of their pre-cold war origins.

When research materials are viewed as infrastructure, Oudshoorn (1990) points out, we can see how they can and do create the very kinds of networks Latour (1987) suggests we study. Oudshoorn examines how Dutch laboratory scientists in reproductive endocrinology had to create networks with both gynecologists and pharmaceutical companies to meet their materials needs. The pharmaceutical companies provided glandular material which they in turn extracted from slaughterhouse remains, while the gynecological clinics provided vast amounts of urine from pregnant women, the "ideal source" of certain female hormones. The gynecological clinics also then served as loci of distribution of "finished" hormonal products as treatments for women. Oudshoorn clarifies how the absence at that time of a specialty in men's (reproductive) health as both source and distribution point limited the development of male hormonal products. Gender bias in science was not only reinforced but metamorphosed in this process. Whether the still emergent specialty of andrology in medicine will change these patterns almost a century later remains to be seen.

Drosophila have been objects of study for most of this century, engendering considerable analysis of their utility. In terms of social studies of science, Allen (1975) led off in this area, and Kohler (1991a, 1991c; 1993a,1993b) has widened the scope and deepened the analysis. Using *Neurospora* as a useful comparison case, his work obviously moves across the categories considered here. Kohler's (1991b) work on the development of moral economies of scientific practice leads us nicely into the domain of vivisection studies.

The Use of Living Materials and Vivisection Movements

Lederer has focused her research on the organization and use of living materials in research and consequential relations with vivisection movements for some years. Her work began with the use of humans in particular lines of research, such as Noguchi's Luetin Experiments (Lederer 1985) and Wile's study of syphilis (Lederer 1984). Lederer (1987) then moved on to study the controversy over animal experimentation in America, 1880–1914, and how this controversy has itself shaped the biomedical research literature in terms of how the work of research is—and is not—(re)presented in the reports (1991b).

Lynch (1988b) brings a sociological perspective to bear on the processes of sacrifice as a ritual practice as part of the laboratory culture in the neurosciences. His concern centers on how the animal body is transformed into "scientific object." Some other recent studies of animals in medical research have been done by Paton (1993[1984]); Rowan (1984) who focuses especially on mice; the volume edited by Rupke (1987) placing vivisection in historical perspective; and Singer's (1990) work on animal liberation. Historian Tansey (1994) has examined experiments on dogs in Britain and the intensification of resistance to the research use of animals that "count" more commonly as pets. Guerrini's (1993) recent work is titled "Animal Tragedies: The Moral Theater of Anatomy, 1660–1750."

Particular Materials and Lines of Research

This has been the most developed of recent foci on research materials. Haraway has followed primate materials into different lines of research including space research, the new physical anthropology, psychology, paleoanthropology, and primatology, including repatriation as conservation (see 1989: chap. 6–15). Lowy (1993) has begun to examine the role of mice in the development of medical oncology in the United States after World War II.

Several papers in Clarke and Fujimura's (1992b) volume on practice in twentieth-century life sciences focused on particular organisms and lines of work. Kimmelman (1992) examined corn as a strategic organism for the organization and display of "vested" interests in agricultural genetics early this century in the United States, where it was agricultural scientists who pioneered genetics research, but were soon displaced. Displaced too were their interests in a more physiological rather than hereditary genetics. This latter theme is also taken up by Mitman and Fausto-Sterling (1992), who ask "Whatever happened to *Planaria*?", which had been a key material in studies of the physiology of inheritance. Both corn and *Planaria* were marginalized vis-à-vis what became the powerful *Drosophila* machine, so that we begin to see the outline of careers of various research materials over time, space, and circumstance. Star (1992a) continues a career or trajectory framework in her examination of the rise and fall of taxidermy in the preservation and (re)presentation of research materials in natural history. In addition, she takes up the intersections of (marked) race and gender in the elaboration of scientific infrastructure and the invisibility of technicians.

The collection of papers recently published in the *Journal of the History of Biology* under the frame "The Right Organism for the Job" is a major contribution to this line of science studies research. The authors by and large are attempting to answer the question why a particular material was right or wrong in a specific historical situation of scientific practice. Right and wrong here are generally constructed via a scientist's-eye view. Lederman and Tolin (1993) take up viruses and add to the usual viral genetics focus a broader view of virology over the past century. Summers (1993) examines how bacteriophage came to be used by the "Phage

Group," arguing that such stories are intrinsically important given that alternative paths and materials are and were usually available.

Zallen (1993) begins by reiterating A.H. Sturdevant's (1971) distinction between "wise" and "lucky" choices of materials. In the latter case, scientific progress was attributed to properties of the organism not known at the time of its selection. Zallen argues for a third position—"inevitable choices"—and finds that certain green algae fit this bill admirably in the study of photosynthesis. (Clause's [1993] work on the Wistar rat published with these was discussed above.)

Kohler (1993a, 1993c) continues his explorations of *Drosophila*, here from an ecological, natural history perspective. He argues that *Drosophila melanogaster* were "just ideal for life in the laboratory" compared to other species of *Drosophila* which were much less cosmopolitan and hence less able to be successful "camp followers" of humans, including following them into the laboratory. Like Kohler, Holmes (1993) takes up one of the "classic materials" in the history of science, the frog in experimental physiology. This project reminded Holmes, along with many of us, of "pithing experiences" in biology classes from long ago. Demonstrating remarkable personal archiving, we are even treated to Holmes's own drawing of his frog for a high school zoology course in 1948!

This ambitious set of papers ends with Burian's (1993) elaboration of how—the many different ways in which—the choice of experimental organism does matter. He asserts that there may be a disjunction between theoretical "rightness" and practical—in practice—rightness. The latter is usually temporary and more or less local or regional.

(Re)Presentation, Symbolism, and Display of Materials

One of the most interesting current lines of research in social studies of science generally and materials in particular focuses on processes and modes of representation and analyses of symbolism and display. Lynch and Woolgar's (1990) edited volume broke new ground in this approach in the life sciences. Lynch (1990) examined the externalized retina in terms of the selection and mathematization of visual documentation. Myers (1990) offers a powerful analysis of the politics of display in a study of the illustrations in E.O. Wilson's *Sociobiology* (see also Latour 1990a). Ruse and Taylor (1991) edited a journal issue on pictorial representation, in biology and the introduction by Taylor and Blum is a most helpful entree.

In *Primate Visions*, Haraway (1989) renders processes of representation, symbolism, and display as central phenomena. In fact, the old Yerkes field station for primate research in Florida that Haraway (1991b, chap. 3) examines is now a tourist attraction, inviting further study. Fox and Lawrence (1989) include some images of research in their study of images and power in medicine since 1840. Lederer (1991a) has analyzed the symbolic representation of dogs as shifting "from pets to subjects to research dog heroes."

A number of people have also begun to examine cinematic attention to what we would consider research materials, or cinematic representations of research using animal materials. Cartwright (1992) looks at what were called "experiments of destruction" as physiology was initially subjected to cinematic inscriptions, such as the death by electrocution of an elephant. She argues that experimentalists regarded the cinematic devices offered up ca. 1895 as improved versions of photographic apparatuses they were already using. Even the earliest films were often used in presentations to the lay public, "extending the senses" (Borell 1986) of scientists and nonscientists alike.

"Laboratory Life on the Silver Screen: Animal Experimentation and the Film Industry in the 1930s" is the title of a recent paper by Lederer (1993), taking up vivisection among other themes. Mitman (1993a) has begun a major project centered on the use of film in animal behavior studies and its relation to popular culture. This includes examining the "private life" of the dolphin as recorded via underwater cameras in display tanks which were constructed to simultaneously serve tourism and research purposes (Mitman 1993b).

Human/Nonhuman/Cyborg: Continuities and Divides

Last, Haraway (1985), Latour (1987, 1988d, this volume), Callon (1986b), Callon and Latour (1992), and others have drawn our attention in social studies of science and technology to the significance of the nonhuman in various practices, asserting the value of attending to nonhumans as actors or actants in heterogeneous situations, including scientific and technological research sites. This contested development is the focus of much concern about issues such as agency (see, e.g., Ashmore et al. 1994).

Life sciences research materials generally fall into the domain of the nonhuman, and some can certainly epitomize a recalcitrance so consequential that the significance of the nonhuman in scientific practice is vividly demonstrated, such as the need for squeeze-boxes for primate monitoring and injection (Clarke 1987). But what counts as human or nonhuman is a constructed boundary and, somewhat like the color line which separate races in segregating situations, the boundary moves—is historical, located, specific, and negotiated (Park 1952). For example, Casper (1994a, 1994b) finds that fetal positions in science and medicine lie at the margins of humanity. If the fetus is a live object for antenatal surgery, it is vividly humanized, granted a dominant personhood. If instead it is no longer alive and is desired tissue for research purposes, it is rapidly shoved over the boundary line and thereby transformed into the nonhuman. Petchesky (1987) analyzes fetal images as part of the deployment of visual culture in current reproductive politics in the Unites States centered on the question of when the fetus becomes human. (See also Maynard-Moody 1984[1979]; Steinbock 1992; and Duden 1993.)

Cyborgs pose yet another set of boundary disputes in social studies of science in terms of "what counts as what," when, where, and so on. Haraway (1991a), Downey et al. (in press), and others (e.g., Hogle 1993; Casper, in press; Clarke, in press b) offer perspectives that will likely complicate notions of the human/nonhuman "divide," rendering it as more conditional and varied, perhaps even episodic. For example, Hogle's work (in press b)on the procurement of human biological materials and their technological transformations and deployments frames the possibility of indeterminacy.

Suggestions for Further Research

A piece of good advice I want to share in thinking about suggestions for further research came from Anselm Strauss, my dissertation advisor. He said, "Study the unstudied," which is how I dared get into the study of research materials in the first place. So what remains unstudied or understudied in the area of research materials? Quite a bit.

In the paper above, I discussed exotic, mundane, medically supplied, on-site colony, and biological supply house materials. Actually, I only pointed to the need for further studies of the development of biological supply houses and to my knowledge, no one has addressed this need to date from a science studies perspective. Here I would argue that the focus should be on the particular houses and whatever they did ("ended up doing" as the organization itself developed over time) rather than just particular research materials. That is, if materials lines were part of a broader supply framework, we need to understand that framework as it would likely have influenced the overall organization of many elements of scientific research and teaching infrastructure. Whether and how the networks of consumers of the products of supply houses expanded from local to regional, national and even international sites is particularly of interest as this involves yet another set of infrastructures—transportation. Studies of the development of lab-based colonies such as that of Evans noted earlier are also needed. I offered above a preliminary chronology of colony development in terms of needs, commitments, and organizational elements; such case studies would establish conditions for expansion, contraction, and even extinction.

Another relatively unstudied phenomenon is the procurement of research materials from surgical and related medical procedures. Here it is likely that a single kind or genre of material will of necessity be the focus of study because such arrangements for procurement are likely to be highly local and individual. Entree into a network of researchers around the country focused on similar problems would probably be the most effective means of data collection. Given the remarkable changes in procurement processes in medicine for various kinds of transplantation over the past two decades (Fox and Swazey 1992; Hogle 1993, 1994, in press a; in press b; Kimbrell 1993), there are likely parallel elaborations in procurement of medical materials for research that will be of considerable interest.

Certainly continued study of processes of procurement of fetal materials for research will be fascinating. We can anticipate that introduction of RU486, the "French abortion pill," will rupture the taken-for-granted organization of the delivery of abortion services in the United States, now in free-standing clinics that could have facilitated the collection of fetal materials. The RU-486 pill can obviously be given in physicians' offices which could reduce the need for free-standing abortion clinics to women beyond the ninth week of pregnancy—about half of all U.S. abortions. The desire to use fetal materials to treat Parkinson's and other diseases and potentially for nonscarring surgeries will be strong, while access to them will be impaired in new ways of particular interest to those in social studies of science (Casper 1994b; Clarke and Montini 1993; Annas and Elias 1989).

Obviously the presentations and symbolic representations of research materials for scientific and for tourist and other popular cultural purposes deserve considerably more study. For example, zoos have shifted by and large from zoological gardens for gentleman scientists to tourist attractions and now, due to the vulnerability of many species to extinction, also serve research and reproductive purposes. Some collections are largely colonies. The intersections of particular species with research and display orientations, including procurement expeditions, would make excellent research topics, as would zoos as organizational phenomena (see, e.g., Bendiner 1981; Bridges 1966; Crandall 1966; Green 1985).

Last and far from least is the need for study of the development of animal technician as a professionalized occupation with specialized training and licensing requirements, almost but not quite completely ignored (e.g., Shapin 1989). The shift from getting janitors not to destroy experiments while cleaning, to having graduate assistants, to having technically expert assistants often more knowledgeable about specific materials than the researcher is quite significant (Clarke 1987). Sources for such research are quite bountiful both in the training texts and in the methods sections of research papers in different fields of the life sciences between the wars and after World War II. This may turn out to be a particularly fruitful area in which to examine the intersections of race and gender with parascientific training (cf. Star 1992a; Strauss, et al. 1964:22).

In sum, while some excellent work on research materials has been done, we have far from exhausted the possibilities or the kinds of payoffs that studies of infrastructural elements such as research materials will yield.

NOTES

1. This paper is reprinted here with permission as originally published in a volume edited by Gerald Geison (1987) with a new epilogue. An earlier version was presented at the meetings of the American Association for the History of Medicine, San Francisco, CA., 1984. I am grateful to Howard S. Becker, Kathy Charmaz, Nan P. Chico, Joan H. Fujimura, Elihu M. Gerson, Marilyn Little, Sheryl Ruzek, Leonard Schatzman, Susan Leigh Star, and Anselm L. Strauss for their support of the project on which this paper was based. Gerald Geison (organizer), Merriley Borell, Jane Maienschein, and other participants in the conference from which this

chapter emerged also provided invaluable assistance. The Special Collections of the University of Chicago, the University of California, Davis, The Chesney Archives of the Johns Hopkins Medical Institutions, the Carnegie Institution of Washington, and the Rockefeller Archives graciously allowed me access to unpublished materials The research has been supported by the Tremont Research Institute, the University of California, San Francisco, and the Rockfeller University.

2. This shift of emphasis took place in most areas of biological, medical, and agricultural research. See, for example, Allen (1978), Clarke (1985), Coleman (1977; 1985), Geison (1979, 1978), Gerson (1982, 1983b), Harvey (1976), Rosenberg (1976,1979a), Rossiter (1979).

3. For some of the leading examples of such developments, see Long (1987), Borell (1987), and Maienschein (1987).

4. For analyses of problems of disciplinary emergence and substantive cases, see Clarke (1985), Geison (1981, 1983), Graham et al. (1983), Kohler (1982), Lemaine et al. (1976), Pauly (1984), Farber (1982a), Rosenberg (1976, 1979a, 1979b).

5. In terms of the history of materials, for example, articles have appeared on the history of cell lines and tissue cultures, but the question of biological materials is largely subordinated to the careers of the individuals who developed them and their institutional settings. See, for example, Bang (1977) and Harvey (1983). Works focused on other matters occasionally discuss biological materials tangentially; see, for example, Coleman (1977), Corner (1981), Latour (1984), Lederer (1984, 1985). On instrumentation in physiology, see Borell (1987).

6. The blurred boundaries between disciplines at the turn of the century were clear to the historical actors themselves. Ross Harrison, for example, once stated that departmental affiliations, what he called "present fortuitous attachments," were often a matter of institutional contingency. See Blake (1980).

7. See, for example, Becker (1982), Bucher and Strauss (1961), Bucher (1962), Freidson (1976), Hughes (1971a), Strauss (1975, 1978b, 1985). My own approach draws on the grounded theory research method as developed by Glaser and Strauss (1967), Glaser (1978), and Strauss (1987).

8. See, for example, the varied perspectives of Gerson (1983a), Latour and Wolgar (1979), Law (1974), Knorr-Cetina and Mulkay (1983), Star (1983, 1985) and Volberg (1983).

9. Robert Frank has noted that materials, instruments, and techniques are commonly "packaged" together in scientific work. A classic example is the hypophysectomized rat, discussed later. For an analysis of

the relation of such packaging to overall scientific research production, see Fujimura (1987).

10. There is considerable debate about whether this was really a shift, or whether it was instead an integration into biology of approaches previously reserved to medical domains, and about whether the changes were revolutionary or evolutionary in character. See Allen (1978, 1979, 1981a, 1981b), Benson (1981, 1985), Churchill (1981), Coleman (1982, 1985), Farber (1982a, 1982b), Maienshein (1981, 1983), Maienschein et al. (1981), and Werdinger (1980). For analyses of the shift in population biology, see Gerson (1982).

11. My chief sources here include the Frank Rattray Lillie Papers [hereafter, the Lillie Papers}, the Zoology Department Papers [hereafter, the Zoology Papers], University of Chicago Archives, and the President's Papers [hereafter, the President's Papers] from Special Collections, University of Chicago Archives; the Embryology Department Papers of the Carnegie Institution of Washington, DC (also held at the Johns Hopkins Medical Institution, Chesney Archives); the Animal Science Department Papers [hereafter, the Animal Science Papers], University of California, Davis, Special Collections; the Rockefeller Foundation Papers, Rockefeller Archives Center, North Tarrytown, NY; and the Hunterian Laboratory Papers, Johns Hopkins Medical Institution, Chesney Archives.

12. Lillie to President Judson, 10 December 1914, p. 3 and part of his proposal, "Department of Zoology; The Chief Need," p. 1, Zoology Papers, box IV, folder 2.

13. Lillie also noted that Princeton University was currently building such a vivarium at a cost of about $25,000; see E. G. Conklin to Lillie, 23 January 1915, Zoology Papers, box IV, folder 2. Such sharing of architectural plans and facilities proposals was common practice in scientific networks.

14. See, for example, the President's Reports to the Regents of the University of California, ca. 1910–1940, sections on Gifts to the University, especially to the Museum of Vertebrate Zoology.

15. The application of statistical methods to materials and colonies is a problem worthy of investigation. See, for example, Kevles (1985) and MacKenzie (1981).

16. Elsewhere (Clarke 1985) I discuss these assumptions of physiological parallelism and argue that they led to systematic research biases that routinely emphasized similarity and deemphasized diversity both within and across species. Cf. Wimsatt (1980).

17. Such collections are expensive to develop and maintain, whether of zoological or botanical specimens. See, for example, Rothschild (1983) and, in botany, Volberg (1983).

18. The secondary literature has little to say about collecting expeditions for morphological specimens or expeditions for live animals for zoos. On morphological expeditions, see Haraway (1984), Attenborough (1980), Bendiner (1981), Bridges (1966), and more popular books such as Buck with Anthony (1930), Durrell (1964), and Webb (1953).

19. I am indebted to Leigh Star for this insight.

20. The issue of access to materials is, of course, not limited to mammalian reproductive science, but rather affects most if not all sciences. For example, one of the major features of both the Naples Zoological Station and the Woods Hole Marine Biological Laboratory was ease of access to plentiful and fresh materials, especially marine invertebrates. In these cases, the scientist went to the site of the materials to do research. See, for example, Allen (1978), Maienschein (1981a, 1983, 1985b, 1985c), and Werdinger (1980).

21. See Becker (1982) for a discussion of the ways in which organizational commitments become conventionalized, and the ultimate consequences of such conventions.

22. See Clarke (1985). I am indebted to Reuben Albaugh, extension agent emeritus, University of California, Davis, for an interview that confirmed this aspect of his work. The Animal Science Papers record expenditures for herds and research. The [Annual] Reports of the Secretary of the University of California, ca. 1890–1940, document the not insignificant income (used for subsequent research) that various agricultural programs generated through services to farmers and ranchers and through sales of animals and products that grew out of such research.

23. The history of materials, record keeping, and the breeding of colony animals evolved into the notion of controlled, standardized materials, perfected models with extensive control of genetic, environmental, and other sources of variation. An early criterion was known age; a later criterion was freedom from certain parasites or pathologies: recently, a host of new standards has been invoked. See, for an early example of an argument for standardization, the Hull Zoological Laboratory Report, President's Report, University of Chicago Archives, 1902:438. See also Yerkes (1916, 1935–1936), and Saver (1960). Also of interest are the Rockefeller Foundation Papers on the National Research Council's work on "Genetic Stocks, 1939–1941," Rockefeller Archives Center, RG1.1, S200D, box 31, folders 1835–1836.

24. Whitman, chair, and faculty members J. M. Coulter, T. R. Williston, Frank R. Lillie, Edwin O. Jordan, and William I. Tower for the biologi-

cal departments, to Martin Ryerson, President of the Board of Trustees. 1 December 1906, President's Papers 1, box 13, folder 12.

25. President's Report, University of Chicago Archives, 1902:437. There seems to have been some misunderstanding of the meaning of the term *biological farm.* Whitman (1902a:509) drew a careful distinction between a farm as a research endeavor and public zoological or botanical gardens. In the President's Papers at the University of Chicago, however, the biological farm proposal was filed under "Animal Husbandry," likely reflecting earlier misunderstanding. See President's Papers 1, box 7, folder 1. The Rockefeller Institute purchased a New Jersey farm for laboratory animal production in 1908, subsequently called "Hell Farm" by antivivisectionists (Lederer 1985:37). Many years later, Geoffrey H. Bourne referred to primate research centers in their breeding capacity as "animal farms." See Bourne (1973:490).

26. Hartman further noted that Corner later used materials from rabbits, monkeys, and occasionally dogs, guinea pigs, and humans in his research on the corpus luteum. Applying cytological criteria, Corner was able " as early as 1945 to distinguish seven stages of pregnancies as reflected in the corpus luteum of the sow"; parallel research on the rat by Joseph Long and Herbert Evans followed six years later.

27. The funds were provided by the Department of Anatomy (Corner 1981:159).

28. Wilson was then a member of Corner's Department of Anatomy at Johns Hopkins (Wilson 1926: 418–32).

29. When abattoirs disappeared from New York City, as zookeeper Crandall noted, they could no longer keep certain animals such as vampire bats for which they had relied on abbattoirs for a regular supply of fresh blood (Crandall with Bridges 1966:21).

30. See Lillie to G. O. Fairweather, 3 October 1918, Lillie Papers, box VII, folder 14. There is routine mention of the stockyards as a source of materials in both the Lillie Papers and the Zoology Papers.

31. Lillie to Wallace Heckman, Faculty Exchange, 14 March 1919, Lillie Papers, box VII folder 15.

32. Zondek's comments on stallion urine as an excellent source of female estrogenic hormones are especially interesting.

33. I am indebted to Andrew Nalbandov, professor emeritus of the University of Illinois, for the interview on which this information is based.

34. This was considerably easier than collecting sperm from many domestic species for artificial insemination. See Herman (1981).

35. For this information, I am indebted to Perry T. Cupps and Hubert Heitman of the University of California, Davis.

36. I have not included some fascinating accounts of access to cadavers at Hopkins and Rochester due to limitations of space and because of their tangential use in reproductive biological research. But see Sabin (1934) and Corner (1981).

37. Letter from Mall to President Woodward, 21 March 1914, Department of Embryology drawers, folder "Embryology Miscellany 1," Carnegie Institution of Washington (hereafter, Embryology file). The Department of Embryology, Carnegie Institution of Washington, established in 1913, was physically located at and was closely associated with Johns Hopkins University and its medical school. See Mall (1903:33–39); and "The Embryological the Carnegie Institution," undated but printed document, Embryology file.

38. See "Data Bearing Upon the Cause of Abortion and Age of the Embryo," undated form bearing Mall's name and the Johns Hopkins Medical School address. Embryology file.

39. "Request . . ." from Mall, June 1913, Embryology file. This document appears to have been mailed to departmental alumni of the medical school.

40. Letter "Dear Doctor," from Evans, undated, Embryology file.

41. For example, the Department of Embryology also collected alligator, monkey, and baboon embryos. For alligators, see letter to President Woodward from George L. Streeter, 4 February 1919, regarding a trip to British Guyana by Albert M. Reese of West Virginia University, Embryology file. Regarding monkey embryology, see Personnel file of Dr. Chester H. Heuser; and, regarding baboon embryology, see Personnel folders of George L. Streeter, letter to O. A. Scherer from Corner, 20 August 1947; Personnel drawers at the Carnegie Institution of Washington.

42. See Lillie's (1917) paper on the freemartin calf for the classic study of fetal endocrine interrelations.

43. Allen and his associates (1930) reported that Corner had recovered the first primate ovum from rhesus (1923), Allen recovered three tubal primate ova (1928), and degenerating uterine ova were recovered by Corner (1923) and Hartman (1924, et al.).

44. There is no mention in their article of informed consent processes in relation to the research.

45. In their article Hertig and Rock do not, in fact, discuss their means of access to the materials; this was likely done in conversation with Hartman, who reported it. See also McLaughlin (1982), especially

chap. 4. McLaughlin notes (p. 61) that Hertig had worked with Streeter at the Carnegie Department of Embryology in 1933 on primate embryology.

46. McLaughlin pursues the informed consent and abortion issues: see pp. 63–64.

47. The results of this research focused on (1) the timing of ovulation in terms of the extrusion of the first polar body; (2) the stage of development of corpora lutea in relation to ovulation; (3) the timing of ovulation in relation to the menstrual cycle and related epithelial changes; and (4) determination of whether ovulation always accompanies menstruation (Allen et al. 1930:73–74).

48. See the Hunterian Laboratory Papers in the Chesney Archives at Johns Hopkins.

49. Ibid. Antivivisectionists often influenced the access of scientists and faculty to these mundane materials, and animal house managers went further and further afield to achieve their quotas.

50. See the Rockefeller Foundation Papers on the National Research Council's work on "Genetic Stocks, 1939–1941," Rockefeller Archives Center, RF RG1.1, S200D, Box 51, folders 1855–6. The Rockefeller Foundation provided extensive support to Charles Stockard, Department of Anatomy, Cornell University Medical School, to develop a colony of genetically standardized dogs.

51. Aside from works on antivivsectionism, these issues have not been subjected to adequate historical research.

52. I am indebted to Merriley Borell for locating this information. Fruit flies were also colonized with considerable success in genetics research. See Allen (1978).

53. The development of colonies of smaller laboratory animals deserves study, especially in relation to the emergence of the biological supplies industry. See, for example, the Hunterian Laboratory Papers, Chesney Archives at Johns Hopkins.

54. For a detailed explanation of Smith's technique, see Agate (1973).

55. McCardy notes (on p. 9) that in his own research he also utilized materials collected and left by Hartman at the Wistar Institute.

56. See also Coghill (1939).

57. I use the term *materials* throughout this chapter to refer to both living and dead animals, both as wholes or parts. Some contemporary scientists have objected to use of the term *materials* in reference to nonhuman primates, and specifically to the term *primate materials.* They are accustomed to using the term *subjects* for the nonhuman primates they

study. In contrast, my focus here is not on the animals per se but rather on their organization in the service of scientific research, as part of the requisite infrastructure. Hence the term *materials* seems appropriate, if not customary.

58. Animals displayed in some zoological gardens were occasionally used for research purposes. See Pocock (1906) and Zuckerman (1930). They were not, however, intentionally organized for research but rather for public observation. Research was usually carried out on them only after their death. Circuses were another source of primate cadavers. See Schultz (1971).

59. Among other such efforts were those of Allen at Washington University, St. Louis, and Gertrude Van Wagenen and John Fulton at Yale. See Backman (1982, 1983), Allen (1926), and (van Wagenen 1950).

60. Regarding the physical sciences, see Reingold (1979). Regarding the life sciences, see, for example, the Grants-In-Aid files of the Rockefeller Foundation, RF RF1.1 S200 B38 F433 and passim, Rockfeller Archives Center. In 1939, requests to the Rockefeller Foundation for support of various colonies to produce standardized materials became so numerous that Warren Weaver of the Foundation requested the assistance of the National Research Council in developing an improved understanding of the needs in this area to which the Foundation could best respond. See the Rockefeller Foundation Papers on the National Research Council's work on "Genetic Stocks, 1939–1941," RF, RG1.1 S200D, Box 31, folders 1853–1856, Rockefeller Archives Center.

61. Foundations and other philanthropies had earlier sponsored buildings and expeditions to collect specimens. See, for example, Haraway (1984).

62. By comparison, Sewall Wright's major colony of guinea pigs at the University of Chicago after 1924/1925 was costly for its day, with an annual feed budget of about $2500–$3000 (Zoology Department Papers, box 11, folder 3). However, such figures did not begin to approach primate colony costs.

63. For example, the head zookeeper of the New York Zoological Park in 1980, Samuel Stacey, was himself the son of the Duke of Wellington's water bailiff in charge of water fauna. As a youth, Sam Stacey had become the Dutchess's "bird boy" and eventually moved onto the London Zoological Gardens as a bird keeper, coming to New York for the same purpose and to train a staff of bird keepers (Crandall 1966:3–4).

64. For an account of a private primate colony in Cuba, see Yerkes (1925).

65. Regarding this shift of focus, see Heuser and Streeter (1941). The Department had been committed for many years to the development of a human embryo collection and related embryological work. Streeter argued that a monkey colony would allow conceptions to be controlled and fetuses to be dated (unlike those of human materials), while a host of other problems could also be pursued. For the full rationale for this colony, see Memorandum to Doctor Merriam from George L. Streeter, 20 April 1925, folder "Embryology-Director, 1913–1935." Department of Embryology files, at the Carnegie Institution in Washington, D.C.

66. Hartman quoted in Vollman (1965:vii).

67. In Vollman's words: "Thus, for the first time, a primate became a laboratory animal." Somehow Corner's earlier work was ignored in this claim, which is odd in that (according to both Corner and Hartman) Hartman's breeding practices were based in part on Corner's research, which had yielded crucial data on the timing of fertility and the menstrual cycle that was helpful in breeding programs (Vollman 1965:vii). Hartman also published on the development of primate colonies. See Hartman (1930, 1945).

68. Quite similar arguments were made regarding federal assumption of sponsorship of major primate colonies. See Eyestone (1966).

69. For a discussion of such commitments and their consequences, see Becker (1982) and Gerson (1976).

MICHAEL LYNCH

6

Laboratory Space and the Technological Complex: An Investigation of Topical Contextures

THE PROBLEM OF "WHERE" THE ACTION OCCURS

If we are able to recognize what kind of space we are in, we find we possess implicit knowledge of how it is customary to behave there.
—Shapin 1988:390

When I began my ethnographic studies of laboratory practices in 1974, I was struck by the fact that these actions were not always transparent to my attempts to observe them. At that time I was attending laboratory sessions in an introductory biology course where students performed exercises with the microscope. I was interested in the social process through which students learned to see with the instruments, and I was given permission to observe and tape-record the sessions.

From a vantage point within the laboratory I saw a group of about twenty students and an instructor dispersed along two rows of benches in the room. Several binocular microscopes were placed on the benches, and students clustered in groups of two or three around each instrument. Supplies of pond water, slides, cover slides, eyedroppers, and stains were dispensed at the end of one of the benches. At the beginning of the period the instructor advised the group on how to prepare the slides. These preliminary instructions outlined procedures to be followed and precautions to be taken, and they included a sketch on the characteristics of animal and plant cells that students were supposed to recognize during the exercise. Each student also possessed a manual of instructions, which was consulted at various times during the exercise.

Following these preliminaries, the students set about gathering the materials, arranging the equipment, and peering through the oculars of their microscopes. While doing so they chatted with one another and with the instructor about what they saw, asked questions on whether they were seeing what they were supposed to see, and solicited instructions on what to do next. What I saw was a spectacle of students handling bits of material and peering into the instruments. I could make sense of their actions and talk only by imagining what it was that they might be seeing "in" the microscope. I too could peer into the microscopes—and on a couple of occasions students enlisted my help, as though I were an adjunct instructor—so it was not as though each student had access to a completely private world or that I was ultimately blocked from understanding what the students might have been talking about.

The problem was an analytic one. Where was the *action* occurring? Or, in Gooding's (1988) terms, what "activity field" did the scene implicate? Standing along the back wall of the laboratory with my tape recorder I witnesses a cacophony of voices and a complex array of body movements and gestures. Had I videotaped this spectacle, I would have obtained a detailed spatiotemporal record of these voices and movements. Such a record would not give access to the visual fields that the students presumably witnessed in their instruments and that provided a frame of reference for their locational expressions (e.g., "The chloroplast is that green thing at the bottom of the field"; "It's out of focus"; "That's a bubble under your cover slide, not a cell"). Indeed, such a record would provide little or no basis for discerning the rich texture of activities within the developing witnessable events in the microscopic field. To analyze such a record would be analogous to analyzing a videotaped recording of a symphony with the sound turned off. Such a record would provide a dense texture of analyzable events, but it would deeply inhibit any effort to discern the actions expressed *in and as* musical events.

I continued to encounter more elaborate versions of this problem in subsequent ethnographic studies of researchers' activities in astronomy and the neurosciences (Lynch 1985a; Garfinkel et al. 1981; Lynch and Edgerton 1988). In each of these studies, although always in a quite different way, I had to discover how to *witness* the spatial and temporal "environments" of laboratory activities, since many recognizably spatial and temporal references were initially opaque, undecidable, ambiguous, and strange. These references did not elaborate an already familiar mundane world; they did not readily attach themselves to the social and institutional identities, architectural environments, and intelligible activities that I took for granted as relevant backgrounds. Beyond questions on the intelligibility of technical references was the more massive fact that laboratory work seldom provides a very interesting spectacle. For long periods of time one or a few individuals would sit silently, tapping at the keyboard of a computer terminal or scribbling notes while viewing data displays. The bodies did not move, the voices were not animated, and an ethnographer's questions were not always honored with polite answers.

One might figure, after Schutz (1962), that these largely silent and immobile scientists were acting in a separate "world of scientific theorizing"—an experiential world remote from the commonplace "world of everyday life"—with a logic all its own associated with a unique disciplinary corpus of knowledge. The multiple realities argument fails, however, to account for how the scientists I observed seemed to treat the references to laboratory events that I found so baffling to be no less ordinary or everyday than commonplace references to persons, places, and things. For them, it seemed, such references were part of the nexus of daily life and not of a separate reality.

Topical Contextures and Equipmental Complexes

The place of laboratory work is not a locale within a unitary physical space, since it is constituted by the actions that *dwell* grammatically within it. By drawing upon Gurwitsch's and Merleau-Ponty's phenomenological writings on perception, and a limited use of Foucault's (1972:31ff.) notion of "discursive formations," I will try to show how spatial grammars are topically tied to complexes of action and equipment ("paradigms" in a particularistic sense of the term).

The *topic* of laboratory space is not primarily concerned with the "naming" or "labeling" of space but with the grammars of spatial concepts associated with particular practices. It would be incorrect to say that any particular application of language *creates* a space of operations; rather, any such application participates in a contexture of activities in which a space is organized. Gurwitsch (1964:106ff.) demonstrates how perceptual space is organized by topical contextures—local orderings of referential details exhibiting visible relations of above/below, next to/separate from, inside/outside, before/behind, aligned with/askew, and so on. These spatial predicates are topically bound to particular constellations of details rather than to an invariant spatial matrix: "By implying, modifying, and qualifying each other, the several appearances of a perceived thing are given as coordinated by virtue of their mutual intrinsic reference to one another" (p. 296).

Merleau-Ponty elaborates how embodied actions are implicated in such scenic contextures. His classic *Phenomenology of Perception* (1962) develops a strange inventory of clinical observations on brain-injured and bodily disabled patients, along with a review of experiments in perceptual psychology. His study of the "shadow limb" phenomenon described by amputees, and of experimental subjects' apprehension and modes of adaptation to optically inverted, reversed, and tilted visual fields, enables him to specify how the "lived-body" with its perceptual and motile capacities constitutively reaches into space-time to establish the terms under which "it" is appropriated. "It is never our objective body that we move, but our phenomenal body, and there is no mystery in that, since our body, as the potentiality of this or that part of the world, surges towards objects to be grasped and perceives them" (Merleau-Ponty 1962:106). For Merleau-Ponty, embodied spatiality is not a "subjective" gloss over a transcendental space or a set of terms to be canceled out by deleting all reference to a perceiving subject. Nor is it an "ideal" space, emanating from a deep intellectual reserve and imposed on a formless chaos. He elaborates how our movements in space establish the predicates under which we encounter things, including their standard modes of orientation, typical facets and fronts, discriminable surfaces and points of entry, boundaries, synesthetic integrity, and—in Gibson's (1986) terms—their "affordances" (how an environment presents itself in accordance with the species capacities and cultivated habits of a living body). Merleau-Ponty draws a contrast between the spatiality of *situation* and the spatiality of *position.* The former is the lived-space through which we operate prereflectively, while the latter is what is sometimes wrongly called "physical" space, a space whose coordinates are seemingly abstracted from merely human forms of perception.

Merleau-Ponty treats the phenomenal body as a transitive condition for perception and action, but he does not consider historical or technological variations in the body's pre-predicative transparency. As Foucault's various researches demonstrate, the spatiality of situation is subject to various historical transformations within a public order of discourse and technology. To account for the transformations of embodied spatiality brought about in technologically (and textually) mediated action, we need to go beyond the perceptual technology of the naked subject. "Power has its principle not so much in a person as in a certain concerted distribution of bodies, surfaces, lights, gazes; in an arrangement whose internal mechanisms produce the relation in which individuals are caught up" (Foucault 1979:202).

The idea that "readable technologies" (Heelan 1983:206) extend embodied perception is of course a familiar one, and it dates back at least to Francis Bacon. It is particularly well developed in Polanyi's discussion of the primitive case of the "probe" (1958:59). Polanyi illuminates how the blind man's stick extends his embodied access to *what* he probes with the end of the stick. What the metaphor of "extension" misses, however, is the transformation of embodied spatialty associated with an instrumental complex. Foucault's regional analyses, with their emphasis on the discontinuities between the spaces brought into play within instrumental complexes, problematize any notion of extension from a naked existential ground. A historically specific "discourse" is irreducibly part of that ground. He also forcibly opposes any suggestion that such a discourse is an organization of thoughts or a network of concepts. An instrumental complex embodies systems of common usage, built environments, and the activities consonant with (though not strictly determined by) those environments.

In this chapter I treat Foucault's historical studies in a very limited way. I do not intend to use particular technological complexes as metaphors for a "dominant discourse" characteristic of a historical *episteme.* Instead, I will discuss topical contextures associated with particular technological interventions. I will suggest that complexes of technology and human actions can take a wide variety of forms, and my interest will be to specify how such complexes help to transform spatial predications. Foucault's (1979:171) explication of panopticism, for instance, is tied initially to an investigation of particular "observatories of human multiplicity":

> Slowly, in the course of the classical age, we see the construction of those "observatories" of human multiplicity for which the history of science has so little good to say. Side by side with the major technology of the telescope, the lens and the light beam, which were an integral part of the new physics and cosmology, there were the minor techniques of multiple and interesting observations, of eyes that must see without being seen; using techniques of subjection and methods of exploitation, an obscure art of light and the visible was secretly preparing a new knowledge of man.

Many of Foucault's interpreters use his explications of prison designs, military ordnance, classroom seating arrangements, and the like, to index an entire historic arrangement of power/knowledge existing in a kind of cognitive master plan. For an ethnomethodological reading, however, Foucault is at his best when explicating the details of texts he appropriates, and he is least convincing when advancing a general theory of power/knowledge (if indeed it can be said that he does advance a general theory). This reading of course directly conflicts with an appreciation of Foucault's mode of analysis as a kind of dark functionalism, portraying an ominous harmony between individual action and social order that leaves no room for an autonomous subject. Ethnomethodologists resist the impulse to view any single document or technological complex as a metaphor for the Enlightenment or modern society. This is not to say that what Foucault delineates cannot be found all over the place, though in accordance with local specifications and on a patchwork topology whose endogenous spaces readily support conflicting theoretical representations. The ethnomethodological move is to treat, for instance, the panopticon as one example among many of an instrumental complex that can coexist with other such complexes in a discontinuous ecology that embodies heterogeneous, and perhaps even contradictory, orderings of knowledge/power. In this respect, ethnomethodology also aligns with the reading of Foucault developed in the "actor-network" approach to the sociology of science and technology (Callon 1986a; Latour 1986, 1987, 1990b).

In what follows I will concentrate on two orders of laboratory "space": opticism and digitality. The paradigm for the former is the lensed instrument and the scrutinizing eye, while the latter is embodied by the play of fingers (digits) on a keyboard instrument. Opticism seems to have taken root during the Renaissance and expanded through the seventeenth century, although it can be traced to far more ancient origins and is still very much with us. Digitality is of course associated with the computer age (Baudrillard 1983), although it is no less characteristic of such earlier activities as playing keyboard instruments and operating a loom (even prior to the Jacquard loom). Neither opticism nor digitality discriminate historical periods, and it would be inaccurate to develop a totalizing picture of scientific knowledge in their terms. I shall argue that opticism coexists and overlaps with digitality, and that it is not displaced by the topical contexture associated with the "information age." Nor does it conflict with digitality, except in a fragmented or piecemeal way. The key loci for these systems are the instrumental complexes that "afford" them: they lose their specificity when treated as historical ideologies divorced from concrete arrangement of bodies, textural surfaces, lines of sight, and fields of technical action.

OPTICISM

Gibson characterizes a familiar view of perception as the "orthodox theory of the retinal image" (1986:58). According to this ancient theory *an image of an object* projects into the back of the eye. "The object, of course, is in the outer world, and the back of the eye is a photoreceptive surface attached to a nerve bundle." The technological device whose operations most clearly express this model is the camera (or, earlier, the camera obscura). The topical contexture that comes into play in the "orthodox theory" integrates the terms and practices of Euclidean geometry, theoretical optics, cartographic and graphic representation, the crafting of instruments, and the artistic techniques of linear perspective (Edgerton 1975; Alpers 1983; Latour 1986, 1990a). More than a representational technology or a specific theory of vision, the "orthodox theory" has given us an *observation language* that is ready to hand and all too easily relied on in epistemology. This is what I shall call *"opticism."*

Gibson (1986:58–59) traces the orthodox theory of vision to Johannes Kepler, although Kepler's was only one of a long line of optical theories tracing back to the early Greeks.

> The germ of the theory as stated by [Kepler] was that everything visible radiates, more particularly that every point on a body can emit rays in all directions. An opaque reflecting surface . . . becomes a collection of radiating point sources. If an eye is present, a small cone of diverging rays enters the pupil from each point source and is caused by the lens to converge to another point on the retina. The diverging and converging rays make what is called a *focused* pencil of rays. The dense set of focus points on the retina constitutes the retinal image. There is a one-to-one projective correspondence between radiating points and focus points.

A key feature of this theory of vision is its integration of a particular representational schema, a kind of mathematical analysis, and particular technological designs:

> This theory of point-to-point correspondence between an object and its image lends itself to mathematical analysis. It can be abstracted to the concepts of projective geometry and can be applied with great success to the design of cameras and projectors, that is, to the making of pictures with light, photography. . . . But this success makes it tempting to believe that the image on the retina falls on a kind of screen and is itself something intended to be looked at, that is, a picture (pp. 59-60).

Gibson (1986: 60) calls this the "little man in the brain" theory of perception and argues that it remains "one of the most seductive fallacies in the history of psychology." So seductive is it, in fact, that it informs the typical design of the visual psychology experiment. The experimenter devises ways to "take a photograph" using the human subject as an embodied camera. The laboratory setup inhibits head and body movement, so that the subject is precluded from using what Gibson calls "ambient" and "ambulatory" vision. These latter concepts include embodied practices of turning an object around in one's hands, and moving within a field in a temporally expanded synesthesia of its various inspectable properties. Although the subject does not make a particularly good camera, he or she can be made to act like one by using what Gibson calls "snapshot" and "aperture" vision. Accordingly, the psychology experiment successfully generates the subject-camera's reactions to ingeniously arranged fields of "stimuli."

Gibson argues that the psychology of vision based on such experiments is erroneous, but for our purposes this "error" constitutes a kind of "effective historical consciousness" (Gadamer 1984:167ff.). The orthodox theory is embodied in technology and in the literary figure of a human actor categorically bound and disciplined in accordance with the technological complex. "A real subjection is born mechanically from a fictitious relation" (Foucault 1979:202). The *perceiver* or *observer* is a figure bound to the material and social environments associated with opticism. The relations between image and observer, the position assumed by the observer in the organized realm of light, and the active or passive role assigned to the observer in relation to the image's construction are all varied within the scheme of opticism. Consider the different positions and spaces assigned to observers, subjects, and audiences in optical demonstrations using lenses and prisms, exhibits of linear perspectival art, and systematic uses of such instruments as the camera obscura, the simple microscope, the Newtonian telescope, the camera lucida, and Daguerre's diorama. This last was a marvelous theater in which massive semitransparent paintings (composed through the use of the camera obscura) were used as stage settings. They were illuminated from different angles to give an illusion of motion and lapse of time. Daguerre's practical experience in this opticist theater was "instrumental" for his coinvention of the daguerreotype (Newhall 1964:14ff.). Like the rigid subjects posing for the early daguerreotyes, the absurd subject in the psychology experiment can also be viewed as a historical actor, since the immobilized head and the skills required of a "good" experimental subject are actual performances in a local social setting. These complex relations are less the makings of a monumental scientific error than the identifying features of an operative topical contexture, a spatiality of situation bound to a technological complex.

A number of mutually supportive elements of opticism can be specified, and in what follows I hope to show that these features can be viewed as "epistemic" conditions of embodied action in particular technological complexes (cf. Alpers 1983:95; Latour 1990b; Lenoir 1989):

1. Ocular vision provides the paradigm of perception and observation.
2. Visual field and viewer's image are clearly distinguished along Cartesian lines (external object–internal image).
3. The viewer's "eye" becomes a singular point or aperture toward which a field is oriented.
4. The field is framed by a window, often represented as the outer edge of the cone of rays linking the field to the eye.
5. The relationship between eye and object (or field) is transacted through a converging arrangement of linear rays. This arrangement integrates the limit forms and axioms of Euclidean geometry with the mechanisms of vision.
6. A transparent lens and/or reflective mirror mediates the linear transfer of rays into (or in some theories out of) the "eye's" image.
7. A point-by-point correspondence obtains between image and object. Note that this correspondence governs even the well-known "defects" of vision, since the biases and distortions are mapped out in reference to the refractory and reflective properties of the bodily instrument.
8. The model of vision supplies a vocabulary and set of topics for a more general epistemology. Discussions and debates about the role of sense data, primary and secondary qualities, signifier-signified relations, and private experience partake of the orthodox theory's opticism.

Linear Perspective and Representational Technologies

The effectiveness of the orthodox theory of perception is particularly evident in the representational conventions it supports, which in turn *document* it with tangible literary imagery. The perspectival and descriptivist art emerging from Italy, Holland, and elsewhere in Europe after the Renaissance is closely connected to the orthodox view of vision. Edgerton's account of the "rediscovery" of linear perspective by Brunelleschi and Alberti in fifteenth-century Florence documents how the skills of the "artisan-engineer" Brunelleschi, combined with the optical theory of Alberti, articulated the new representational conventions. Where Brunelleschi invented a mirror device for perspectival painting, Alberti later articulated the principles through which the lines of sight embedded in the design and operation of the device could be exposed and rectified in a more direct mathematical operation. Painting became a kind of embodied mathematics, trading in dots and marks that concretely approximated geometric limit forms. The canvas a painter composed was simultaneously a "plane of signs," an approximately mathematical space, a mundane workplace, and a graph where a grid of lines link up "like threads in a cloth" (Edgerton 1975:80).

Opticism coordinates spaces and practices: optical instruments, representational technologies, a theory of optics, Euclidean geometry, cartography, and (as Edgerton [1975:37] indicates) the practical arts of measurement employed in the marketplace. Its precepts—the fixed point, convergence of rays, hyperrealism, and point-by-point correspondences between object and image—topicalize an epistemology by providing an account of the mechanisms of vision, an account of their truth, and proposals for the rectification of their errors and limits (cf. Bacon [1620] 1960).

The devices of linear perspective, as well as the representational technologies of the light microscope, the refracting and reflecting telescope, and, in its own way, the visual psychology experiment, all embody the topical contexture of opticism. Opticism supplies a body of tools and operations for adjudicating the correspondence between image and object. "Errors," "inaccuracies," and so forth can be diagnosed to appropriate levels of precision within the technological complex. Similarly, if we ask whether a microscope delivers a "true" image of its object, terms and devices are available for discriminating and suppressing artifacts while "bringing out" a thematic contexture of veridical details within the field (Hacking 1983:186ff.; Lynch 1985a:81ff., 1985b).

While the problems associated with opticism for the practicing photographer, painter, or microscopist may legitimately be addressed through technical innovation, a different order of problem arises when an opticist observation language inflects the terms of a general epistemology or semiotics. So for instance when philosophers' ideal-typical examples of perception, observation, representation, and the like are couched exclusively in opticist terms, a particular topical contexture gets elevated to a transcendental account of all conditions of knowledge (or, in psychology, all perception). Barnes (1977:2), for instance, critiques what he calls the "contemplative" conception of knowledge, a conception marked by its opticist imagery.

> It describes knowledge as the product of isolated individuals. And it assumes that the individuals intrude minimally

> between reality and its representation: they apprehend reality *passively*, and, as it were, let it speak for itself . . . learning and knowledge generation are thought of in terms of visual apprehension, and verbal knowledge by analogy with pictorial representation.

Barnes further explicates this "analogy between learning and passive visual apprehension":

> We talk of understanding as "seeing," or "seeing clearly"; we are happy to talk of valid descriptions giving us a "true picture." Similarly, we are able to characterize inadequate knowledge as "coloured," "distorted," "blind to relevant facts," and so on. The overall visual metaphor is a resource with which we produce accounts of the generation and character of truth and error. And in many ways these accounts serve us well (p. 2).

To this familiar epistemological picture it can be added that the "observer" confronts a static monocular spectacle. Both observer and object remain frozen, while epistemology "reflects" upon their relationship. The terms of linear perspective remain in force even when the laws for reconciling multiple "viewpoints" are held to be problematic. Similarly for Cartesian semiotic systems: words (usually names, sometimes other nouns and verbs) correspond to objects; signifiers to signifieds; symbols to meanings. Words are mapped onto a sensory manifold as though enunciated by a talking eye, and the analyst's job is to discern the laws of refraction or reflection that transact that relation.

Barnes (1977:2) conducts a kind of immanent critique of the "contemplative" account in order to demonstrate that "knowledge is not produced by passively perceiving individuals, but by interacting social groups engaged in particular activities." He remarks that the contemplative account often maintains a resilient hold on even those who would abandon it: "the associated pictorial metaphor for knowledge is so pervasive, intuitively attractive and, indeed, valuable as an explanatory resource, that it can be difficult in practice to structure one's thought independently of it" (pp. 3–4). Rather than abandon this "pictorial metaphor" altogether, Barnes works within its framework to attack the particular assumption of a "passive" relation to visual reality. Instead of simply reflecting reality, "when a representation conveys knowledge or information about, say, an object, it is by classifying it, making it an instance of or mere kinds of entity recognized by the culture whose resources are drawn upon" (p. 5). Maps, diagrams, and even photographs "remain constructs for use in activity" (p. 9).

Barnes's argument effectively carries forward the debate between representational realism and social constructivism. However, as Tibbetts (1988:120) points out, Barnes's arguments are developed from examples that are "physically static and spatially fixed: circuit boards, built environments, musculature and elevations." While these have their place in a number of practical activities, this "picturing model" has limited applicability to cases where Tibbetts asserts there are "no spatially defined *objects* to represent (e.g., airfoil lift-drag coefficients, meteorological phenomena, unemployment rates, or demographic changes)." Tibbets's argument is partly based on Hacking's (1983:167) suggestion that observation is "overrated." By "observation," Hacking means a sort of "naked perception" or "raw observation" of the object, distinct from efforts "to get some bit of equipment to exhibit phenomena in a reliable way." An opticist account of observation can, however, be compatible with Hacking's focus on getting equipment to work. In lab vernacular, it is commonplace to speak of "making an observation" when setting up a complex of optical and electronic equipment to record data. Depending upon the topical contextures brought into play by the design, arrangement, and use of the equipment, the complex can produce an automated opticism. The challenge Tibbetts raises is not that observation is overrated, but that many common forms of observation are not readily subsumed to an oculocentric picture.

Although Barnes criticizes oculocentrism, he operates within the topical contexture of opticism to argue for an inversion of the "contemplative" account's representational scheme. This retains some of the framework of opticism, if only for the sake of argument. In a fashion, Barnes's debate with representational realism is a modern form of archaic controversies between proponents of "extromission" and "intromission" theories of optics (Edgerton 1975:67). These ancient debates concerned the directionality of the "rays" between perceiver's eye and object. The extromissionists believed that the eye emitted a spirit of "fire," which projected through the visual cone to activate the object's visibility; the intromissionists argued for a movement from the object to the eye. Although the intromissionists eventually emerged victorious, the debate moved within and served to refine the geometric account of optics. Of course Barnes makes no claim about substantive projections, and he abandons the individualistic account of "observation" in favor of a sociological theory of knowledge. What gets lost in his critique of the opticist picture is the local adequacy of its use. If we figure that the "contemplative conception of knowledge" is, in Garfinkel's terms, a "production account," its general epistemic inadequacies are beside the point. Instead, such an account is situated in a set of actions performed in accord with the various optical knowledge-production machines: a disciplinary compliance on the part of the subjects in those systems. The photographic subject, the viewer of perspectival art, and the user of the optical microscope all learn to inhabit a static field (or to construct one if it is not initially at hand), to assume the fixed pose, to steady the eye, and to align with a set of coordinates that define the *posit*. The passive subject of the contemplative approach becomes an embodied relation and thus a subject that locally validates the opticist picture.

DIGITALITY

In his review of the orthodox theory of perception, Gibson (1986:61) discusses a more recently developed variant of the opticist doctrine that takes account of the retina's mediation of the image transmitted to the homunculus:

> Even the more sophisticated theory that the retinal image is transmitted as signals in the fibers of the optic nerve has the lurking implication of a little man in the brain. For these signals must be in code and therefore have to be decoded; signals are messages, and messages have to be interpreted. In both theories the eye sends, the nerve transmits, and the mind or spirit receives. Both theories carry the implication of a mind that is separate from a body.

Gibson's objections remain in force, since this semiotic version of the orthodox theory of vision retains the Cartesian picture of a mental image to be brought into point-by-point correspondence with an external world. In this case the homunculus does not simply view the eye's mirror image of the world, it deciphers a psychophysical code inscribed by luminous "messages" interacting with retinal sensitivities to color, line, edge, position, orientation, and other abstract parameters. Although Gibson treats it as incidental to the continued dominance of the orthodox theory, he notes that this "more sophisticated" technological metaphor shifts emphasis from "a camera that forms and delivers an image" to "a keyboard that can be struck by fingers of light" (p. 61). Gibson makes nothing more of this shift, as he argues that neither metaphor sufficiently accounts for the realities of embodied perception. Although I agree with Gibson that the *keyboard* metaphor does indeed retain the Cartesian picture so central to opticism, I will argue that it brings into play a topical contexture and an order of practical relations that center around the *digit* rather than around the ocular-centric image of the realm of light. Again, as in the discussion of opticism, I will emphasize implications of digitality that do not turn on the correctness of the visual psychology or epistemology developed in its terms. Instead, I will argue that digitality is an embodied relation with its own forms of practical efficacy.

Asymmetric Alternates

It would not be correct to say that Gibson "confuses" digitality and opticism, since this would assume an epistemic discontinuity between them. As Gibson suggests, visual psychologists insinuate the keyboard within the operative terms of opticism, and by doing so they participate in a broader movement in twentieth-century cognitive science—one that uses the computer as both a research tool and a source of analogies for investigating mental activities. Although cognitive science has undoubtedly invested "mind" and "the mental" with a novel set of predicates, these are largely organized within the mentalist and Cartesian assumptions of the older psychology. So, for instance, many of the arguments Wittgenstein (1953) launched against psychologism apply rather precisely to its latest incarnations in cognitivist psychology and philosophy of mind (Coulter 1983, 1989). Accordingly we can read Gibson to be suggesting that the digital imagery of contemporary psychophysics *simulates* or reiterates the opticist framework. What this implies, then, is that when we compare opticism and digitality, we are not dealing with incommensurable or discontinuous "discursive formations" but with what Garfinkel has called "asymmetric alternates" (Garfinkel et al. 1989).

What this means for the present discussion is that digital modes of image production and image processing can be said to subsume the representational figures associated with opticism.

Moreover, it can be argued (in a Whiggish sort of way) that the representational imagery of the earlier era *always was and already was* digital in composition. Thus Gombrich's (1960) explication of "illusionist" art demonstrates how the *trompe l'oeil* decomposes into painterly glossing practices through which realistic detail arises from an artful arrangement of color patches. And, as discussed above, Alberti's theory of linear perspective provided for the painting's surface to be composed of significant dots and points. Although these features are placed in relation to the idealized lines, points, and mathematical relations of Euclidean geometry, we can now say that they already express an atomism operating beneath the sensible threshold of the image in the realm of light. Similarly, photographic realism can be resolved into a microcomposition of grains in the emulsion. Digitality can thus be said to *undermine* the operative terms of opticism, but not vice versa. Digitality does not necessarily bring about the destruction or replacement of opticism, since it can just as readily *underpin* or simulate the operative conditions of opticism. The keyboard-retina metaphor can be viewed as just such a simulative use of an asymmetric relation. The retinal keyboard plays the spatial tunes of a classical opticism; but once the keyboard is separated from the orthodox theory of the retinal image and topicalized as a facility in its own right, a distinct contexture can be opened up.

The Topicalization of Digitality

As with opticism, we can think of the topical contexture of digitality as covering a representational art, a technological complex, an embodied "subject," a practical mathematics, and an epistemology. As with opticism, certain historical associations come into play, although again I will suppress any suggestion of discrete periods or stages and will focus on local relations in a technological complex. Digital relations can be distinctively associated with modernist themes: atomistic fragmentation, decentering, and arbitrariness. Although these themes identify a topical contexture proper to digitalized action, they have no necessary association with its technological complex. This turns on the question of topicalization.

Even if digitality has always been "present" beneath the operations of descriptivist art, what happens when it becomes topicalized? Consider the difference between classical representational art and the various innovations in painting developed in the late nineteenth century. The pointillist, divisionist, and impressionist art of the period made an issue of the composition of a painting by color patches. Instead of effacing the artist's work in an attempt to reproduce the camera obscura or camera lucida image, a *point* was made of the compositional details: points, dots, and brushstrokes were exposed and exaggerated in size and systematic form—that is, made obvious from normal viewing distance. Works such as Mondrian's ginger jars selectively exposed the visual impact of different compositions; perspectival features were disrupted and played with to the point of decentering or even canceling the oculocentric subject of linear perspective. While making a "commentary" on the structural supports about which previous art had been so reticent, the impressionists and modernists changed the terms under which art was composed and appreciated.

The embodied site of digitality is the hand—the artist's hand, the fingers on the keyboard (Sudnow 1979)—a hand whose systematic movements in a technological complex are topically bound to be a "digital" space. Digital space is an explicitly simulated or constructed space, built up from series of repeating digits or cellular units that act simultaneously as codes and substantive details. A summary of features includes the following:

1. *"Pixelated" space*: Digital space is mathematized in terms of atomic details rather than of an idealized linear matrix. "Detail" is reduced to a unitary code of "bits." The pixels, or picture elements, on a video monitor, for instance, are uniform and equally spaced, with each element capable of expressing a range of values. (A video screen is an array of pixels, and each pixel can be illuminated with a gray scale level, or a combination of red, green, and blue.)
2. *Arbitrary code*: The atomic details are emptied of content, in the sense that they express an open range of values. The atomic code does not resemble anything, although it can be made to simulate or mimic an image.
3. *Manipulable details*: The elements of the "keyboard" have no inherent tie to an object; they are places for the fingers to strike in the course of an expressive action.
4. *Diachronic* (syntagmatic) organization: A composition or image is not a static field reflecting its object but a spatial product of a temporal scan. Keyboard action is sequential and syntagmatic, with an order emerging from the play of signs in a series, rather

than from a point-by-point correspondence between image and object.

5. *An equivalence of "qualities" and "quantities"*: Digitality obliterates the classic distinction between primary and secondary qualities (Hacker 1987). In the linear perspectival drawing there is a distinct separation between form and content, between the geometric arrangement of the framework and the objects to be represented within that framework. In pixelated space detail is finite, limited by the array of pixels "beneath" which there is no finer order of detail. Euclidean mathematical rays transmitting a "quality" are replaced by a code that is simultaneously a mathematical unit and a sensory datum.

The artisan-engineers of digitality work at computer terminals in a wide variety of scientific and other fields. Unlike the artisan-engineers at the dawn of the Renaissance, they do not confront a transcendental space infused with religious mystery. Linear perspective evokes an expansive and luminous space within which mundane objects are situated and ordered. It is very Christian (and Platonic as well), in the sense of implying a deep and abstract realm that pervades appearance and gives it form (Edgerton 1975: chap. 2). Pixelated space is manipulatory; there is no necessary implication of a transcendental order within which mundane reality manifests. Instead, the "world" breaks down into arbitrary bits of information that allow its composition and recomposition to be arranged by a *user*. If a sense of a mysterious and infinite universe is to be retained, its retention is owed to a hands-on accomplishment.

The Spatiality of Situation: Digital Image Processing[1]

The Image Processing Facility (IPF) at the Harvard-Smithsonian Center for Astrophysics is a "center of calculation" (Latour 1987) whose major items of equipment are a couple of large computers acting as "host system" and "slave system" for running interactive image-processing technology. The facility also houses several stand-alone image-processing units along with photographic and film-editing equipment, a digitizing unit, a hardcopier, and various associated electronic devices. Also on hand are racks of magnetic tape storing digital data from orbiting and ground-based observatories, along with manuals and documentation for various software programs used by astronomers at the center.

The image-processing units are arrayed in several different rooms at the IPF. One unit, for instance, is housed in an office-sized room alongside another identical unit. The room also contains a hardcopier and a digitizing camera. Arrays of electronic and photographic equipment and attachments are strewn about the room. Each image-processing unit is connected to the mainframe computers, and it includes a computer terminal with a large high-resolution monitor connected to a "track-ball"—a device operating like a computer "mouse" that can be rolled by hand to change the composition of an image on the monitor. Associated with the monitor is a separate keyboard connected to a green glowing "touch screen," which displays an array of touch-sensitive boxes with names for various subroutines printed on them (see figure 6.1).

CHANNEL	ZOOM	ROAM	GR. ROAM	PATCH	8 BIT	16 BIT	HISTGM
PROFILE	LABEL	GR. OFF	GR.COLOR	PALETTE	SLICE	OFM	HISTSCALE
DIG.	FREEZE	CIRCLE	BOX	SMOOTH	TABLET	SPLITSCR	HIST. EQ.
LUT	LINEAR	LOG	POS.	NEG.	NO WRAP	WRAP	BAR
IFM	CHAN. C	CHAN D.	LINEAR	LOG	*2	/2	BAR
FLIP							
CURSOR	XLINK	YLINK	XYLINK	#LEVELS	CONTOUR	TR.COLOR	G.ERASE
HELP	MENU	IDLE	128 INIT	MATRIX	BLINK	MOVIE	IM.ERASE

Figure 6.1
Touch Screen Layout

At different times during a typical day or evening at the IPF, several different astronomers and image-processing assistants can be seen "interacting" with any of the several image-processing units. Usually they work alone at the machine, although occasionally two or three colleagues will gather round while a demonstration is made, a problem is addressed, or a joint project is accomplished. Often there is not much to see or hear within the "physical space" of the IPF other than sporadic tapping at keyboards while practitioners quietly stare at screens and shuffle through notebooks, occasionally murmuring to each other about different input values or complaining about the slowness of the machine's response.

Despite the ostensive lack of embodied action and communication, plenty of action and communication takes place. To describe "where the action is" (Goffman 1967), however, we need to identify the action in its proper space: the keyboards, display monitors, and touch screen. Without going into detail on the improvisational use of these facilities, we can gain a preliminary appreciation of the topical contexture of activities by taking an inventory of image-processing "functions." The boxes on the touch screen provide identifying labels for some of these, and they and others are elaborated in software manuals for image-processing programs (such as AIPS, the Astronomical Image Processing System).

Functions are identified with names, printed on the touch screen, such as *channel, zoom, roam, patch, histogram, profile, label, palette, slice, freeze, box, smooth,* and *tablet.* Each of these names for the visible result of an action can be translated into a verb for the action itself, so that *patch* is usable as a name for a delimited 9 by 15-pixel array (see figure 6.2) as well as a verb describing the routine through which the user delimits such an array with a box cursor. A *patch* can be *zoomed* to enlarge the display of pixels, or *profiled* by graphing the intensities along a particular cross-section of the patch. These functional terms can also be placed in the grammatical role of *agents* (cf. Callon 1986b; Latour 1987, 1988b), as in the following account of *smooth* in the IPF manual:

> Performs a sliding-window average to smooth the contents of a refresh memory channel. SMOOTH averages the intensities of a 2x2 array of pixels and writes the result in the lower left hand pixel. Then the array is moved one pixel position and the process is repeated, until all pixels in the refresh memory have been replaced. SMOOTH is typically used prior to applying LEVELS and CONTOUR to generate simpler contours by reducing slope and complexity of intensity gradients in an image.

Smooth is variously an agency, an action, and a visual result through which, for instance, an intensity map of a supernova remnant is normalized. The value for each pixel is averaged with those immediately adjacent to it, so that the overall effect is to reduce extreme variations and create an image with less jagged or discontinuous contours.

Y/X:	6	7	8	9	10	11	12	13	14
17:	3	3	3	4	3	4	3	2	2
16:	1	5	4	7	8	9	5	3	4
15:	3	8	7	8	11	9	7	5	4
14:	5	10	13	17	21	14	13	8	6
13:	14	22	34	44	49	41	28	17	8
12:	23	50	104	159	168	127	73	36	18
11:	36	120	324	517	523	355	169	69	28
10:	55	203	613	1016	1030	666	285	97	36
9:	50	196	566	947	955	628	274	94	33
8:	32	98	253	414	428	303	150	59	28
7:	21	39	81	120	128	103	58	28	16
6:	12	18	29	38	41	38	27	18	11
5:	5	10	14	17	17	17	12	10	6
4:	6	8	8	10	12	10	10	10	5
3:	4	5	5	6	8	6	5	5	3

Figure 6.2
Array of Numbers

Many of the functions identified on the touch screen can be intuitively recognized: *zoom* is little different from viewing a photograph through a magnifying glass; *palette* enables the different pixel intensities to be "pseudocolored" by selecting a table for transforming intensity values into specific color or gray-scale levels in a kind of "paint-by-numbers" operation. Some of the functions are more "technical," but many have to do with working on a picture to isolate a section, sharpen the contrast, move the picture up or down, rotate it, label its features, or convert it to a contour map or a graph. The functions seem to duplicate work that might otherwise be done in a photography developing lab or a graphics workshop. Keyboard and touch-screen operations displayed on the video monitor replace the classical toolbox of scissor, paper, rule, paintbrush, and bench. A bricoleur's vocabulary is displaced into the digital system, as *palette, paintbrush,* and *slice* become electronic operations.

If we limit our attention to optical astronomy (astronomy using visible wavelengths of the electromagnetic spectrum), it initially seems to be the case that digital detectors and digital image processors act as stand-ins for familiar optical equipment and the associated representational technologies for processing photos. A digital camera includes a CCD (charge couple device) detector—a charged plate composed of a grid of light-sensitive cells. The rectilinear array of light-sensitive elements replaces the irregular (although more finely textured) smear of chemical grains on the photographic plate. In some ways the digital camera is held to be inferior, since its CCD detector does not resolve as fine an order of detail as a good photograph, and its advantages are described primarily in opticist terms. Unlike a photographic plate, the CCD detector is said to respond to the objective field in a linear way; its response to incident light is more efficient, and it responds to radiation over a wider range of wavelengths (Janesick and Blouke 1987). Accordingly, the digital camera can be viewed as a technological extension of the operative terms of opticism: Gibson's keyboard played by fingers of light, enabling a set of "telescoped acts" (in G. H. Mead's terms) to take place between keyboard and image. But, as indicated above ("Asymmetric Alternates"), digitality can just as readily undermine as underpin the opticist contexture. In either case it infiltrates and takes possession of opticism. But how do we disclose this difference?

"A Picture Is an Array of Numbers"

A Boston University astronomer describes a picture displayed on a screen. The picture, he tells us, is disrupted by periodic lines, which he attributes to an electronic artifact:

> Over here is an array of numbers that has intensity coming out, and it has space in the X direction and space in the Y direction. So we said if we consider that as the sum of a lot of waves, uhm, there's a very high frequency wave here. All these lines are fast Fourier components . . . in each row . . .

and so uh there are, there are periodicities in that picture that are very high frequency.

Note the complex of predicates: "an array of numbers," "intensity coming out," "space in the X . . . Y direction," "the sum of a lot of waves," "fast Fourier components . . . in each row." In a second-order analytic project, we could try to sort these terms into their proper domains by, for instance, assigning some to the objective intensities (in the sky), some to the visible residues of the electronic equipment, some to the format of a graph, and some to a mathematical function "expressed" through the technological display. But this would be to reiterate abstractly what the astronomers have already accomplished in the singular space of the instrumental complex. The perspicuous reference, for our purposes, is "an array of numbers."

On a separate occasion, this astronomer explains to his visitors that "a picture is an array of numbers." He does *not* say, "a picture can be *represented* as an array of numbers." What are the implications of this identity function? Consider figure 6.2—reproducing the array of numbers in a "patch" an astronomer called up on the IPC monitor to represent the intensities of incident light across a region of sky including a thematic object. Background has been set to zero. Looking at the variable numbers across the rectangular array we can "see" a region of relatively high intensity in the central area of the patch (the peak of which is located at the four adjacent pixels reading 1016, 1030, 947, 955). The intensity then spreads out in a roughly symmetrical pattern. In this case, the figure was identified as a possible QSO (quasistellar object).

Using commonplace functions, the astronomer can instantaneously turn this array into a histogram or a false-colored map with variable contour intervals and contrasting hues. Alternatively, intensities can be represented with more or less continuous gray-scale levels or with monochromatic false-color palettes. Further operations can be used to blot out backgrounds, so that numbers less than, say, fifty would appear as a uniform dark background against which the higher-intensity regions stand out as a bounded "source" object. By setting a color palette composed of contrasting "false colors" to correspond to numerical intervals within the high intensity region, a "figure" with a distinctive visual composition can be constructed. So, for instance, a typical false-color palette uses white for the "peak" contrasts, and dark blue for the "background," with yellow, red, green, and light blue set at even intervals in between these extremes. Figure 6.3 exhibits a similar translation of the array of numbers from the central region of figure 6.2 (columns 6–14, rows 6–13). Since colors are not available for use here, a "palette" of black and white patterns is used to resolve the numerical intervals into a map.

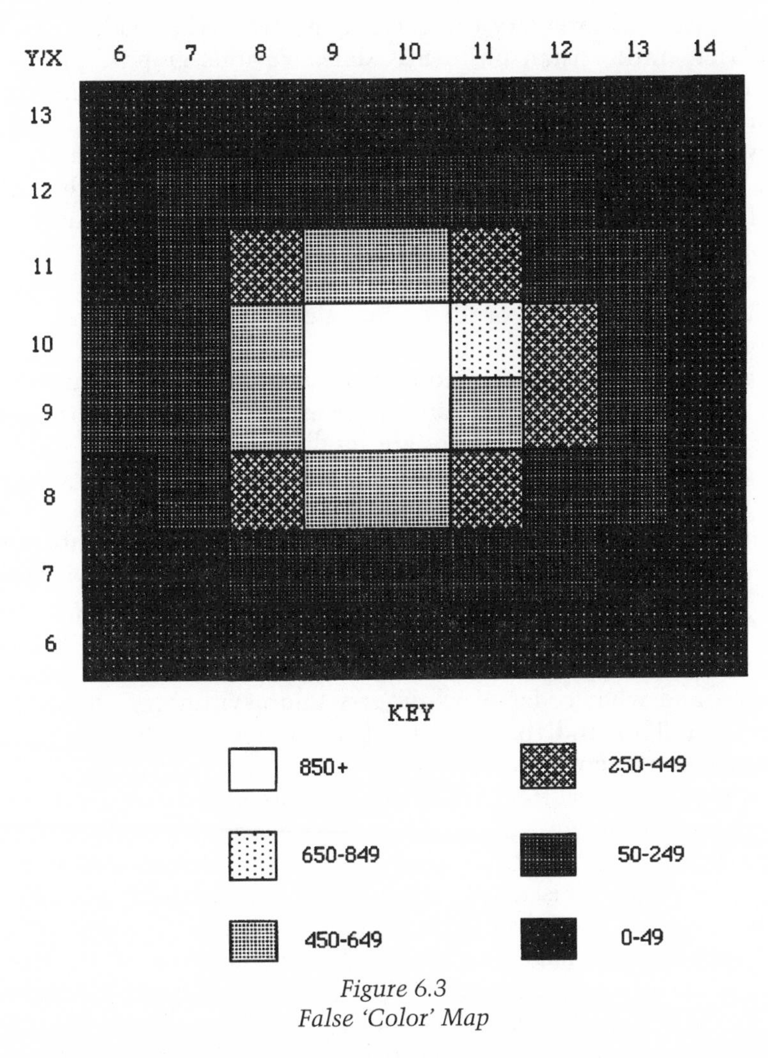

Figure 6.3
False 'Color' Map

Such a figure is but one of an indefinite series of compositions that can be arranged on the screen and preserved for later analysis. Changing the interval corresponding to the palette resolves different contours. If, for instance, the background range was expanded from 0–50 to 0–250, the visible figure would constrict around its edges. Similarly, setting the "peak" level above 1,000 would constrict the brightest area from four pixels to two pixels. Adding more levels to the palette would differentiate the figure into finer contour lines, and with more differentiation at the lower levels of intensity the background would be made to look "noisier" rather than uniformly dark. Various graphic profiles along the rows and columns can be constructed through different manipulations of the palette. Depending on where the breaks between intervals are set, the figure can be made to appear more or less symmetrical. This is to say nothing of the operations through which the "raw data" are analyzed and processed to subtract "noise" from the apparent "signal." So, for instance, the "numbers" in figure 6.2 can be added to those recorded in an entire series of pictures taken of the "same" field, so that "random" variations are canceled out from one picture to the next while recurrent patterns accumulate more sharply against the background.

The medium of operations is the array of numbers, although the constituents of this array are modified when various operations are performed on the picture. In the above example a figure is brought out against a background. The false-color palette and the various cutoff points are calibrated to isolate or bracket a coherent "object" while preparing it for further analysis. A sense of what the picture shows guides the project: a sense of how many pixels the object's profile should cover; what intensities and intensity gradients are appropriate for a "point source" versus a nebular object; and what order of symmetry and asymmetry should be expected. This and the various other operations—for example, sharpening the contours and resolution of the image—are guided by a variety of practical and theoretical assumptions: about the operations of the equipment, about the sensitivities of the detecting system, about singular conditions at the observatory at the time the picture was taken, about the thermodynamic properties of light scattering, and so forth. In brief, the picture is progressively composed and shaped to "get the phenomenon out of the data" and, relatedly, to "make the thing look like what it is" (Garfinkel et al. 1989). But note that the "phenomenon" is never "reflected" in any single composition, whether represented as numerals, colors, gray scales, or contour lines. Nor, strictly speaking, does it correspond to any particular array of numbers once such operations as *smooth* and various other editing functions are used to remove "cosmetic defects" (Lynch and Edgerton 1988). These latter functions change an array of numbers from a raw data configuration to a processed image. In IPC vernacular, the expression *raw data* refers to the data frame initially transferred from a detector to recording tape. The data frame is not treated as a pristine reflection of "reality" but as the residue from a confused field where electronic noise, detector defects, ambient radiation, and cloudy skies mingle indiscriminately with the signal from a source object. The processed image is often considered the more accurate and "natural" rendering (cf. Yoxen 1987:282). The relation to the phenomenon is not accomplished through a synchronic "snapshot," nor is it impressed "wholesale" on the instrumental retina.

Reality As an Ethical Simulation

We might characterize the "true" digitally processed image as a *sincere picture-sentence* rather than as a reflection of a preexisting object. *Simulation,* as previously mentioned, is a key feature of digitality (Baudrillard 1983:103ff.); but simulation is subject to numerous conventions and constraints. Consider, for instance, the significance of *color* in digital image processing. On the one hand, the technology provides for the use of "false color"—that is, color as a code for any of a variety of values. In the above example, a palette is used as part of a mapping process to separate figure from ground and to resolve contours in numerically measured levels of intensity.

Among the properties color can be used to code are, odd as it may seem, *color.* One astronomer, for instance, specialized in using digital detection and image-processing equipment to make "realistic" and "very slightly color enhanced" pictures for popular and semipopular articles and shows. These pictures were also significant as research objects, since color differences could be used to index different sources of emission in a galaxy. This astronomer would use three different color filters to take three separate exposures, with the digital camera mounted on a telescope. One shot would be taken with a blue filter, another with green, and the third with near-infrared. Each of these would be stored as an array of numbers. At the IPF he would then reassign the "original" color function to the blue and green data arrays, and would assign "red" to the infrared intensities. He would then combine the three images, the result being a color-enhanced image (because of the infrared-red conversion). In principle, he could have enhanced the colors to a much greater extent than he did, but he limited himself to about a 15 percent enhancement because, as he explained it, he did not want to "cheat" (as, he said, NASA did with *Voyager's* planetary images). In his words, "It's important to be kind of honest about what the colors are, because we don't want people to have an exaggerated view of what the colors are." (For a related discussion on the "true" color of the planets, see Young 1985.)

This is a fairly trivial instance of "ethical simulation," but it is sufficient to alert us to the conventional limitations on the arbitrariness of the "code." Color becomes iconic when used for color enhancement or when signifying intensity or red shift. That is, the code is selected intuitively to suggest properties the object *should* have. In this case the astronomer expressed a sense of public responsibility to produce "kind of honest" representations and, while doing so, acknowledged the innumerable possibilities of doing otherwise. Note that doing otherwise is not necessarily to be placed in the category of fraud, since—as astronomers who work in other regions of the spectrum aver—"these things have no true color anyway." At this stage of the game, there seems not to be a very tight sense of what a picture should look like. So, for instance, one astronomer said in an interview that virtually anything goes, as long as the "look up table" (correspondence between palette and intensities) is published; at the same time, however, he indicated that using colors to "tease out" slight variations in signal against backgrounds can be controversial. This is because the breaks between colors can visually suggest categorical distinctions, phenomenal edges, and interior structures, rather than minor fluctuations within a noisy medium (Lynch and Edgerton 1988). The responsibility, it seems, is to retain a semblance of a "sensical" universe through a kind of nostalgic recovery of the normative properties things "should" have.

Epistemological Nostalgia

Epistemological nostalgia has to do with the way astronomers use digital image processing to evoke the "classic" themes of opticism. Practitioners freely admit to using digitally processed pictures for such rhetorical purposes as dazzling naive audiences with false-color or color-enhanced pictures, and conveying a "qualitative" appreciation of the mathematized things astronomers study in order to persuade Congress to support NASA astronomy projects. In some instances—such as the "slightly enhanced" color compositions discussed above—astronomers compose digital images to evoke photographic realism, and while doing so they exploit the "mystique" associated with photography and related forms of "mechanical reproduction" (Barthes 1981; Benjamin 1969). As Bastide (1990) argues in her sophisticated treatment on the semiotics of scientific images, photographs enable the viewer to grasp the "facts" at a glance: they economize textual persuasion and confer a hyperrealistic materiality on the objects portrayed.[2] Bastide goes so far as to declare the persuasive power of photographs to be superior to that of numbers, digital "bits," and language. What she misses about the "bit," however, is that even while reproducing the conventions of photographic realism, practitioners of digital image processing become attuned to a mathematical assemblage. While a bit in isolation is simply a binary number, the array of bits *in use* constitutes a topical contexture, and for all practical purposes the bit *infiltrates* the photograph and disposes of its details within a different topical contexture—the contexture of the keyboard, digitablity.

The real-time work of digital image processing involves a play at the keyboard, where images on the monitor are continuously recomposed by changing the palette, using touch-screen routines, plugging in parameters, and trying out different software manipulations. A picture on the screen provides less of a stable ground for observation and analysis than a moment in an open series of eidetic variations. A complex series of adjustments and modifications of an image, or of a series of images combined into a unitary display, enables researchers, in one astronomer's words, to "*see the physics.*" "The physics" is not immediately identified with any stable photographic array; instead, it is located in and through the lateral play of compositions, readings, and calculations. The *play* is brought to a conclusion whenever practitioners attempt to select a picture for an article or slide show. At that point, the task becomes one of composing a picture to *illustrate* or exemplify a finished argument or set of findings. Such pictures are selected to show concisely what will be said about them, and they can be adjusted to fit their captions. A split thus emerges between the practitioner's local historical grasp of the image as a moment in an eidetic flux *versus* a treatment of the image as a "pure deictic" reference (Barthes 1981:4) conveying an appreciation of the things discussed in an article. There is no rupture, however, since this split between fluctuating digital contextures and fixed opticist rendering is scaled over by a digital simulation of photographic realism. The properties earlier ascribed to "digitality"—pixelated space, arbitrary code, manipulable details, syntagmatic organization, and an equivalence between qualities and quantities—are made to subserve a smooth depiction of a bounded object within a luminous field viewed from a fixed standpoint. What the viewer sees at a glance is thus a pieced-together assemblage, dutifully composed to evoke the gestalt properties of an opticist contexture. In contrast to the pointillist or impressionist artists, who topicalized the compositional elements of painting, thereby exploding the established conventions of pictorial realism, astronomers tend not to bring out fully digitality's distinctive contexture. As a consequence they act as members of a community of the "wise" (Goffman 1963), sharing a secret understanding of nonapparent qualities while putting on a front for the sake of prevailing standards of taste and decorum.

CONCLUSION

In this chapter I have argued for a revised understanding of the "place" in which scientific work occurs. This place of work is not confined by laboratory walls or inscribed upon a floor plan. Although the arrangements indicated by a floor plan are certainly not irrelevant to the routine activities conducted on that floor, they provide a rather spare rendering of such activities. Although walls, entrances, hallways, and the like have the advantage that they are readily described in diagrams and they do not disappear when the persons inhabiting them go away, this does not necessarily give them a more fundamental or "objective" role in activities than those less obvious contextures of space, technique, and language that disappear when relevant activities come to a close. Although the latter are essentially bound to activities, this does not make them "subjective," nor does it imply that they are mere effects of a skeletal architectural machinery. The topical contextures of "opticism" and "digitality" that I have sketched here are not readily tagged to a material base; they generate material for analysis at the same time that they inhabit a material ground. The elucidate physical space while endowing the "physical" with distinctive configurations.

If we are to treat the architecture and lived-space of the laboratory as a place of knowledge, we need to view walls, room dimensions, partitions, enclosures, and restrictions on entry as surface features of the phenomenal fields investigated by the scientists inhabiting such a "place." This does not make floors and walls irrelevant, but it suggests that when various practitioners construct and tear down walls, assign equipment to places and practitioners to equipment, manage computer files, set up precautions against contamination, and so forth, they are not obeying generalized laws so much as effecting local rearrangements within a specific topical field. What I have suggested here is that by trying to understand the space of knowledge we confront an ecology of local spaces integrated with disciplinary practices.

The central theme of "knowledge" also needs to be reexamined. Throughout this chapter I have focused on contextures of practice and equipment, rather than on ideas or even theories. The point of treating "knowledge" in this way is made clear by histories of science that develop detailed accounts of experimental demonstrations and representational systems (Edge and Mulkay 1976; Rudwick 1976; Shapin and Schaffer 1985; Galison 1987, 1989; Gooding 1986, 1988; Pickering 1988). These and many other studies show that what counts in particular disciplinary contexts as convincing arguments or persuasive results is inexorably tied to the practical devices through which phenomena become *accountable.* This contrasts to a theory-centric epistemology that represents scientific work as an implementation of ideas or a grounding of propositions. Sociological studies of contemporary scientific activities (Latour and Woolgar 1986 [1979]; Collins 1985; Woolgar 1988b; Amann and Knorr-Cetina 1988; Gerson and Star 1988) have similarly elaborated on the practices through which representational devices are constructed and linked together in laboratory work. An upshot of these studies is that the concept of "knowledge" may no longer be adequate to handle the "components" of scientific activity. Scientists construct and use instruments, modify specimen materials, write articles and make pictures, and build organizations. This praxis is not limited to "intellectual" or "cognitive" operations, since it takes place in material environments while at the same time reconstructing those environments. I have argued, however, contrary to historical materialism, that scientific praxis is bound up in locally organized topical contextures. Ethnomethodology is often criticized for eschewing a panoramic view of history and society in favor of an excessive attention to detail.

While I plead guilty to all such accusations and more, the policy recommended here is to forget the eye in the sky and pay more attention to the local "order at all points" (Sacks 1984).

NOTES

1. The discussion in this section is based on an ongoing study in collaboration with Samuel Edgerton on digital image processing in astronomy (Lynch and Edgerton 1988). The study examines contemporary astronomers' use of image-processing technology at two Boston area astrophysics facilities.

2. This was not always the case, as can be gathered from the troubles experienced in the portrait studios that proliferated shortly after the invention of the daguerreotype. The pose became a key point of contention in the transaction between photographer and subject. Newhall (1964:21) quotes the following passage from the satirical journal *Le Charivari* to illustrate this problem: "You want to make a portrait of your wife. You fix her head in a temporary iron collar to get the indispensable immobility. . . . You point the lens of the camera at her face, and when you take the portrait it doesn't represent your wife: it is her parrot, her watering pot, or worse."

Despite their novelty, and popular expectations of an almost magical realism, daguerreotypes were compared unfavorably to the more "natural" portraits enabled by oil painting. These complaints were emblematic of the contingencies involved in forging an operative "natural representation," and they subsided with further technological innovations in the camera, the studio, and the techniques of the pose. The family photograph eventually assimilated the classic conventions of portraiture, to the point that the snapshot is now seen at a glance *as* its subject.

JIM JOHNSON
(a.k.a. BRUNO LATOUR)

7

Mixing Humans and Nonhumans Together: The Sociology of a Door-Closer[1]

The most liberal sociologist often discriminates against nonhumans. Ready to study the most bizarre, exotic, or convoluted social behavior, he or she balks at studying nuclear plants, robots, or pills. Although sociology is expert at dealing with human groupings, when it comes to nonhumans, it is less sure of itself. The temptation is to leave the nonhuman to the care of technologists or to study the impact of black-boxed techniques upon the evolution of social groups. In spite of the works of Marx or Lewis Mumford and the more recent development of a sociology of techniques (MacKenzie and Wacjman 1985b; Bijker et al. 1986; Winner 1986; Latour 1987), sociologists still feel estranged when they fall upon the bizarre associations of humans with nonhumans. Part of their uneasiness has to do with the technicalities of complex objects and with the absence of a convenient vocabulary allowing them to move freely from studying associations of humans to associations of nonhumans. In this chapter I want to contribute to the reinsertion of nonhumans into the mainstream of American sociology by examining an extremely simple technique and offering a coherent vocabulary that could be applied to more complex imbroglios of humans and nonhumans.

REINVENTING THE DOOR

On a freezing day in February, posted on the door of the sociology department at Walla Walla University, Washington, could be seen a small handwritten notice: "The door-closer is on strike. For God's sake, keep the door closed." This fusion of labor relations, religion, advertisement, semiotics, and technique in one single insignificant fact is exactly the sort of thing I want to help describe. As a technologist teaching in an engineering school in Columbus, Ohio, I want to challenge some of the assumptions sociologists often hold about the "social context" of machines.

Walls are a nice invention, but if there were no holes in them, there would be no way to get in or out; they would be mausoleums or tombs. The problem is that, if you make holes in the walls, anything and anyone can get in and out (bears, visitors, dust, rats, noise). So architects invented this hybrid: a hole-wall, often called a "door," which, although common enough, has always struck me as a miracle of technology. The cleverness of the invention hinges upon the hinge-pin: instead of driving a hole through walls with a sledgehammer or a pick, you simply gently push the door (I have supposed here that the lock has not been invented; this would overcomplicate the already highly complex story of this door). Furthermore, and here is the real trick, once you have passed through the door, you do not have to find trowel and cement to rebuild the wall you have just destroyed; you simply push the door gently back (I ignore for now the added complication of the "pull" and "push" signs).

So, to size up the work done by hinges, you simply have to imagine that every time you want to get in or out of the building you have to do the same work as a prisoner trying to escape or a gangster trying to rob a bank, plus the work of those who rebuild either the prison's or the bank's walls.

If you do not want to imagine people destroying walls and rebuilding them every time they wish to leave or enter a building, then imagine the work that would have to be done in order to keep inside or to keep outside all the things and people that, left to themselves, would go the wrong way. As Maxwell could have said, imagine his demon working *without* a door. Anything could escape from or penetrate into the department, and there would soon be complete equilibrium between the depressing and noisy surrounding area and the inside of the building. Techniques are always involved when asymmetry or irreversibility is the goal; it might appear that doors are a striking counterexample since they maintain the hole-wall in a reversible state, but the allusion to Maxwell's demon clearly shows that such is not the case. The reversible door is the only way to irreversibly trap inside a differential accumulation of warm sociologists, knowledge, papers, and also, alas, paperwork; the hinged door allows a selection of what gets in and what gets out so as to locally increase order or information. If you let the drafts get inside, the drafts will never get outside to the publishers.

Now, draw two columns (if I am not allowed to give orders to the reader of this volume, then take it as a piece of strongly worded advice). In the right column, list the work people would have to do if they had no door; in the left column write down the gentle pushing (or pulling) they have to do in order to fulfill the same tasks. Compare the two columns; the enormous effort on the right is balanced by the little one on the left, and this thanks to hinges. I will define this transformation of a major effort into a minor one by the word *translation* or *delegation*; I will say that we have delegated (or translated or displaced or shifted out) to the hinge the work of reversibly solving the hole-wall dilemma. Calling on a sociologist friend, I do not have to do this work nor even to think about it; it was delegated by the carpenter to a character, the hinge, that I will call a nonhuman (notice that I did not say "inhuman"). I simply enter the Department of Sociology. As a more general descriptive rule, every time you want to know what a nonhuman does, simply imagine what other humans or other nonhumans would have to do were this character not present. This imaginary substitution exactly sizes up the role, or function, of this little figure.

Before going on, let me cash out one of the side benefits of this table: in effect, we have drawn a scale balance where tiny efforts balance out mighty weights. The scale we drew (at least the one that you drew if you have obeyed my orders—I mean, followed my advice) reproduces the very leverage allowed by hinges. That the small be made stronger than the large is a very moral story indeed (think of David and Goliath). By the same token, this is also, since at least Archimedes' days, a very good definition of a lever and of power: the minimum you need to hold and deploy astutely in order to produce the maximum effect. Am I alluding to machines or to Syracuse's King? I don't know, and it does not matter since the King and Archimedes fused the two "minimaxes" into one single story told by Plutarch: the defense of Syracuse. I contend that this reversal of forces is what sociologists should look at in order to understand the "social construction" of techniques and not at a hypothetical social context they are not equipped to grasp. This little point having been made, let me go on with the story (we will understand later why I do not really need your permission to go on and why, nevertheless, you are free not to go on, although only *relatively* so).

DELEGATING TO HUMANS

There is a problem with doors. Visitors push them to get in or pull on them to get out (or vice versa), but then the door remains open. That is, instead of the door you have a gaping hole in the wall through which, for instance, cold rushes in and heat rushes out. Of course, you could imagine that people living in the building or visiting the Department of Sociology would be a well-disciplined lot (after all, sociologists are meticulous people). They will learn to close the door behind them and re-transform the momentary hole into a well-sealed wall. The problem is that discipline is not the main characteristic of people. Are they going to be so well behaved? Closing a door would appear to be a simple enough piece of know-how once hinges have been invented; but, considering the amount of work, innovations, sign-posts, recriminations that go on endlessly everywhere to keep them closed (at least in Northern regions), it seems to be rather poorly disseminated.

This is where the age-old choice, so well analyzed by Mumford (1966), is offered to you: either to discipline the people or to *substitute* for the unreliable people another *delegated human character* whose only function is to open and close the door. This is called a groom or a porter (from the French word for door) or a gatekeeper, or a janitor, or a concierge, or a turnkey, or a jailer. The advantage is that you now have to discipline only one human and may safely leave the others to their erratic behavior. No matter who these others are and where they come from, the porter will always take care of the door. A nonhuman (the hinges) plus a human (the porter) have solved the hole-wall dilemma.

Solved? Not quite. First of all, if the department pays for a human porter, they will have no money left to buy coffee or books or to invite eminent foreigners to give lectures. If they give a poor little boy other duties besides that of porter, then he will not be present most of the time, and the damned door will stay open. Even if they had money to keep him there, we are now faced with a problem that two hundred years of capitalism has not completely solved: how to discipline a youngster to reliably fulfill a boring and underpaid duty. Although there is now only one human to be disciplined instead of hundreds (in practice only dozens because Walla Walla is rather difficult to locate), the weak point of the tactic is now revealed: if this one lad is unreliable then the whole chain breaks down. If he falls asleep on the job or goes walkabout, there will be no appeal; the damned door will stay open (remember that locking it is no solution since this would turn it into a wall, and then providing every visitor with the right key is an impossible task). Of course, the little rat may be punished or even flogged. But imagine the headlines: "Sociologists of science flog porter from poor working-class background." And what if he is black, which might very well be the case, given the low pay and widespread racism? No, disciplining a porter is an enormous and costly task that only Hilton Hotels can tackle, and that for other reasons that have nothing to do with keeping the door properly closed.

If we compare the work of disciplining the porter with the work he substitutes for, according to the list defined above, we see that this delegated character has the effect opposite to that of the hinge. A simple task, forcing people to close the door, is now performed at an incredible cost; the minimum effect is obtained with maximum spending and spanking. We also notice, when drawing the two lists, an interesting difference. In the first relationship (hinges vis-à-vis work of many people), you not only had a reversal of forces (the lever allows gentle manipulations to heavy weights) but also a reversal of *time*. Once the hinges are in place, nothing more has to be done apart from maintenance (oiling them from time to time). In the second set of relations (porter's work versus many people's work), not only do you fail to reverse the forces, but you also fail to modify the time schedule. Nothing can be done to prevent the porter who has been reliable for two months from failing on the sixty-second day; at this point it is not maintenance work that has to be done, but the same work as on the first day—apart from the few habits that you might have been able to *incorporate* into his body. Although they appear to be two similar delegations, the first one is concentrated in time, whereas the other is continuous; more exactly, the first one creates a clear-cut distinction between production and maintenance, whereas in the other the distinction between training and keeping in operation is either fuzzy or nil. The first one evokes the past perfect ("once hinges had been installed"); the second the present tense ("when the porter is at his post"). There is a built-in inertia in the first that is largely lacking in the second. A profound temporal shift takes place when nonhumans are appealed to: Time is folded.

DISCIPLINING THE DOOR-CLOSER

It is at this point that you have this relatively new choice: either to discipline the people or to substitute for the unreliable humans a delegated nonhuman character whose only function is to open and close the door. This is called a door-closer or a "groom." The advantage is that you now have to discipline only one nonhuman and may safely leave the others (bellboys included) to their erratic behavior. No matter who they are and where they come from—polite or rude, quick or slow, friends or foes—the nonhuman groom will always take care of the door in any weather and at any time of the day. A nonhuman (hinges) plus another nonhuman (groom) have solved the hole-wall dilemma.

Solved? Well, not quite. Here comes the deskilling question so dear to social historians of technology: thousands of human porters have been put on the dole by their nonhuman brethren. Have they been replaced? This depends on the kind of action that has been translated or delegated to them. In other words, when humans are displaced and deskilled, nonhumans have to be upgraded and reskilled. This is not an easy task, as we shall now see.

We have all experienced having a door with a powerful spring mechanism slam in our face. For sure, springs do the job of replacing porters, but they play the role of a very rude, uneducated porter who obviously prefers the wall version of the door to its hole version. They simply slam the door shut. The interesting thing with such impolite doors is this: if they slam shut so violently, it means that you, the visitor, *have* to be very quick in passing through and that you *should* not be at someone else's heels; otherwise your nose will get shorter and bloody. An unskilled nonhuman groom thus presupposes a skilled human user. It is always a trade-off. I will call, after Madeleine Akrich (1987), the behavior imposed back onto the human by nonhuman delegates *prescription*. How can these prescriptions be brought out? By replacing them by strings of sentences (usually in the imperative) that are uttered (silently and continuously) by the mechanisms for the benefit of those who are mechanized: Do this, Do that, Behave this way, Don't go that way. Such sentences look very much like a programming language. This substitution of words for silence can be made in the analyst's thought experiments, but also by instruction booklets or explicitly in any training session through the voice of a demonstrator or instructor or teacher. The military are especially good at shouting them out through the mouthpiece of human instructors who delegate back to themselves the task of explaining, in the rifle's name, the characteristics of the rifle's ideal user. As Akrich notes, prescription is the moral and ethical dimension of mechanisms. In spite of the constant weeping of moralists, no human is as relentlessly moral as a machine, especially if it is (she is, he is, they are) as "user friendly" as my computer.

The results of such distributions of skills between humans and nonhumans is well known: members of the Department of Sociology will safely pass through the slamming door at a good distance from one another; visitors, unaware of the *local cultural condition*, will crowd through the door and will get bloody noses. This story is of the same form as that about the buses loaded with poor blacks that could not pass under driveways leading to Manhattan parks (Winner 1985). So, inventors get back to their drawing board and try to imagine a nonhuman character that will not prescribe the same rare local cultural skills to its human users. A weak spring might appear to be a good solution. Such is not the case because it would substitute for another type of very unskilled and undecided porter who is never sure about the door's (or his own) status: Is it a hole or a wall? Am I a closer or an opener? If it is both at once, you can forget about the heat. In computer parlance, a door is an OR, not an AND *gate*.

I am a great fan of hinges, but I must confess that I admire hydraulic door-closers much more, especially the old copperplated heavy one that slowly closed the main door of our house in Columbus, Ohio. I am enchanted by the addition to the spring of a hydraulic piston which easily draws up the energy of those who open the door and retains it, then gives it back slowly with a subtle variety of implacable firmness that one could expect from a well-trained butler. Especially clever is its way of extracting energy from each and every unwilling, unwitting passerby. My military friends at the academy call such a clever extraction an "obligatory passage point," which is a very fitting name for a door—no matter what you feel, think, or do, you have to leave a bit of your energy, literally, at the door. This is as clever as a toll booth.

This does not quite solve all problems, though. To be sure the hydraulic door-closer does not bang the noses of those who are not aware of local conditions, so its prescriptions may be said to be less restrictive. But it still leaves aside segments of human populations. Neither my little nephews nor my grandmother could get in unaided because our groom needed the force of an able-bodied person to accumulate enough energy to close the door. To use the classic Langdon Winners motto (1985), because of their prescriptions these doors *discriminate* against very little and very old persons. Also, if there is no way to keep them open for good, they discriminate against furniture removers and in general everyone with packages, which usually means, in our late capitalist society, working- or lower-middle class employees. (Who, even coming from a higher stratum, has not been cornered by an automated butler when you had your hands full of packages?) There are solutions though: the groom's delegation may be written off (usually by blocking its arm) or, more prosaically, its delegated action may be opposed by a foot (salesmen are said to be expert at this). The foot may in turn be delegated to a carpet or anything that keeps the butler in check (although I am always amazed by the number of objects that fail this trial of force, and I have very often seen the door I just wedged open politely closing when I turned my back on it).

As a technologist, I could claim that, provided you put aside maintenance and the few sectors of population that are discriminated against, the groom does its job well, closing the door behind you constantly, firmly, and slowly. It shows in its humble way how three rows of delegated nonhuman actants (hinges, springs, and hydraulic pistons) replace, 90 percent of the time, either an undisciplined bellboy who is never there when needed or, for the general public, the program instructions that have to do with remembering-to-close-the-door-when-it-is-cold. The hinge plus the groom is the technologist's dream of efficient action, at least it was until the sad day when I saw the note posted on the Walla Walla sociology department's door with which I started this article: "The door-closer is on strike." So not only have we been able to delegate the act of closing the door from the human to the nonhuman, we have also been able to delegate the little rat's lack of discipline (and maybe the union that goes with it). On strike. Fancy that! Nonhumans stopping work and claiming what? Pension payments? Time off? Landscaped offices? Yet it is no use being indignant because it is very true that nonhumans are not so reliable that the irreversibility we would like to grant them is complete. We did not want ever to have to think about this door again—apart from regularly scheduled routine maintenance (which is another way of saying that we did not have to bother about it)—and here we are, worrying again about how to keep the door closed and drafts outside.

What is interesting in the note on the door is the humor of attributing a human character to a failure that is usually considered "purely technical." This humor, however, is more profound than the synonymous notice they could have posted: "The door-closer is not working." I constantly talk with my computer, who answers back; I am sure you swear at your old car; we are constantly granting mysterious faculties to gremlins inside every conceivable home appliance, not to mention cracks in the concrete belt of our nuclear plants. Yet, this behavior is considered by moralists, I mean sociologists, as a scandalous breach of natural barriers. When you write that a groom is "on strike," this is only seen as a "projection," as they say, of a human behavior onto a nonhuman cold technical object, one by nature impervious to any feeling. They call such a projection anthropomorphism, which for them is a sin akin to zoophily but much worse.

It is this sort of moralizing that is so irritating for technologists because the automatic groom is already anthropomorphic through and through. *Anthropos* and *morphos* together mean either what has human shape or what gives shape to humans. Well the groom is indeed anthropomorphic, and in three senses: first, it has been made by humans, it is a construction; second, it substitutes for the actions of people, and is a delegate that permanently occupies the position of a human; and third, it shapes human action by prescribing back what sort of people should pass through the door. And yet some would forbid us to ascribe feelings to this thoroughly anthropomorphic creature, to delegate labor relations, to "project"—that is to say, to translate—*other* human properties to the groom. What of those many other innovations that have endowed much more sophisticated doors with the ability to see you arrive in advance (electronic eyes), or to ask for your identity (electronic passes), or to slam shut—or open—in case of danger? But anyway, who are you, you the sociologists, to decide forever the real and final shape of humans, to trace with confidence the boundary between what is a "real" delegation and what is a "mere" projection, to sort out forever and without due inquiry the three different kinds of anthropomorphism I listed above? Are we not shaped by nonhuman grooms, although, I admit, only a very little bit? Are they not our brethren? Do they not deserve consideration? With your self-serving and self-righteous social problems, you always plead against machines and for deskilled workers; are you aware of *your* discriminatory biases? You discriminate between the human and the nonhuman. I do not hold this bias but see only actors—some human, some nonhuman, some skilled—that exchange their properties.

So the note posted on the door is an accurate one. It gives a humorous but exact rendering of the groom's behavior: it is not working; it is on strike (notice that the word *strike* is also an anthropomorphism carried from the nonhuman repertoire to the human one, which proves again that the divide is untenable). What happens is that sociologists confuse the dichotomy human/nonhuman with another one: *figurative/nonfigurative*. If I say that Hamlet is the figuration of "depression among the aristocratic class," I move from a personal figure to a less personal one (class). If I say that Hamlet stands for doom and gloom, I use less figurative entities; and if I claim that he represents Western civilization, I use nonfigurative abstractions. Still, they all are equally actants, that is to say entities that *do* things, either in Shakespeare's artful plays or in the commentator's more tedious tomes. The choice of granting actants figurativity or not is left entirely to the authors. It is exactly the same for techniques. We engineers are the authors of these subtle plots or *scenariis*, as Madeleine Akrich (1987) calls them, of dozens of delegated and interlocking characters so few people know how to appreciate. The label *inhuman* applied to techniques simply overlooks translation mechanisms and the many choices that exist for figuring or de-figuring, personifying or abstracting, embodying or disembodying actors.

For instance, on the highway the other day, I slowed down because there was a guy in a yellow suit and a red helmet waving a red flag. Well, the guy's moves were so regular and he was located so dangerously and had such a pale although smiling face that, when I passed by, I recognized it to be a machine (it failed the Turing test, a cognitivist would say). Not only was the red flag delegated, not only was the arm waving the flag also delegated, but the body appearance was also added to the machine. We engineers could move much further in the direction of figuration, although at a cost; we could have given him/her (careful here, no sexual discrimination of robots) electronic eyes to wave only when there is a car approaching or regulated the move so that it is faster when cars do not obey. Also we could have added—why not?—a furious stare or a recognizable face like a mask of President Reagan, which would have certainly slowed drivers down very efficiently. But we could also have moved the other way, to a *less* figurative delegation; the flag by itself could have done the job. And why a flag? Why not simply a sign: "Work in progress"? And why a sign at all? Drivers, if they are circumspect, disciplined, and watchful will see for themselves that there is work in progress and will slow down.

The *enunciator* (a general word for the author of a text or for the mechanics who devised the machine) is free to place or not a representation of himself or herself in the script (texts or machines). The engineer may delegate or not in the flag-mover a shape that is similar to himself/herself. This is exactly the same operation as the one I did in pretending that the author of this article was a hardcore technologist from Columbus, Ohio. If I say, "we, the technologists," I propose a picture of the author-of-the-text that has only a vague relation with the author-in-the-flesh, in the same way as the engineer delegates in the flag-mover a picture of him/her that bears little resemblance to him/her. But it would have been perfectly possible for me and for the mechanics to position no figurated character at all as the author *in* the scripts *of* our scripts (in semiotic parlance, there would be no narrator). I would just have had to say things like "recent developments in sociology of science have shown that" instead of "I," and the mechanics would simply have had to take out the dummy worker and replace it by cranks and pulleys.

APPEALING TO GODS

Here comes the most interesting and saddest lesson of the note posted on the door: people are not circumspect, disciplined, and watchful, especially not Walla Walla drivers after happy hour on Friday night. Well, that's exactly the point that the note made: "The door-closer is on strike. *For God's sake*, keep the door closed." In our societies, there are two systems of appeal: nonhuman and superhuman, that is machines and gods. This note indicates how desperate its frozen and anonymous authors were (I have never been able to trace them back and to honor them as they deserved). They first relied on the inner morality and common sense of humans. This failed; the door was always left open. Then they appealed to what we technologists consider the supreme court of appeal, that is, to a nonhuman who regularly and conveniently does the job in place of unfaithful humans. To our shame, we must confess that it also failed after a while. The door was again always left open. How poignant their line of thought is! They moved up and backward to the oldest and firmest court of appeal there is, there was, and ever will be. If human and nonhuman have failed, certainly God will not deceive them. I am ashamed to say that, when I crossed the hallway this fatal February day, the door *was* open. Do not accuse God, though, because the note did not appeal directly to Him (I know I should have added "Her" for affirmative action reasons, but I wonder how theologians would react). God is not accessible without mediators. The anonymous authors knew their catechisms well, so instead of asking for a direct miracle (God Himself/Herself holding the door firmly closed or doing so through the mediation of an angel, as has happened on several occasions, for instance when Paul was delivered from his prison), they appeal to the respect for God in human hearts. This was their mistake. In our secular times, this is no longer enough.

Nowadays nothing seems to do the job of disciplining men and women and forcing them simply to close doors in cold weather. It is a similar despair that pushed the road engineer to add a Golem to the red flag to force drivers to beware—although the only way to slow drivers is still a good traffic jam. You seem to always need more and more of these figurated delegates aligned in rows. It is the same with delegates as with drugs: you start with the soft ones and end by shooting up. There is an inflation for delegated characters too. After a while they weaken. In the old days it might have been enough just to have a door for people to know how to close it. But then, the embodied skills somehow disappeared; people had to be reminded of their training. Still, the simple inscription "Keep the door closed" might have been sufficient in the good old days. But you know people; they no longer pay attention to such notices and need to be reminded by stronger devices. It is then that you install automatic grooms, since electric shocks are not as acceptable for men as for cows. In the old times, when quality was still good, it might have been enough just to oil it from time to time, but nowadays even automatisms go on strike.

It is not, however, that the movement is always from softer to harder devices, that is, from an autonomous body of knowledge to force through the intermediary situation of worded injunctions, as the Walla Walla door would suggest. It also goes the other way. Although the deskilling thesis appears to be the general case (always go from intrasomatic to extrasomatic skills; never rely on undisciplined people, but always on safe, delegated nonhumans), this is far from true. For instance, red lights are usually respected, at least when they are sophisticated enough to integrate traffic flow through sensors. The delegated police officer standing there day and night is respected even though it has no whistles, gloved hands, and body to enforce this respect. Imagined collisions with the other cars or with the absent police are enough to keep drivers and cars in check. The thought experiment What would happen if the delegated character was not there? is the same as the one I recommended above to size up its function. The same incorporation from written injunction to body skills is at work with car user manuals. No one, I guess, will cast more than a cursory glance at the manual before starting the engine. There is a large body of skills that we have now so well embodied or incorporated that the mediations of the written instructions are useless. From extrasomatic they have become intrasomatic. Incorporation in human or in nonhuman bodies is also left to the authors/engineers.

OFFERING A COHERENT VOCABULARY

It is because humans, nonhumans, and even angels are never sufficient in themselves and because there is no one direction going from one type of delegation to the other, that it is so useless to impose a priori divisions between which skills are human and which ones are not human, which characters are personified and which remain abstract, which delegation is forbidden and which is permissible, which type of delegation is stronger or more durable than the other. In place of these many cumbersome distinctions why not take up a few simple descriptive tools?

Following Madeleine Akrich's lead, we will speak only in terms of *scripts* or scenes or scenarios played by human or nonhuman actors, which may be either figurative or nonfigurative. Humans are not necessarily figurative; for instance, you are not allowed to take the highway police officer as an individual chum. He/she is the representative of authority, and if he/she is really dumb, he/she will reject any individualizing efforts from you, like smiles, jokes, bribes, or fits of anger. He/she will fully play the administrative *machinery*.

Following Akrich, I will call the retrieval of the script from the situation *description*. These descriptions are always in words and appear very much like semiotic commentaries on a text or like a programming language. They define actors, endow them with competencies and make them do things, and evaluate the sanction of these actions very much like the narrative program of semioticians.

Although most of the scripts are in practice silent either because they are intra- or extrasomatic, the written descriptions are not an artifact of the analyst (technologist, sociologist, or semiotician) because there exist many states of affairs in which they are *explicitly* uttered. The gradient going from intrasomatic to extrasomatic skills through discourse is never fully stabilized and allows many entries revealing the process of translation. I have already listed several entries: user manuals, instruction, demonstration or drilling situations (in this case a human or a speech-synthesizer speaks out the user manual), practical thought experiments (What would happen if instead of the red light a policeman were there?). To this should be added the innovator's workshop where most of the objects to be devised are still at the stage of projects committed to paper (If we had a device doing this and that, we could then do this and that); market analysis in which consumers are confronted with the new device; and, naturally, the training situation studied by anthropologists where people faced with a foreign device talk to themselves while trying out various combinations (What will happen if I attach this lead here to the mains?). The analyst has to capture these situations in order to write down the scripts. The analyst makes a thought experiment by comparing presence/absence tables and collating all the actions done by actants: If I take this one away, this and that other action will be modified.

I will call the translation of any script from one repertoire to a more durable one *transcription* or *inscription* or encoding. Translation does not have here only its linguistic meaning but also the religious one, "translation of the remains of St. Christel," and the artistic one, "translating the feelings of Calder into bronze." This definition does not imply that the direction always goes from soft bodies to hard machines, but simply that it goes from a provisional less reliable one to a longer lasting, more faithful one. For instance, the embodiment in cultural tradition of the user manual of a car is a transcription, but so is the replacement of a policeman by a traffic light. One goes from machines to bodies, whereas the other goes the other way. Specialists of robotics have very much abandoned the pipe dream of total automation; they learned the hard way that many skills are better delegated to humans than to nonhumans, whereas others may be moved away from incompetent humans.

I will call *prescription* whatever a scene presupposes from its *transcribed* actors and authors (this is very much like "role expectation" in sociology, except that it may be inscribed or encoded in the machine). For instance, a Renaissance Italian painting is designed to be viewed from a specific angle of view prescribed by the vanishing lines, exactly like a traffic light expects that its users will watch it from the street and not sideways. In the same way they presuppose a user, traffic lights presuppose that there is someone who has regulated the lights so that they have a regular rhythm. When the mechanism is stuck it is very amusing to see how long it takes drivers to decide that the traffic light is no longer mastered by a reliable author. "User input" in programming language is another very telling example of this inscription in the automatism of a living character whose behavior is both free and predetermined.

This inscription of author and users in the scene is very much the same as that of a text. I already showed how the author of this article was ascribed (wrongly) to be a technologist in Ohio. It is the same for the reader. I have many times used "you" and even "you sociologists." If you remember well, I even ordered you to draw up a table (or advised you to do so). I also asked your permission to go on with the story. In doing so, I built up an *inscribed reader* to whom I prescribed qualities and behavior as surely as the traffic light or the painting prepared a position for those looking at them. Did you *subscribe* to this definition of yourself? Or worse, is there anyone at all to read this text and occupy the position prepared for the reader? This question is a source of constant difficulties for those who do not grasp the basics of semiotics. Nothing in a given scene can prevent the inscribed user or reader from behaving differently from what was expected (nothing, that is, until the next paragraph). The reader-in-the-flesh may totally ignore my definition of him or her. The user of the traffic light may well cross on the red. Even visitors to the Department of Sociology may never show up because Walla Walla is too far away, *in spite of* the fact that their behavior and trajectory have been perfectly anticipated by the groom. As for the computer user input, the cursor might flash forever without the user being there or knowing what to do. There might be an enormous gap between the prescribed user and the user-in-the-flesh, a difference as big as the one between the "I" of a novel and the novelist. It is exactly this difference that so much upset the authors of the anonymous appeal posted on the door. It is because they could not discipline people with words, notes, and grooms, that they had to appeal to God. On another occasion, however, the gap between the two may be nil: the prescribed user is so well anticipated, so carefully nested inside the scenes, so exactly dovetailed, that it does what is expected. To stay within the same etymological root, I would be tempted to call the way actors (human or nonhuman) tend to extirpate themselves from the prescribed behavior *des-inscription* and the way they accept or happily acquiesce to their lot *subscription*.

The problem with scenes is that they are usually well prepared for anticipating users or readers who are at close quarters. For instance, the groom is quite good in its anticipation that people will push the door open and give it the energy to re-close it. It is very bad at doing anything to help people arrive there. After fifty centimeters, it is helpless and cannot act, for example, to bring people to Washington State. Still, no scene is prepared without a preconceived idea of what sort of actors will come to occupy the prescribed positions. This is why I said that, although *you* were free not to go on with this chapter, *you* were only "relatively" so. Why? Because I now you are hard-working, serious American sociologists, reading a serious volume on sociology of science and technology. So, I can safely bet that I have a good chance of having you read the chapter thoroughly! So my injunction "read the chapter up to the end, you sociologist" is not very risky. I will call *pre-inscription* all the work that has to be done upstream of the scene and all the things assimilated by an actor (human or nonhuman) before coming to the scene as a user or as an author. For instance, how to drive a car is basically pre-inscribed in any (Western) youth years before he or she comes to passing the driving license test; hydraulic pistons were also pre-inscribed for slowly giving back the energy gathered years before innovators brought them to bear on automated grooms. Engineers can bet on this predetermination when they draw up their prescriptions. This is what Gerson and his colleagues call "articulation work" (Fujimura 1987, Gerson and Star, 1986). A lovely example of efforts at pre-inscription is provided by Orson Welles in *Citizen Kane,* where the hero not only bought a theater for his singing wife to be applauded in, but also bought the journals that were to do the reviews, bought off the art critics themselves, and paid the audience to show up—all to on avail, since the wife eventually quit. Humans and nonhumans are very, very undisciplined no matter what you do and how many predeterminations you are able to control upstream of the action.

Drawing a side conclusion in passing, we can call *sociologism* the claim that, given the competence and pre-inscription of human users and authors, you can read out the scripts nonhuman actors have to play; and *technologism* the symmetric claim that, given the competence and pre-inscription of the nonhuman actors, you can easily read out and deduce the behavior prescribed to authors and users. From now on, these two absurdities will, I hope, disappear from the scene, since the actors at any point may be human or nonhuman and since the displacement (or translation, or transcription) makes the easy reading-out of one repertoire into the next impossible. The bizarre idea that society might be made up of human relations is a mirror image of the other no less bizarre idea that techniques might be made up of nonhuman relations. We deal with characters, delegates, representatives, or, more nicely, lieutenants (from the French *lieu tenant,* i.e., holding the place of, for, someone else); some figurative, others nonfigurative; some human, others nonhuman; some competent, others incompetent. You want to cut through this rich diversity of delegates and artificially create two heaps of refuse: "society" on one side and "technology" on the other? That's your privilege, but I have a less messy task in mind.

A scene, a text, an automatism can do a lot of things to their prescribed users at close range, but most of the effect finally ascribed to them depends on a range of other setups being aligned. For instance, the groom closes the door only if there are people reaching the sociology department of Walla Walla. These people arrive in front of the door only if they have found maps and only if there are roads leading to it; and, of course, people will start bothering about reading the maps, getting to Washington State, and pushing the door open only if they are convinced that the department is worth visiting. I will call this *gradient* of aligned setups that endow actors with the pre-inscribed competencies to find its users a *chreod* (a "necessary path" in the biologist Waddington's Greek): people effortlessly flow through the door, and the groom, hundreds of times a day, re-closes the door—when it is not stuck. The result of such an alignment of setups is to decrease the number of occasions in which words are used; most of the actions become silent, familiar, incorporated (in human or in nonhuman bodies)—making the analyst's job so much harder. Even the classic debates about freedom, determination, predetermination, brute force, or efficient will—debates that are the twentieth-century version of seventeenth-century discussions on grace—will be slowly eroded away. (Since *you* have reached this point, it means I was right in saying earlier that you were not at all free to stop reading the chapter. Positioning myself cleverly along a chreod, and adding a few other tricks of my own, I led you *here* . . . or did I? Maybe you skipped most of it; maybe you did not understand a word of it, oh you undisciplined American sociologist readers!)

There is one loose end in my story: Why did the little (automatic) rat go on strike? The answer to this is the same as for the question earlier of why few people show up in Walla Walla. It is not because a piece of behavior is prescribed by an inscription that the predetermined characters will show up on time and do the job expected of them. This is true of humans, but it is truer of nonhumans. In this case the hydraulic piston did its job, but not the spring that collaborated with it. Any of the words above may be used to describe a setup at any level and not only at the simple one I chose for the sake of clarity. It does not have to be limited to the case where a human deals with a series of nonhuman delegates; it can also be true of relations among nonhumans. In other words, when we get into a more complicated lash-up than the groom, we do not have to stop doing sociology; we go on studying "role expectation," behavior, social relations. The nonfigurative character of the actors should not intimidate us.

THE LIEUTENANTS OF OUR SOCIETIES

I used the story of the door-closer to make a nonhuman delegate familiar to the ears and eyes of sociologists. I also used reflexively the semiotic of a story to explain the relations between inscription, prescription, pre-inscription, and chreods. There is, however, a crucial difference between texts and machines that I have to point out. Machines are lieutenants; they hold the places of the roles delegated to them, but this way of shifting is very difficult from other types (Latour 1988b).

In storytelling, one calls *shifting out* any displacement of a character either to another space or to another time or to another character. If I tell you "Millikan entered the aula," I translate the present setting—you and me—and shift it to another space, another time, and to other characters (Millikan and his audience). "I," the enunciator, may decide to appear or to disappear or to be represented by a narrator who tells the story ("That day, I was sitting on the upper row of the aula"); "I" may also decide to position you and any reader inside the story ("Had you been there, you would have been convinced by Millikan's experiments"). There is no limit to the number of shiftings out a story may be built with. For instance, "I" may well stage a dialogue inside the aula between two characters who are telling a story about what happened at the Academy of Science in Washington, D.C. In that case, the aula is the place *from which* narrators shift out to tell a story about the Academy, and they may or may not *shift back in* the aula to resume the first story about Millikan. "I" may also *shift in* the entire series of nested stories to close mine and come back to the situation I started from: you and me. All these displacements are well known in literature departments and make up the craft of talented writers.

No matter how clever and crafty are our novelists, they are no match for engineers. Engineers constantly shift out characters in other spaces and other times, devise positions for human and nonhuman users, break down competencies that they then redistribute to many different actants, build complicated narrative programs and subprograms that are evaluated and judged. Unfortunately, there are many more literary critics than there are technologists and the subtle beauties of techno-social imbroglios escape the attention of the literate public. One of the reasons for this lack of concern may be the peculiar nature of the shifting-out that generates machines and devices. Instead of sending the listener of a story into another world, the technical shifting-out inscribes the words into another matter. Instead of allowing the reader of the story to be at the same time away (in the story's frame of reference) and here (in his or her armchair), the technical shifting-out forces a choice between frames of reference. Instead of allowing enunciators and enunciatees a sort of simultaneous presence and communication with other actors, technics allow both of them to ignore the delegated actors and to walk away without even feeling their presence.

To understand this difference in the two directions of shifting out, let us venture out once more onto a Columbus highway. For the umpteenth time I have screamed to Robin, "Don't sit in the middle of the rear seat; if I brake too hard, you're dead." In an auto shop further along the highway I come across a device *made for* tired-and-angry-parents-driving-cars-with-kids-between-two-and-five-(that is, too old for a baby seat and not old enough for a seat belt)-and-from-small-families-(that is, without other persons to hold them safely)-and-having-cars-with-two-separated-front-seats-and-head-rests. It is a small market but nicely analyzed by these Japanese folks and, given the price, it surely pays off handsomely. This description of myself and the small category into which I am happy to *subscribe* is *transcribed* in the device—a steel bar with strong attachments to the head rests—and in the advertisement on the outside of the box. It is also *pre-inscribed* in about the only place where I could have realized that I needed it, the highway. Making a short story already too long, I no longer scream at Robin and I no longer try to foolishly stop him with my extended right arm: he firmly holds the bar that protects him—or so I believe—against my braking. I have delegated the continuous injunction of my voice and extension of my right arm (with diminishing results, as we know from Fechner's law) to a reinforced, padded steel bar. Of course, I had to make two detours: one to my wallet, the second to my toolbox. Thirty bucks and five minutes later I had fixed the device (after making sense of the instructions encoded with Japanese ideograms). The detour plus the translation of words and extended arm to steel is a shifting out to be sure, but not of the same type as that of a story. The steel bar has now taken over my competence as far as keeping my son at arm's length is concerned.

If, in our societies, there are thousands of such lieutenants to which we have delegated competencies, it means that what defines our social relations is, for the most part, prescribed back to us by nonhumans. Knowledge, morality, craft, force, sociability are not properties of humans but of humans *accompanied by* their retinue of delegated characters. Since each of those delegates ties together part of our social world, it means that studying social relations without the nonhumans is impossible or adapted only to complex primate societies like those of baboons (Strum and Latour

1987). One of the tasks of sociology is to do for the masses of nonhumans that make up our modern societies what it did so well for the masses of ordinary and despised humans that make up our society. To the people and ordinary folk should now be added the lively, fascinating, and honorable ordinary mechanism. If the concepts, habits, and preferred fields of sociologists have to be modified a bit to accommodate these new masses, it is a small price to pay.

NOTES

1. The author-in-the-text is Jim Johnson, technologist in Columbus, Ohio, who went to Walla Walla University, whereas the author-in-the-flesh is Bruno Latour, sociologist, from Paris, France, who never went to Columbus nor to Walla Walla University. The distance between the two is great but similar to that between Steven Jobs, the inventor of Macintosh, and the figurative nonhuman character who/which says, "welcome to Macintosh" when you switch on your computer. The reason for this use of pseudonym was the opinion of the editors that no American sociologist is willing to read things that refer to specific places and times that are not American. Thus I inscribed in my text American scenes so as to decrease the gap between the prescribed reader and the pre-inscribed one. [*Editors' note*: Since we believed these locations to be unimportant to Bruno Latour's argument, we urged him to remove specific place references that might have been unfamiliar to U.S. readers and thus possibly distracting. His solution seems to have proven our point.]

2. To the shame of our trade, it is a art historian, Michael Baxandall (1985), who offers the most precise description of a technical artifact (a Scottish Iron Bridge), and who shows in most detail the basic distinctions between delegated actors which remain silent (blade-boxed) and the rich series of mediators who remain *present* in a work of art.

PART 3: WORKPLACE ECOLOGIES

JOHN LAW
MICHEL CALLON

8

Engineering and Sociology in a Military Aircraft Project: A Network Analysis of Technological Change[1]

There is an old rule of sociological method, unfortunately more honored in the breach than the observance, that if we want to understand social life then we need to follow the actors wherever they may lead us (Latour 1987). We should, or so this dictum suggests, avoid imposing our own views about what is right or wrong, or true and false. We should especially avoid assuming that those we study are less rational or have a weaker grasp on reality than we ourselves. This rule of method, then, asks us to take seriously the beliefs, projects, and resources of those whom we wish to understand. It suggests that an analysis of social life depends upon such understanding, and it implies that we make best sociological progress when we are sociologically humble.

Applied to technology and its place in society, this rule of method leads to some interesting findings. For instance, it turns out that, when we look at what technologists actually *do*, we find that they pay scant regard to distinctions between technology on the one hand and society, economy, politics, and the rest on the other. The disciplinary distinctions so dear to social scientists seem to be irrelevant to engineers. Thus, when they work, they are typically involved in designing and building projects that have *both* technical *and* social content and implications. For this reason if no other, it is clear that the study of technological innovation is important to sociology. Engineers are not just people who sit in drawing offices and design machines; they are also, willy nilly, social activists who design societies or social institutions to fit those machines. Technical manuals or designs for nuclear power stations imply conclusions about the proper structure of society, the nature of social roles, and how these roles should be distributed (see Winner 1986; Latour 1992b). Engineers were practical sociologists long before the discipline of sociology was invented.

This suggestion and the methodological principle upon which it rests lead to a conclusion that is counterintuitive for many sociologists. This is that we must study not only the social but also the *technical* features of the engineer's work; in other words, we have to understand the *content* of engineering work because it is in this content that the technical and the social are simultaneously shaped. Any attempt to separate the social and the nonsocial not only breaks the original methodological principle of following the technologist. It is also, quite simply, impossible, because the social runs throughout the technical and thus cannot be separated from it. We cannot, and should not wish to, avoid the technical.

If we are to study the work of the *engineer-sociologists* (Callon 1987a) in our midst, then we need to press our methodological adage one stage further. Specifically, it is important to *avoid taking sides* in cases of controversy or failure. We have to be agnostic about the prospects of success for any engineering project, and in particular we must avoid assuming that the fate of projects is written into them from the outset. To take sides is, of course, to abandon the original methodological adage. But it is also to run the risk of assuming that success (or failure) was preordained. It makes it difficult to detect the contingencies that lie behind the possibly complex processes by which success was achieved (or thwarted). Our approach stresses that an achievement is precisely that, an achievement; and it suggests that the causes of failure require analysis and cannot be deduced from the fact of that failure.

In what follows we use the notion of *network* to talk about the interconnected character of the social and technical. We use this notion in a way that differs quite fundamentally from standard usage in sociology. Thus we are not primarily concerned with mapping interactions between individuals. Rather, in conformity with the methodological commitment to follow the actors no matter how they act, we are concerned to map the way in which they *define and distribute roles, and mobilize or invent others to play these roles*. Such roles may be social, political, technical, or bureaucratic in character; the objects that are mobilized to fill them are also heterogeneous and may take the form of people, organizations, machines, or scientific findings. A network metaphor is thus a way of underlining the simultaneously social and technical character of technological innovation. It is a metaphor for the interconnected heterogeneity that underlies socio-technical engineering (Law 1992a).

In this paper we analyze a British military aircraft project in these terms. The aircraft in question, the TSR 2, was conceived in the late 1950s as a light bomber and reconnaissance machine and was intended to serve in much the same role as the General Dynamics F 111, which was developed in the United States at roughly the same time. We trace how this aircraft was conceived, designed, and developed, consider some of the difficulties that were encountered along the way, and describe the process that led to its ultimate cancellation in 1965. Our aim is not to advance a particular theory about technical change. It is rather to recommend the method of social analysis that we have described above, to illustrate the suggestion that the social is both technical and human in character, and to propose a vocabulary for the analysis of social and technical engineering, and to describe some of the tactics that are used by engineer-sociologists as they seek to bring their projects to fruition. Our object, then, is to trace the interconnections built up by technologists as they propose projects and then seek the resources required to bring these project to fruition.

A SOCIO-TECHNICAL SCENARIO

In the Royal Air Force (RAF) there is a section of the headquarters Air Staff called the Operational Requirement Branch

(ORB) that tries to anticipate the future needs of the RAF and propose specifications for possible aircraft that might fulfill those needs. In 1957 the ORB was faced with a serious problem. The British government stated as a formal defense policy that:

> Having regard to the high performance and potentialities of [existing] . . . bombers and the likely progress of ballistic rockets and missile defence, the Government have decided not to go on with the development of a supersonic manned bomber . . . (Ministry of Defence 1957).

Nor was this all, for it went on to say that, in the age of missiles, there would be no need for manned fighters. The defense statement thus raised the question as to whether the RAF had any future as a piloted force. White the RAF might continue to perform transport functions, it looked as if the future of combat in the air lay with missiles rather than with aircraft. RAF culture and the careers of RAF officers rested on flying, however. Thus, the problem for the ORB was to find a role for a combat aircraft that was neither a strategic bomber nor a fighter.

The British defense situation suggested several possibilities. The first was for an all-weather tactical strike aircraft. This aircraft would be capable of operating deep within Russian-controlled European airspace in the event of war and destroying important pinpoint targets such as bridges or railway lines. A second possibility was for a reconnaissance aircraft with sophisticated electronic equipment that would be capable of reporting battlefield conditions back to commanders. This, too, would require an all-weather capability. A third possibility was for a light bomber to be used in Britain's (then considerable) empire, either for protection from hostile powers (such as Indonesia) or for counterinsurgency. The ORB thought it important that all these roles be filled, and filled quite urgently. The urgency arose because the survival of existing aircraft in the airspace of a sophisticated enemy using radar and antiaircraft missiles was increasingly questionable. The capacity to fly very fast at treetop heights was rapidly becoming essential in order to avoid radar detection and subsequent destruction.

There were, then, three possible aircraft types that might be built. However, this posed a further problem for the ORB because in the late 1950s defense expenditure was being severely restricted. Thus the Treasury, which was responsible for these budget cuts, tended to doubt the need for any aircraft at all. In addition, the Royal Navy, which was in competition with the RAF for scarce resources, was hostile to the production of a new RAF aircraft. The Navy was purchasing a smaller aircraft called the Buccaneer and was anxious to persuade the RAF to buy this aircraft because this would simultaneously cut unit costs for the Navy and reduce the size of the overall procurement bill for the RAF.

It was clear to the ORB that it would have to choose between the three possible aircraft discussed above or, alternatively, combine them into a single aircraft. It was also clear that, if the RAF was to have its own distinctive aircraft, it would be necessary to specify this in a way that differentiated it from the smaller Royal Navy Buccaneer. During 1957 there were tough negotiations within government about the future requirements of the RAF. In the end it was agreed that an operational requirement for a large versatile aircraft should be issued to industry. Accordingly, in September 1957, a seven-page booklet called *General Operational Requirement 339 (GOR)* was issued that specified a large combined tactical strike and reconnaissance aircraft for use both in Europe and East of Suez (Gunston 1974; Williams et al. 1969; Wood 1975). Nine airframe manufacturers were asked to produce ideas and designs for an aircraft that would meet this specification by January 1958.

GOR 339 can be seen as a solution to an interconnected set of political, bureaucratic, and strategic problems. As we have indicated, problems were posed for the RAF by the Treasury, with its insistence on economies, and the Ministry of Defence, with its missile-oriented defense policy. To satisfy all factions, a single aircraft that was neither a fighter nor a strategic bomber was required. The situation was further complicated by competition from the Royal Navy. Thus, a decision was made to specify a large, two-engined aircraft with highly sophisticated terrain-following radar and a pinpoint bomb delivery system—a set of requirements that placed the aircraft in a class quite apart from the rival Buccaneer. Strategic problems were posed by the Soviet Union and its allies, Britain's continuing imperial role, theories about the course of any future hostilities in Europe, and a series of advances in air warfare techniques and capabilities. Together these suggested that the aircraft would have to be capable of high Mach speeds at altitude and transonic speeds at treetop height and have a large operational radius, a long ferry range for quick deployment between trouble spots, short takeoff and landing (STOL), and some highly complex navigational, reconnaissance, and weapons delivery systems.

The ORB was thus operating within a network of other actors, actors which together posed certain problems for the RAF. *GOR 339* was what we will call a "socio-technical scenario." A socio-technical scenario is a plausible proposal for a revised network of both social and technical roles that does not rest on an a priori distinction between human beings and machines. *GOR 339* was thus both a proposal for a machine or, more correctly, a weapons system that might be built *and* a theory about how the political, bureaucratic, and strategic world could be made to look five or ten years later. The proposed machine was part of the social theory of the ORB; thus its size, shape, and specification reflected ORB notions about the network of intentions, powers, and capabilities of relevant national and international actors. But the social theory was also part of the proposed technical solution of the ORB; plausible roles for the various bureaucratic and strategic actors were implied in the TSR proposal. In the sense discussed above, then, the ORB may be seen as an engineer-sociologist.

INDUSTRY'S SOCIO-TECHNICAL SCENARIOS

GOR 339 represented a putative socio-technical network. It defined and distributed roles to actors from the Soviet Union through the Royal Navy to the industrial firms whose participation would be necessary if the aircraft were ever to exist. But these actors were not only social. Many of them were technical in character. The putative aircraft was expected to perform in a particular way. Its components—for instance, its radars, engines, and navigational systems—were similarly allocated roles, albeit rather vague and general in *GOR 339*. But *GOR 339* was only a scenario—the design for an ideal world—and it is easy to design ideal worlds. The problems arise when it becomes necessary to mobilize or create the actors that will play these parts. The RAF knew that they would face both technical and bureaucratic battles. Technically, *GOR 339* specified an aircraft that was considerably beyond the state of the art. At the same time, there were, as we have seen, several powerful bureaucratic opponents to the TSR aircraft. A stepwise approach would be necessary. The RAF would start by mobilizing those who were most ready to play the role allocated to them in the scenario: the aircraft industry, which was hungry for work. Then, armed with the plans and suggestions from industry, it would return to government with a case that it hoped would be more solid and, most crucially, would persuade the defense chiefs and the Treasury to allocate funds to the project.

Thus *GOR 339* was really a fishing trip, a search for ideas and designs, and the British aircraft industry was eager to provide them. In all, there were nine submissions (Gardner 1981:25). Here we will mention only three. Vickers, a well-known engineering and aircraft firm whose aircraft interests were absorbed into the British Aircraft Corporation in 1960, offered two "type 571" suggestions. One was for a small single-engined aircraft that was relatively cheap but diverged considerably from the aircraft in *GOR 339*. The other conformed closely to the specifications in *GOR 339*, and, like its small cousin, represented a "weapons systems" approach to design with an integrated approach to airframe, engines, equipment, and weapons (Wood 1975:156). Although this was a departure from traditional methods of British military aircraft procurement in which airframes were designed, built, and tested first, and weapons and equipment were added afterwards, the approach was consistent with Ministry of Defence thinking and was well received.

Nevertheless, though the general philosophy of the submission was clear, well articulated, and closely argued, Vickers was not able to do all the necessary design work and wanted to go into partnership with another major airframe firm, English Electric. However, English Electric had made its own submission, codenamed the P17A, which was a detailed aerodynamic and airframe design for a large delta-winged Mach 2 strike bomber with twin engines and two seats (Hastings 1966:30; Williams et al. 1969:18; and Wood 1975:155). Though the P17A met many of the specifications of *GOR 339*, it lacked all-weather avionics and a STOL capability (Williams et al. 1969:18). English Electric countered the latter deficiency by arguing that STOL was not the most urgent requirement but suggested that this could be provided at a later date by a STOL platform that would lift, launch, and recover the P17A in the air (Hastings 1966:29; Williams et al. 1969:18; Wood 1975:155). Advanced avionics could be introduced when these became available.

The proposals from the firms, like *GOR 339*, were socio-technical scenarios; they represented proposals about the structure and distribution of both technical and social roles. Furthermore, they differed substantially. The small Vickers proposal was for a quite different type of aircraft and was, we can assume, based on the assumption that the RAF would find it impossible to mobilize the

Treasury to support a larger and more expensive project. Indeed, it attracted immediate Treasury support but had the disadvantage, from the point of view of the RAF, of not meeting *GOR 339* requirements and being much closer to the Buccaneer in size. If the smaller Vickers proposal was acceptable to the RAF, then it was not clear why the Buccaneer should be rejected. Unsurprisingly, the Air Staff found this proposal wanting and insisted on a large, long-range, twin-engined machine that would offer greater reliability.

The larger Vickers proposal rested on the alternative assumption: that Treasury support for an expensive project could indeed be mobilized. Whether this was plausible was not yet clear. Nevertheless, the large 571 submission was particularly attractive to the ORB, the RAF, sections of the Ministry of Defence, and the Ministry of Supply (later named the Ministry of Aviation), the contracting government department that handled aircraft procurement on behalf of the Ministry of Defence (for further details, see Law 1988). This was because it coincided closely both to the ORB scenario with its large twin-engined VTOL (Vertical Takeoff and Landing) or STOL (Short Takeoff and Landing) aircraft and to the fashionable weapons systems approach. In addition, the Ministry of Defence was impressed by the integrated design philosophy advocated by the company and concluded that the latter had the management capacity to control and integrate a complex project (Wood 1975:158; Gardner 1981:33).

However, the ministry was also impressed by the English Electric scenario. Many of the technical features of this submission had been worked out in considerable detail. There had, for instance, been substantial aerodynamic work, and it was thought to be "a first class design" (Wood 1975:155). The design was the product of wide experience with supersonic aircraft. Against this, however, it required revision of parts of the *GOR 339* scenario. STOL took a back seat, as did all-weather avionics. The English Electric scenario thus reworked some of the thinking behind *GOR 339* and argued that these requirements were not important in the short run; it was a low-altitude, supersonic, deep-strike capability that was crucial. However, this revision was not received with favor by the Air Staff, which liked much of the English Electric design but preferred the Vickers weapons systems approach. In addition, though contact between the two firms had been limited (with English Electric contractually tied to another firm that would provide the VTOL lifting platform), Vickers had indicated its wish to have English Electric as its partner. Accordingly, the Air Staff came to the conclusion that a combination of the Vickers type 571 and the English Electric P17A collaboratively built by the two firms would offer the best possible solution.

At this stage, then, the socio-technical scenario was much clearer. A network of design elements for the aircraft was in place and there seemed to be little doubt that the aircraft was, indeed, technically feasible. In addition, a social theory in the guise of a management philosophy had been elaborated, and the firms that would be involved had been informally selected. All that remained was to mobilize the actors whose support would be needed in order to realize this network of roles. The crucial decision was taken in June 1958 when there was a meeting of the important Defence Policy Requirements Committee (Gardner 1981:32), the body that allocated priorities and thus funds to different defense projects. At this meeting pressure from the Treasury and the Royal Navy for a simpler or cheaper aircraft was fought off and formal approval for the TSR 2 project was granted.

GLOBAL NETWORKS, LOCAL NETWORKS, AND NEGOTIATION SPACES

At this stage, many of the major actors had been mobilized and were in place. The various bureaucracies in government had been persuaded to cooperate, or at least their hostility had been checked, and industry was preparing itself. A *global network* of actors had been built. For the time being the support of these actors could be assumed and the protagonists of the project could turn their attention to other matters. The character of this support is interesting. In effect, the actors in the global network had agreed (or so, at least, it appeared) to grant the project managers a degree of autonomy. Such actors would not concern themselves with the detailed development of the project, and neither would they interfere with its internal running. In return for offering financial support, they were seeking limited and specific returns: periodic accounts of progress and the assurance that, five or six years later, the TSR 2 would be in production and going into service with the RAF. At this point, then, the project came properly into being. The managers had been granted an area of relative autonomy by actors in the global network; they had been granted what we will call a *negotiation space* in order to build a *local network*. (The concept of negotiation space is developed more fully in Callon and Law 1989; compare, also, with Gerson's 1976 analysis of sovereignty).

The local network was to be simultaneously social and technical. Thus the two firms, which were combined in 1960 into the British Aircraft Corporation (BAC), set about the difficult task of building a local network of designers, designs, production teams, management, and subcontractors that would bring the TSR 2 into being within time and budget. The first step was to integrate and take control of two quite separate industrial organizations and designs. The designers had previously worked as two teams some two hundred miles apart; but though the process of settling down to collaborative work was difficult, it was generally successful (Beamont 1968:137, 1980:134; Williams et al. 1969:47), and a joint team of fifty designers was undertaking a detailed study of the technical problems raised by *GOR 339* by the early months of 1959. After this joint study a division of labor was evolved that reflected the relative skills of the two teams: the Vickers designers worked on systems including cost-effectiveness and weapons, while the English Electric team worked on aerodynamics (Wood 1975:164).

Just as difficult was the question of integrating the technical components of the scenario into a consistent local network in order to ensure that the aircraft would perform in the way demanded. As an example, consider the different requirements suggested by the necessity for supersonic flight on the one hand, and a STOL capability on the other. High-speed flight suggested the need for small, thin, swept-back wings—all features of the P17A. By contrast, a STOL capability suggested the need for a low wing loading; in other words, the wings should be large. It also suggested that they should be relatively thick, long, and should not be swept back. The variable geometry option preferred by the designers of the F 111 was not considered. Instead, the team wrestled with the different requirements and eventually resolved them into a single design by (1) providing for very large flaps that effectively increased the thickness of the wings at low speeds, thus increasing lift; (2) forcing high-pressure air over the flaps in order to further increase lift at low airspeeds and prevent stalling; and (3) increasing the thrust-to-weight ratio by specifying two extremely powerful engines, which made it possible to achieve take-off speed in a relatively short distance (Gunston 1974:46; Williams et al. 1969:25, 39; Wood 1975:165).

This design decision was important. Given the operational requirement, many other decisions—for instance about the number of engines, the moving surfaces, the undercarriage and integral fuel tanks—were foreclosed. Moreover, a number of new design difficulties arose. For instance, there was controversy between members of the two teams about the location of the engines. The Vickers team had assumed that these would be slung beneath the wings; but the English Electric designers, wanting thin, uncluttered wings, suggested that they should be located in the fuselage. The Vickers engineers were unenthusiastic about this and pointed out that, since the fuselage was, in effect, a fuel tank, this decision would produce a real fire risk. In the end, however, the English Electric designers carried the day.

We cannot consider the details of the design process here. However, it is worth emphasizing that, rather than being purely technical, it was thoroughly and throughout informed by social considerations. Thus the design of the wings was not simply a function of the theory and practice of aerofoil design but was also influenced by the (socially given) requirement for a long-range aircraft that they also be designed as fuel tanks. Furthermore, their shape and orientation were partially dictated by the necessity of providing a reasonably smooth ride for the pilot and the navigator in the dense and turbulent air found just above ground level.

However, it was not always the social that defined the technical. Thus by 1961 the designers had concluded that the proposed aircraft was going to be too heavy to achieve a take-off roll of five hundred yards. They sought and were given permission to build an aircraft that would take off instead from half runways and rough strips (Gunston 1974:41). In this instance, then, the technical features of the network reshaped their strategic and bureaucratic neighbors.

MANAGEMENT: AN OBLIGATORY POINT OF PASSAGE?

We have argued that the TSR 2 project started with a scenario, the description of an ideal social and technical world, and we have traced how actors were constructed or mobilized in order to realize that scenario. In particular, we have shown how the proponents of the project mobilized the actors in a global network and sought to create a relatively autonomous negotiation space where a local socio-technical network might be designed and brought into being without constant interference from outside. In addition, we implied that all transactions between the local and the global networks, between the project and outsiders, would pass through the project's management. We implied that the management of the project would act as an *obligatory point of passage* for all contacts between the global and the local networks. Certainly, this was the approach that lay behind British government thinking on weapons systems procurement:

> Since the failure of only one link could make a weapons system ineffective, the ideal would be that complete responsibility for coordinating the various components of the system should rest with one individual, the designer of the aircraft (Ministry of Defence and Ministry of Supply 1955:9).

Though the document in question went on to note that experience had shown "that this is not completely attainable," the basic approach rested on the assumption that autonomy for a negotiation space could only be achieved by ensuring that the designer was, indeed, an obligatory point of passage between the two networks. In practice, the future course of the TSR 2 project hinged on these issues and specifically on two interrelated factors: first, the capacity of the BAC (and the relevant government departments) to prevent outside interference; and second, the ability of the BAC to mobilize the elements, technical and social, that were needed in order to create a viable local network.

Despite the formal adoption of a weapons systems philosophy, neither the BAC nor the Ministry of Supply was able, together or separately, to impose itself as such an obligatory point of passage. Thus, though Vickers/BAC was appointed prime contractor (Hastings 1966:35; Williams et al. 1969:22), in practice the project was controlled by a complex and diffuse series of committees on which a range of different agencies, including the hostile Treasury, was represented. The description of this labyrinth of committees is beyond the scope of this chapter, but the result was that no one agency, at least of all the BAC, was in a position to control the process of elaborating a network free from outside interference.

This situation was manifest in a number of ways. Thus there were sometimes considerable delays in specifying and authorizing equipment, and many of the decisions taken were relatively cost insensitive. For instance, the specification required a digital navigational computer, and the subcontract for this was given to a company called Elliott Automaton. Elliott rapidly concluded that the only way the equipment could be provided in time was by purchase of a basic computer from North American Autonetics. The Ministry of Supply balked at this because it had previously sponsored basic British research on airborne digital computing. Eventually it was persuaded otherwise, but, in view of the cost and complexity of the U.S. equipment, the Treasury insisted that the decision should be reviewed after a year. The result was both delay and increased cost (Hastings 1966:160).

Again, the Air Staff tended to make decisions without reference to the BAC. The problem here was that the RAF continued to develop its ideas about the ideal performance and capabilities of the TSR 2. This tendency to upgrade specifications was encouraged by the fact that contractors would often talk directly to the Air Staff and government departments. Sometimes such discussions would lead to changes in the specification of equipment the characteristics of which had already been fixed (or so the BAC thought). In addition, many of the most important contracts were awarded directly by government; the contract for the engines is a case in point. The design team took the unanimous view that this should be awarded to Rolls Royce. However, the Ministry of Supply had other views, apparently deriving from its concern to pursue a policy of industrial merger, and awarded the contract to another firm, Bristol Siddeley Engines (BSE), despite this recommendation (Clarke 1965:77; Gardner 1981:29; Gunston 1974:41; Williams et al. 1969:21). In practice, the BAC controlled only about 30 percent of the project expenditure (Gunston 1974:67; Hastings 1966:40).

If neither the BAC nor the Ministry of Supply was able to impose itself as an obligatory point of passage between the local and the global networks, both encountered equally serious problems in mobilizing the necessary social and technical elements to build the local network. There was, at least in government, a view that management was slipshod (Hastings 1966:157; Williams et al. 1969:54). The most spectacular troubles concerned the engines (Law 1992b). Neither government nor BSE seems to have known what they were letting themselves in for when the contract was awarded. The government appears to have specified the engines in very general terms, and it was at first thought that their development would be a fairly straightforward manner of upgrading an existing type (Williams et al. 1969:27, 52). This was not the case. The engine that was developed had a much greater thrust than its predecessor and operated at much higher temperatures and pressures. This led to a series of problems and delays. The first of these appeared when the new engine was proved on the test bed in the autumn of 1962. Under the new and more severe conditions, the turbine blades, which had been cast, were too brittle and sheered off. It was necessary to replace them with forged blades at considerable cost in both time and money (Hastings 1966:42; Gardner 1981:104).

However, the most serious problems concerning the engine appeared only late in the development process. After the engine had been proved for over four hundred hours on the test bed (Hastings 1966:43), it was installed in late 1963 beneath a Vulcan bomber for further testing. On December 3 this aircraft was taxiing during ground tests at the BSE works when the engine blew up, "depositing," as Wood (1975:174) reports it, "a large portion of smouldering remains outside the windows of the company press office." The aircraft was reduced to burning wreckage, and although the crew was saved, a fire tender was consumed by the flames (Gunston 1974:56). At first it was not clear what had happened. BSE hypothesized that the problem might be due to stress and ordered that the thickness of the low-pressure shaft be doubled. However, further tests led to additional unpredictable and unexplained explosions. Finally, in the summer of 1964 BSE concluded that the problem indeed lay with the low-pressure shaft. In the original unmodified engine this had turned on three bearings. However, the design team had become concerned that the middle bearing might overheat, so it had been removed. Then, in order to provide the shaft with sufficient rigidity, it had been increased in diameter (Beamont 1968:139; Hastings 1966:43; Wood 1975;174). At a particular speed, the shaft started to resonate like a bell, and disintegration quickly followed. Even with a diagnosis at hand, however, a solution was going to cost further time and money.

THE COLLAPSE OF THE NEGOTIATION SPACE

In the last section we traced how the autonomy of the negotiation space was eroded because no one agency was able to impose itself as an obligatory point of passage between the global and the local networks. We also considered some of the problems that arose as the contractor sought to build an effective local network. These failures, when combined with unregulated seepage between the two networks, led to increasing skepticism by the actors in the global network, even though they were partly to blame. In particular, the failures led to concern about costs and time. In the present section we trace how these concerns developed into clearly political concerns.

The relative autonomy of a negotiation space is normally granted only for a time and is subject to satisfactory performance. The concerns of the actors in the global network we have described centered around performance, cost, and time. The RAF had been promised that the TSR 2 would be available for squadron service by 1965, but it was clear, with the engines still unproved in the middle of 1964, that this deadline could not be met. The Ministry of Defence had likewise been promised a vital weapon with which to fight a war in Europe or the Commonwealth by 1965. This was not going to be available. The Treasury had been promised a relatively cheap and versatile aircraft, but by 1963 the estimated cost of the aircraft had nearly doubled. The Navy, which had been hostile from the outset, saw the project swallowing up more and more of the procurement budget. By 1963, then, many of the actors in the global network were restless, and it was clear to all those involved that the project was in deep trouble.

However, though these difficulties were serious, they did not necessarily mean that the project was doomed. If the skeptical actors could be kept in place and obliged to provide the necessary resources, then the project would continue. Funds from the Treasury, expertise and support from the RAF, political support from parts of the Ministry of Defence, and specialist services from a range of government agencies would allow it to carry on. In fact, the RAF, though not necessarily the whole of the Ministry of Defence, remained a somewhat lukewarm supporter of the project and, with the government committed, it was not possible for the Treasury, the Navy, or indeed, the hostile sections of the Ministry of Defence to stop the project. Accordingly, the funds continued to flow. However, armed with the knowledge gained from participation in the cat's cradle of management committees, the skeptics in the global network were in a strong position to undermine the project by indirect means. This involved taking the fight into a wider arena and mobilizing new actors.

The most important of these was the Labour Party. During the period we have been discussing, the Conservative Party, traditionally the more "hawkish" of the two major British parties, had been continuously in power. However, the Labour Party was riding high in the opinion polls, and a general election was due by October of 1964 at the latest. The Labour Party had expressed skepticism about such "prestige projects" as Concorde and TSR 2 and had promised to review them if it was returned to power in 1964. Whispering in the corridors of power, talk in the press, and a series of admissions from the Ministries of Aviation and Defence about delays and escalating costs thus led to the TSR 2 becoming an object of political controversy from 1963 onward. This process was reinforced by a highly controversial setback to the project: the failure to persuade the Australian government to purchase the TSR 2 for the Royal Australian Air Force. In a blaze of publicity, the Australians opted instead for the General Dynamics F 111.

The political arguments ranged far and wide, and many of these concerned the escalating costs and delays that, the Labour opposition argued, had led to the Australian decision to buy the F111. This charge was angrily rejected by the government, which claimed that the constant carping of British critics had led the Australians to doubt whether the aircraft would ever be produced (*Times* 1963a). Other critics suggested that the aircraft had become too expensive for its role and too expensive to be risked in combat (*Times* 1964a). Further political disagreements centered around the role of the aircraft, which had, in government thinking, been widened to include a strategic nuclear specification. This shift attracted criticism both from sections of the press, which felt that the aircraft was neither fish nor fowl, and the left wing of the Labour Party, which was committed to a policy of unilateral nuclear disarmament. Yet others, including the official Labour Party defense spokesperson, concluded that the "strategic bonus" did not so much represent a change in the specification of the aircraft as an attempt by the government to persuade its backbenchers of the soundness of its United States-dependent Polaris nuclear defense policy (*Times* 1963b). Finally, there was also controversy about the continued delays in the first test flight. Labour claimed at the beginning of 1964 that the BAC had "been given an order that it must get the TSR 2 off the ground before the election, and that [this] was a priority" (*Times* 1964b).

Thus by the autumn of 1964 the project was at a crucial point. On the one hand, most of the local network had finally been built, and the aircraft was almost ready for its maiden flight, albeit very much behind schedule and over budget. On the other hand, opposition to the project in the global network was no longer confined to such insiders as the Treasury and the Royal Navy. The dispute was now in the public arena, with firm support from the Conservative government and much criticism (though no commitment to cancel) from an important actor that was new to the TSR 2 scene, the Labour Party. The future of the project thus depended on two factors. On the one hand, it was important to demonstrate the technical competence of the project, and the best way to do this was to have a successful first flight. This would reinforce the position of those in the global network who wished to see the project through. On the other hand, the outcome of the general election was also vital. Conservative success would assure the future of the project. Labour victory would call it into question by redistributing crucial roles and reshaping the network of actors operating within government.

The first flight took place just eighteen days before the general election. The test pilot (Beamont 1968:144) has described the rather subdued group of engineers, technicians, managers, and RAF personnel who assembled before the flight. Most knew, as the large crowd beyond the perimeter wire did not, of the potentially lethal nature of the engine problem, and they knew that, although its cause had been diagnosed, it had not yet been cured. In fact, the flight was highly successful, the aircraft handled well, and there was no hint of the destructive resonance that had plagued the engines. Deep in the election battle, the Conservative prime minister described it as "a splendid achievement" (Beamont 1968:151). The aircraft was then grounded for several months in order to modify the engines and tackle vibration problems.

On October 15 the general election took place. The result was close, and it was not until the following day that it became clear that the Labour Party had been returned to power with a tiny majority. Beset by economic problems, it quickly ordered a detailed scrutiny of the various military aircraft projects and started a review of the proper future shape and size of the aircraft industry (Campbell 1983:79). Discussions within the new government were long and difficult. In February the new Prime Minister, Harold Wilson, told Parliament that the future of the TSR 2 would depend upon four factors: (1) a technical assessment of the aircraft and its alternatives; (2) the fact that though the overseas purchase of an alternative aircraft would save £250 million, this would also involve considerable expenditure; (3) the future shape of the aircraft industry, and the possible unemployment that would result; and (4) the nature of the terms that could be negotiated with the BAC.

At the beginning of April the spokespersons for the principal actors in the newly restructured global network—the Cabinet

Ministers responsible for departments of government—met to make a final decision. They considered three courses of action: to continue with the TSR 2, to cancel it and put nothing in its place, or to cancel it and replace it with the similar F 111 (Crossman 1975:191; Wilson 1971:90). The Treasury remained hostile to the TSR 2 and accordingly sought cancellation. Though it was concerned that a large purchase of an alternative U.S. aircraft such as the F 111 would impose severe costs, it was prepared to accept that an option for the purchase of this aircraft should be taken out on the understanding that this did not imply a firm commitment. The Ministry of Defence was also in favor of cancellation on cost grounds, and it was joined by those such as the Navy that favored the claims of other services and projects (Hastings 1966:68, 70). The Minister of Defence was in favor of an F 111 purchase, but there was some uncertainty whether Britain really needed either type of aircraft in view of its diminishing world role (Williams et al. 1969:31). He was thus happy to take out an option on the American aircraft rather than placing a firm order.

The position of the Minister of Defence probably in part reflected a shift in view within the RAF. The combination of delay and cost overrun, together with a much tougher policy of economies introduced by the new Minister of Defence, had convinced the Air Staff that it was most unlikely that there would be a full run of 150 TSR 2s, and this had led to doubt about whether it would be possible to risk such a small number of expensive aircraft in conventional warfare. For some officers, this pointed to the desirability of acquiring larger numbers of cheaper aircraft that might be more flexibly deployed. In addition, though the technical problems of the TSR 2 appeared to be soluble, its delivery date was still at least three years away. Since the F 111 was designed to essentially the same specifications and was already in production, the RAF found this quite an attractive alternative (Reed and Williams 1971:181).

The Ministry of Aviation was concerned that a decision to scrap the TSR 2 would seriously reduce the future capacity of the British aircraft industry to mount advanced military projects. It tended to favor cancellation combined with the purchase of a lower-performance British substitute. However, most ministers, including the Minister of Aviation, believed that the industry was much too large for a medium-sized nation. The real problem was that there was not yet a policy in place about the future shape and size of the British aviation industry. Even so, the TSR 2 was costing about £1 million a week, and further delay in cancelling did not, on balance, seem justified.

The government was concerned that cancellation would lead to unemployment. With the party's tiny majority in Parliament, ministers were anxious not to court unnecessary unpopularity. Against this, however, ministers felt that the resultant unemployment would mostly be temporary, that many of those working on the TSR 2 would quickly be absorbed by other projects or firms.

Nevertheless, the decision was by no means clear-cut; there was no overall cabinet majority for any of the three options (Wilson 1971:90). A number of ministers—mainly, it seems, those who were not directly involved—wanted to postpone cancellation until a long-term defense policy was in place (Crossman 1975:190). Overall, however, those who wanted to maintain the project were outnumbered by those in favor of cancellation, with or without the F 111 option. The vagueness of the latter commitment ultimately made it possible for these two groups to reconcile their differences. The cancellation was announced on April 6, 1965. With this announcement there was immediate withdrawal of funding for further work, and the local network was dissolved overnight. Many were thrown out of work; the TSR 2s in production were scrapped, and the three prototypes were grounded. As an ironic footnote it should be added that the U.S. F 111 was never actually purchased.

DISCUSSION

In this paper we have used a network vocabulary to describe the rise and fall of a major technological project because this vocabulary is equally applicable both to social and technical phenomena. We have used this vocabulary because it is neutral; it does not distinguish on a priori grounds between the technical and the social. This is useful, indeed essential, when we study innovation because seemingly technical innovations usually have profound social consequences, and social innovations almost always imply technical change.

We have tried to encapsulate this awesome promiscuity in a phrase by talking of technologists as "engineer-sociologists." We might equally well have described them as "heterogeneous engineers" (Law 1987:113). At any rate, it is clear that engineers pay scant attention to the divisions that most of us detect when we separate the technical from the social, then subdivide the social into the economic, the political, the sociological, and the rest. Innovation is a seamless web (Hughes 1986), the social is rebuilt alongside and interpenetrates the technical, and we need a neutral, matter-of-fact, nondisciplinary vocabulary to describe it. In our description of the TSR 2 project we have thus mixed the technical and the social not in order to be difficult, but because heterogeneity was the name of the game. If at times our description sounds as if it is drawn from a technical manual or, indeed, from political science, we make no apology, because the project went through different phases, some of which were more technical in character, while others were more political. Overall, however, the development of the TSR 2 project is just as incomprehensible in the absence of technical understanding as it would be without a willingness to chart the links between social actors.

We have thus sought to emphasize how the development of the project was contingent. Neither success nor failure was written into the project at the outset. The aircraft, when it flew, flew well. The pilots were delighted with the way it handled, and there is every reason to suppose that it would have been a military success had it actually entered service. Similarly, early versions of the subsequently successful F 111 were underpowered, prone to stalling, suffered excessive drag, and were subject to structural failure (Coulam 1977).

But if technology alone does not explain the cancellation of TSR 2, then neither was this preordained by politics. As we have shown, the Labour Cabinet reached its decision only with the greatest difficulty. Indeed, one source tells us that the Prime Minister was personally opposed to cancellation (Crossman 1975:191) but for unknown reasons did not make this clear to his colleagues. Had he chosen to do so, it is possible that the decision would have gone the other way. Again, if we may mention another political might-have-been, the Labour government came to power with a majority of only five. Had a handful of citizens voted otherwise, the story might have turned out quite differently, as it might had the engines not suffered from bell-resonance and held up the project for a year.

Hypotheticals, however, should not hold our attention for too long. The crucial point is that the contingency of the process can be seen only if we are prepared to deal with both the social *and* the technical evenhandedly. The socio-technical processes that interest us are interactive and emergent; their course and fate are not easily predicted. Yet they are crucially important, for they shape and reshape the modern world.

Explanations of social and technical change must avoid three traps. Two of these take the form of reductionisms. Social reductionism, the doctrine that relatively stable social categories can explain technical change, and technological reductionism, the converse view, that technological change automatically shapes society, are both one-sided, incomplete, and misleading (MacKenzie and Wajcman 1985a). But even if the social and the technical are both taken to be important, there is a third trap to avoid. This is the notion that the technical and the social evolve as a result of separate processes and only subsequently interact. By contrast, our aim has been to suggest that they are *jointly* created in a single process. For this reason, we have not distinguished between technical content on the one hand and social context on the other. Context and content are similar in that *both* are social and technical. Furthermore, we have tried to show that *context is internalized in the object* or, in the language we have used here, the local network contains the global network (see Callon 1987; Law 1987).

If we want to avoid both forms of reductionism and the kind of interactive approach that arises from compartmentalization of the social from the technical, the search for explanation should be conducted in a rather different way. A possibility that we have started to explore in this paper is to seek out regularities in heterogeneous networks. Thus, we have tried to identify strategies that generate relatively stable networks of socio-technical objects that therefore, for a time, exert a disproportionate influence on those around them. Here we have pointed to one strategy for creating such stability, that of generating a negotiation space. Within a negotiation space it is possible to make mistakes in private, it is possible to experiment and, if all goes well, it is possible to create relatively durable socio-technical combinations. That this failed in the case of the TSR 2 merely underlines the fact that many of the innovations that appear in the modern world emerge fully fledged from places that are both relatively private and autonomous. Our conclusion is that privacy should be treated as fundamental technology of power.

NOTE

1. We are grateful to the Nuffield Foundation for the award of a Social Science Research Fellowship to John Law that has made it possible to undertake research on the TSR 2 project.

JOAN H. FUJIMURA

9

Ecologies of Action: Recombining Genes, Molecularizing Cancer, and Transforming Biology

Harold Varmus, I and our numerous colleagues have been privileged to assist as a despised idea became the ruler over a new realm. The notion that genetic changes are important in the genesis of cancer has met strenuous resistance over the years. But not that notion has gained ascendancy..
—J. Michael Bishop, "Retroviruses and Oncogenes II," Nobel Lecture, 8 December 1989

In the late 1970s and early 1980s in the United States, participants from many different lines of work came to agreement on how best to study cancer. The molecular biological approach gained an increasing proportion of cancer research commitments. A new representation of cancer called the proto-oncogene theory constructed by a few molecular biologists and tumor virologists became accepted as a fact by other researchers, sponsors, suppliers, and diverse participants in the cancer research arena. Cancer became defined as a disease of the DNA. In the 1990s proto-oncogene research continues to be the state of the art in cancer research and additionally defines research projects on normal growth and development, embryology, human gene therapy, and evolutionary theory.

My (hi)story of the development of proto-oncogene research is the story of how the definition of the situation constructed by several molecular biologists and tumor virologists became the accepted definition.[1] Science, like other human activities, is not determined by social interests. Techno-scientific products are instead the outcomes of what I call "ecologies of action."[2] I began to try to understand the ecologies of action involved in the proto-oncogene story by borrowing Everett Hughes's (1971a) view of the workplace as "where peoples meet." Hughes asked how members of different social worlds cooperate in the workplace while holding different "definitions of the situation."[3] His answer was that work gets done in these places only through conflict, struggle, and negotiations over the set of conventions that should guide action and interaction at this meeting of worlds. Hughes's question was framed around the situations in factories in 1940s and 1950s Chicago when new immigrant groups from diverse cultures came together and were obliged to work together. In my case, cancer research is the workplace, the "space" where members of different worlds meet, and the negotiations are about how one should approach solving the problem of cancer. This meeting of different social worlds created a new social world, the hybrid world of proto-oncogene research. In this process, the diverse social worlds involved in the interactions also were changed.

However, there is a difference between the two situations. In the case of cancer research, participants from different social worlds were not forced to work together. Until the early 1980s when oncogene funding commitments attracted many new and established researchers to the field, scientists and medical researchers in different lines of research could have continued in their own trajectories without engaging in or using the findings of new proto-oncogene research. Recent histories and sociologies of science have also shown that agreement and consensus are not taken-for-granted events in science; they are not events based on reality's guiding hand. The question of how proto-oncogene research became "the ruler over a new realm" is the puzzle of this paper. I present an account of the growth of proto-oncogene research and the acceptance of proto-oncogenes as facts in terms of changes in work practices across laboratories.

Callon, Latour, and Law have framed an actor-network theory that discusses science in terms of a field of battle where whose "fact" is more "factual" depends on whether actors succeed in enrolling allies much as leaders enroll armies and armories.[4] A major strategy used by scientists in fact-making is to translate others' interests into their own interests. Callon and Latour conceived of the movement from finding to fact as the enrolling and disciplining of many heterogeneous actors into a network of stable linkages.[5] They proposed several strategies, including translating others' interests into their own interests by making one's laboratory an "obligatory passage point," by which such enrolling and disciplining of actors and finally stabilizing of linkages occur. More generally, translation is the way in which certain actors gain control or power over the way society and nature are organized, by which "a few obtain the right to express and to represent the many silent actors of the social and natural worlds they have mobilized" (Callon 1986a:224). However, as Latour (1987) pointed out, others are simultaneously trying to enroll allies behind their fact. In this view, science is a battle over which fact, that is, which network, will become and remain stabilized.

The story I present here is also one of enrollment, but it is framed in less warlike and individualist goal-oriented terms. Instead, it is framed in terms of work, practice, and ecologies of action. I focus on how people manage to get their work done in science. I agree with Callon and Latour that there are empire builders in science as there are in other human activities. However, many aggressive empire builders do not succeed, and the products of many less aggressive participants are elevated to fact status without vigorous promotion. I focus on the linking of practices, work routines, and theory in diverse situations and across time to tell a story of a proto-oncogene bandwagon. Instead of isolated, goal-oriented action, mine is a story of collective work across different scientific enterprises. Latour (1987) argues that making one's laboratory a "center of authority" is one strategy used by scientists to win allies. I present an account of a package of theory and methods that became a *distributed* center of authority, or in the words of J. Michael Bishop, "the ruler over a new realm." Through the proto-oncogene theory and molecular genetic technologies, players in the worlds of science transformed practices in many different research laboratories and communities, as well as those of their sponsors and suppliers. These transformations produced a new ecology.

CRAFTING A THEORY-METHODS PACKAGE

> By the end of the 1960s the first phase of expansion of molecular biology was over. Some of the key issues, like the nature of the genetic code, had been sorted out, which made the selection of the next set of comparable problems rather difficult. What questions could be as important? The early 1970s were also a time of some anxiety in science. Research budgets were starting to fall in real terms for the first time since the war. Money was available in the field of cancer research, but all the approaches to that problem looked formidably difficult.
>
> Then a series of technical discoveries came together that have completely altered the landscape of possibility in molecular biology. What became practicable, for the first time, was the controlled manipulation of pieces of genetic material. Researchers could snip out sections of DNA from one organism and transfer them to another. One could cut genes out of one cell and splice them into another. The technical term for this activity is "recombinant DNA research" because what is involved is the controlled "recombination" of sections of DNA, the hereditary material. . . . The implications of this for many areas of research were staggering. For example, one could now think about splicing new bits of DNA into tumor viruses, or taking them out with precision, to see how that modified their impact on the cells they infected (Yoxen 1983:30–31).

As recombinant DNA technolgoies came on the scene in the early to mid-1970s, researchers began immediately to think of using them in cancer research. These technologies quickly moved into diverse social worlds and transformed cancer biology. Molecular biologists and cancer researchers framed the problem of cancer in molecular genetic terms and created a new theory, a new line of research, and connections to many other biological problems.

However, I do not assume that the rush of research activity on the molecular biology of cancer in the 1980s and in particular proto-oncogene research is the result of simple technology transfer. I portray the development and maintenance of this research as coconstructed with a flexible set of conventions for action in many different situations. By coconstruction, I mean that the development and maintenance of this research and the package of conventions gave form to and modified each other as time went on. This set of conventions provided broad outlines for action and recreated research problems in laboratories across the United States. In addition to recombinant DNA technologies, the package also included a theoretical model for explaining cancer, the proto-oncogene theory. Oncogene proponents theorized that cancer is caused by normal cellular genes called "proto-oncogenes" that somehow go awry and turn into cancer genes, loosely called "oncogenes." Recombinant DNA and other molecular genetic technologies were the methods used to create and to test the theory.

Elements critical to the story of the growth of oncogene research include both the construction and dissemination of this "theory-methods package" by oncogene researchers and the situations and practices of their colleagues, members of other lines of research, funding agencies, and suppliers who adopted or supported the package. This package captured the interests of many scientific workers in both academic and private settings. Variations on the work of early oncogene researchers were constructed by investigators in biological and clinical research laboratories and other work sites with different local conditions. The collective commitments of resources established the conditions for gaining new adherents (scientists, laboratories, organizations, and research tools) through a "snowball effect." By 1984, the line of research had grown to the point where it was sustained by its own momentum and infrastructure of commitments.[6] It was what many of my respondents called a "bandwagon."

The development of proto-oncogene research was a consequence as well as a part of the creation of collective commitments made to molecular biological approaches to understanding cancer as well as normal growth and development and other biological processes since the mid-1970s in the United States. Participants engaged in both oncogene and the general molecular biological approach to cancer included people, organizations, and tools from many social worlds. These worlds include cancer research, molecular biology, the new field of bioinformatics, university administrations, commercial biotechnology and biological supply industries, science funding agencies, and the United States Congress.

Proto-oncogene research protocols and molecular genetic technologies have transformed practices in many laboratories and lines of research and consequently the kinds of knowledge they produce. Science is work and scientific information is constructed in organizational contexts.[7] Changing conventionalized work organizations and practices involves convincing and persuading, teaching and learning, meshing and adjusting, and redefining and reconstructing. *Conceptual change* in science, in turn, is achieved through these individual and collective changes in the way scientists organize their work. In the following sections, I present an account of the activities, contexts, and processes through which proto-oncogenes were constructed and, in the process, changed the world.

CRAFTING PROTO-ONCOGENES AS A UNIFYING METAPHOR

In the late 1970s, according to scientists, the development of recombinant DNA techniques provided a means by which researchers could enter the eukaryotic cell nucleus and actually *intervene* and *manipulate* nucleic acids to study their activities in ways never before possible. As these technologies were imported into diverse laboratories, scientists tinkered with them to make them work in these new situations with new contingencies. Researchers used the techniques to examine at the molecular level questions previously approached at the cellular and organismal levels. While some molecular biologists worked to improve and extend recombinant DNA technologies (that is, creating new techniques and materials as well as making others easier, faster, more efficient, and less expensive), others explored the structure of DNA, and still others applied the techniques to long-standing questions about function in many biological subdisciplines. Molecular biologists regarded recombinant DNA technologies and eukaryotic cell genes as the keys to the previously locked doors in normal differentiation and development, cell proliferation, cancer (especially human cancer), and even evolution. However, the remarkable news of the 1980s was the proto-oncogene theory that appeared to provide at least partial answers to many of these long-standing biological problems, including those previously unrelated to cancer. More significantly, the theory was represented as a way of linking all of these problems through a *unified* explanation of biological functions and perhaps even to intervention (that is, *reconstructing nature* via gene therapy).

Back in the late 1970s, several tumor virology and molecular biology laboratories were engaged in two separate lines of research on the molecular mechanisms of cancer causation. They claimed that they had found a class of genes in the normal cell that could be triggered to transform the normal cell into a cancer cell. They posited several types of triggering mechanisms. Although there were many debates about specifics, these tumor virologists and molecular biologists framed the proto-oncogene theory in a way that they claimed encompassed and unified many other areas of cancer research. They claimed that further investigation using the oncogene framework would produce explanations at the molecular level for problems previously pursued in classical genetics, as well as on chemical, radiation, hormonal, and viral theories of cancer. Proto-oncogene theory proponents claimed to have developed a molecular explanation for many different types (classifications) and causes (causal explanations) of cancer. The ultimate claim was that their research might lead to a common cure, a "magic bullet" for cancer. They also proposed connections between their theory and other work in the molecular biology of normal growth and differentiation. At the time (and at present), the proto-oncogene theory was the only coherent and widely accepted theory for activities at the molecular level in cancer causation.[8]

According to the scientists I interviewed, a unifying theory appeals to them because of a quality they call "elegance." An elegant theory or fact is one that can explain many disparate observations with relative precision. However, claiming that one's theory unifies many lines of research does not mean that others will agree with the claim and rush to pursue experiments based on the theory. By widening our focus to follow both unifiers and the unified, we can construct a story that also includes the perspectives and practices of participants other than the original oncogene proponents.

Between Cancer and Evolution: Proto-Oncogenes

During the 1960s and early 1970s tumor virologists extended their research on viral oncogenes to develop the concept of normal cellular genes as causes of cancer by borrowing and using the concept of gene conservation as developed by evolutionary biologists. Tumor virologists reported that they had found specific "cancer" genes, which they named "viral oncogenes," in the viruses that transformed cultured cells and caused tumors in laboratory animals. This experimental work was done using traditional methods and technologies of virology and molecular biology to investigate RNA tumor viruses.[9] As other laboratories joined in this line of research and explored other viruses, they reported discoveries of more viral oncogenes. However, after a decade of research, many investigators and administrators concluded that these viral oncogenes caused cancer only in vitro and in laboratory animals. No naturally occurring tumors in animal and human populations were credited to viral oncogenes.[10]

In 1976 J. Michael Bishop, Harold T. Varmus, and their colleagues at the University of California, San Francisco, announced they had found a normal cellular gene sequence in various normal cells of several avian species that was very similar in structure to the chicken viral oncogene, called *src* (Stehelin et al. 1976).[11] Two years later, after constructing a radioactive probe for their viral oncogene (*src* RNA), they reported that they had also discovered complementary DNA sequences related to the *src* viral oncogene in normal cells in many different vertebrate species from fish to primates, including humans (Spector et al. 1978). Further, the transduced gene sequence is not necessary for viral reproduction. Another significant difference was the lack of introns in the viral DNA (translated from the viral RNA), whereas the normal *src* DNA had these intervening sequences. Bishop and Varmus and their collaborators suggested that the viral gene causing cancer in animals was transduced from normal cellular genes by the virus; that is, the virus took part of the cellular gene (probably the messenger RNA version without the introns) and made it part of its own genetic structure. Based on their research and that of others, Bishop and Varmus proposed that some qualitative alteration (through point mutation, amplification, chromosomal translocation) of this normal cellular gene may play an important role as a cause of human cancer. Before these experiments, decades of efforts to link viruses to human cancer had been unsuccessful.

With the results of these new experiments, the UCSF team constructed a theory that turned an earlier hypothesis on its head. According to Bishop (1982, 1989) and Varmus (1989b), they planned their experiments to test the virogene-oncogene hypothesis proposed by viral oncogene researchers (Huebner and Todaro 1969; Todaro and Huebner 1972). The difference between earlier theories and the Bishop and Varmus theory was the latter's conjecture that the gene was *originally* part of the cell's normal genome, rather than a viral gene recently implanted by viruses into the normal cell (viral infection). They hypothesized that this gene became part of the cell's genome early in the organism's evolutionary history and thereafter was retained as genetic baggage.

After their initial experiments, Bishop and Varmus instead proposed that the viral oncogenes came from normal cellular genes rather than vice versa. Bishop and Varmus proposed that the gene that caused normal cells to become cancer cells was part of the cell's normal genetic endowment. Their proposal used arguments based on "evolutionary logic." Since the gene was reportedly found in fish, which are evolutionarily quite ancient, the gene must have been conserved through half a billion years of evolution. Bishop and Varmus constructed *a two-way relationship* between their research and evolutionary theory. They are not simply drawing on evolutionary arguments. They also injected their theories, inscriptions, and materials into the wealth of research, debates, and controversy in evolutionary biology.[12]

> Transduction by retroviruses is the only tangible means by which vertebrate genes have been mobilized and transferred from one animal to another without the intervention of an experimentalist. How does this transduction occur? What might its details tell us of the mechanisms of recombination in vertebrate organisms? What does it reflect of the potential plasticity of the eukaryotic genome? Can it transpose genetic loci other than cellular oncogenes? *Has it figured in the course of evolution?* How large is its role in natural as opposed to experimental carcinogenesis? These are ambitious questions, yet the means to answer most of them appear to be at hand (Bishop 1983:347–48; emphasis added).[13]

Subsequent research on oncogenes in fact presumed the evolutionary relationship as the basis for the role of proto-oncogenes in cancer causation.

Molecular Biological Oncogenes and Tumor Virological Oncogenes

Associating viral oncogenes with molecular oncogenes was also critical to the growth of oncogene research.[14] In 1978, soon after the announcements by tumor virologists Bishop and Varmus, a few molecular biology laboratories began to study cancer using recombinant DNA technologies, especially gene transfection techniques, and soon reported that they had found cancer genes similar to Bishop and Varmus's proto-oncogenes. In one set of experiments, molecular biologists in Robert Weinberg's laboratory at the Whitehead Institute, Massachusetts Institute of Technology, first exposed "normal" mouse cells to DNA from mouse cells that had been transformed by chemical carcinogens.[15] The outcome, as reported by the researchers, was the transformation of the "normal" cells into cancer cells. They then extracted DNA from these transformed cells and used them to transform normal cells. The results were used to claim that the "normal" cells were transformed by the transfected (in plain terms, transferred) DNA and that the transfected DNA therefore contained active cancer genes (Shih et al. 1979). These and other researchers then applied this gene transfection assay (procedure) as well as secondary DNA transformations to human tumor cell lines and primary tumors. After cloning and isolating the genes, they claimed that human DNA molecules similarly were the transforming agents (Der et al. 1982; Santos et al. 1982; Parada et al. 1982). Weinberg (1983:127A) concluded from the experimental outcomes that "the information for being a tumor cell [was] transferred from one [mammalian] cell to another by DNA molecules."

> The successful isolation of transforming DNA in three laboratories [of Barbacid, Cooper, and Weinberg] by three different methods directly associated transforming activity with discrete segments of DNA. No longer was it necessary to speak vaguely of "transforming principles." Each process of molecular cloning had yielded a single DNA segment carrying a single gene with a definable structure. These cloned genes had potent biological activity. . . . The transforming activity previously attributed to the tumor-cell DNA as a whole could now be assigned to a single gene. It was an oncogene: a cancer gene (p. 130).

During this time, many molecular biologists were attempting to clone transfected cancer genes, but only a few laboratories had the skills and tools to do so. These few laboratories, especially those of Weinberg at the Whitehead Institute, Michael Wigler at the Cold Spring Harbor Laboratory, and Mariano Barbacid at the National Cancer Institute in Washington, D.C., raced to be the first to clone transfected human cancer genes. Researchers in Weinberg's laboratory eventually managed to clone the EJ human bladder cancer gene, only to find that it was the same gene (called T24 human bladder cancer gene) cloned a month earlier by researchers in Wigler's laboratory (Goldfarb et al. 1982; Shimizu et al. 1983) and a by Barbacid's laboratory (Der et al. 1982). More significantly, soon after cloning the EJ gene, a graduate student in Weinberg's laboratory conducted homology searches for sequences similar to the gene and found that it was similar to the *ras* (rat sarcoma) viral oncogene (Parada et al. 1982). According to Angier (1988:110), Weinberg was not happy about this finding, because it meant that instead of finding a new cancer gene via other methods, Weinberg had found another example of the human cell transforming activity of a so-called viral oncogene.[16]

> A month after Chiaho Shih secured his EJ phage, Luis Parada would disclose to the Weinberg lab what neither Bob nor anybody else in the laboratory wanted to hear. . . . The human cancer genes were not new genes, but the same old genes they'd been working with for years. They were the genes that Weinberg had tried so desperately to put behind him.

After subsequent research, Weinberg's laboratory reported that a single point mutation had caused this single normal *ras* gene to become a cancer-causing gene (Tabin et al. 1982).[17] In 1982, Weinberg proposed that his transfected oncogenes were of a class with the oncogenes reported by tumor virologists Bishop and Varmus.

> A second question concerns the relation of these oncogenes to those which have been appropriated from the cellular genome by retroviruses and used to form chimeric viral-host genomes. The most well known of these genes is the avian sarcoma virus *src* gene, the paradigm of a class of more than a dozen separate cellular sequences. Do these two classes of oncogenes, those from spontaneous tumors and those affiliated with retroviruses, overlap with one another or do they represent mutually exclusive sets? Although the answer to this is not yet at hand, it will be forthcoming, since many of the sequence probes required to address this question are already in hand (Weinberg 1982:136).

Weinberg (1982:135; emphasis added) reported that while "the study of the molecular biology of cancer has *until recently* been the domain of tumor virologists," it now was also the domain of molecular biologists. In 1983, he and his associates (Land et al., 1983:391) claimed to have confirmed this equivalence between these sets of oncogenes.

> Two independent lines of work, each pursuing cellular oncogenes, have converged over the last several years. Initially,

> the two research areas confronted problems that were ostensibly unconnected. The first focused on the mechanisms by which a variety of animal retroviruses were able to transform infected cells and induce tumors in their own host species. The other, using procedures of gene transfer, investigated the molecular mechanisms responsible for tumors of nonviral origin, such as those human tumors traceable to chemical causes. *We now realize that common molecular determinants may be responsible for tumors of both classes.* These determinants, the cellular oncogenes, constitute a functionally heterogeneous group of genes, members of which may cooperate with one another in order to achieve the transformation of cells (emphasis added).

Bishop (1982:92) tentatively supported the claims of Weinberg and his associates.

> Weinberg and Cooper have evidently found a way of transferring active cancer genes from one cell to another. They have evidence that different cancer genes are active in different types of tumors, and so it seems likely that their approach should appreciably expand the repertory of cancer genes available for study. None of the cancer genes uncovered to date by Weinberg and Cooper is identical with any known oncogene. Yet it is clearly possible that there is only one large family of cellular oncogenes. If that is so, the study of retroviruses and the procedures developed by Weinberg and Cooper should eventually begin to draw common samples from that single pool.

Indeed, the list of oncogenes as constructed by review articles expanded to include the genes studied by Weinberg and Cooper. In 1984 the list included over twenty oncogenes (including viral oncogenes) and their homologous proto-oncogenes. By 1989 the list had expanded to sixty. According to Bishop (1989:11), "*the definition of proto-oncogene had now become more expansive, subsuming any gene with the potential for conversion to an [activated] oncogene—by the hand of nature in the cell, or by the hand of the experimentalist in the test tube.*" (emphasis added)

To summarize, a few molecular biologists constructed an equivalence between their cancer-causing genes and the proto-oncogenes of tumor virologists. They argued that their cancer-causing genes were in the same class of cancer-causing genes reported by tumor virologists. This representation expanded the category of proto-oncogenes to include genes that had been transformed by chemicals reported to be carcinogens in volumes of previous studies on cells, on whole organisms, and especially on humans. The work in Weinberg's laboratory links carcinogenesis studies, human cancer, and oncogenes. His was one of the first laboratories to claim that they had found cellular oncogenes in human tumor cell lines. This simultaneously provided a new link between Bishop and Varmus's oncogene and carcinogenesis studies. As researchers embraced one another's work, the concept of a normal gene as a cancer-causing gene became more stable.

Associations with Developmental Biology

Normal growth and development are research problems that form the basis of developmental biology. This has been, and remains, an established and popular field of biological research. At the time of the initial announcements by Bishop and Varmus, they proposed that their "normal" proto-oncogene had something to do with cell division. Later, as researchers in the molecular biology and biochemistry on normal growth and development began proposing the existence of growth factor genes based on research on growth factor protein, they tied their work both theoretically and concretely to Bishop and Varmus's work on oncogenes. For example, Russell Doolittle, a molecular evolutionist and chemist at the University of California at San Diego, and Michael Waterfield of the Imperial Cancer Research Fund in London both reported that a partial sequence of platelet-derived growth factor (PDGF) was nearly identical to that deduced for the protein product of the *sis* oncogene of simian sarcoma virus (Doolittle et al. 1983; Waterfield et al. 1983). In 1984, Waterfield's laboratory reported that they had found that the epidermal growth factor (EFG) receptor protein was identical to an oncogene's (*erbB*) protein product studied by the Varmus and Bishop group.

This association between normal growth factors and proto-oncogenes provided an evolutionarily acceptable explanation for finding that potentially cancer-causing genes were conserved through time, as Bishop argued in a 1983 review article and in a 1984 interview.

> The logic of evolution would not permit the survival of solely noxious genes. Powerful selective forces must have been at

> work to assure the conservation of proto-oncogenes throughout the diversification of metazoan phyla. Yet we know nothing of why these genes have been conserved, only that they are expressed in a variety of tissues and at various points during growth and development, that they are likely to represent a diverse set of biochemical functions, and that they may have all originated from one or a very few founder genes. Perhaps the proteins these genes encode are components of an interdigitating network that *controls the growth of individual cells during the course of differentiation.* We are badly in need of genetic tools to approach these issues, *tools that may be forthcoming from the discovery of proto-oncogenes* in Drosophila and nematodes (Bishop 1983:347–48; emphasis added).

> And it took us a while to convince people that [these genes] might have a different purpose in the normal body. . . . But if something went wrong with them, they would become cancer genes as they were in the virus (Bishop interview).

Bishop expanded the number of research problems in his laboratory from the study of one viral oncogene to studies of several viral oncogenes and their related proto-oncogenes. By 1984 some members of the laboratory were asking questions regarding the normal functions of the proto-oncogenes in developmental biology.

> My laboratory doesn't much resemble what it was ten years ago. . . . [How has it changed and why?] The work's evolved in response to progress in the field. You get one problem solved, and you move on to something new that presents itself. A number of people in my laboratory are explicitly interested in normal growth and development. They're here because we believe that the cellular genes we study are probably involved in normal growth and development. And I wasn't studying cellular genes involved in normal growth and development fifteen years ago. . . . There is a conceptual and probably mechanistic connection between cancer and development. But I'm not a developmental biologist, and I haven't read seriously in the field. There are people in my laboratory who will probably become developmental biologists as they fashion their own careers (Bishop interview).

Thus, research on normal growth and development, evolutionary biology, and cancer became part of the constitution of the proto-oncogene theory.

"Back-Translating" Proto-Oncogenes to Viral Oncogenes to the Virus Cancer Program

National Cancer Institute administrators supported and promoted the proto-oncogene theory for several reasons. *Their* sponsors were Congress and the public it represented, including other scientists. The proto-oncogene theory provided them with both the justification for past research investments in the Virus Cancer Program (VCP) and with a product to present to Congress.

In 1964, the NCI focused on the role of viruses in human cancer etiology through a special, heavily funded Virus Cancer Program, initiated as a contract program in 1964. Many virologists and molecular biologists were funded by the NCI through this program both in the 1960s and in the 1970s under its new form mandated by National Cancer Act of 1971. Their research focused on what are now called DNA tumor viruses and retroviruses (or RNA tumor viruses). Both the act and its viral research component became controversial and much maligned efforts in the 1970s (Rettig 1977). Controversy raged over both the contractual basis for dispensing research funds and the huge sums of money concentrated on the virus cancer program, that is, on what was considered by many at that time to be an unsubstantiated and unlikely view that viruses caused human cancer. In 1974 the *ad hoc* Zinder committee submitted an extremely critical report to the National Cancer Advisory Board (NCAB), which oversees the work of the entire NCI. After twenty years of research, tumor virologists reported finding many viral oncogenes in experimental animals and a few in naturally occurring animal populations. However, no viral oncogenes were reported discovered in human tumors, and no viruses had been linked to human cancer.[18] In 1980, NCI leaders decided to break the VCP up and integrate the pieces into other NCI programs. This overhaul had cumulative negative effects on viral cancer research funds.

In this highly politicized situation in the early 1980s, several administrators and scientists used the proposed role of proto-oncogenes in causing human cancer to justify past investments in viral oncology. Bishop (1982:92), for example, states:

> The study of viruses far removed from human concerns has brought to light powerful tools for the study of human disease. Tumor virology has survived its failure to find abundant viral agents of human cancer. The issue now is not whether viruses cause human tumors (as perhaps they may, on occasion) but rather how much can be learned from tumor virology about the mechanisms by which human tumors arise.

Vincent T. DeVita Jr. (1984:1–5), then director of the National Cancer Institute, viewed the research similarly.

> Recent discoveries of retrovirus oncogenes and their human homologs make it reasonable for one to state that few areas of research have been so fruitful. We are closer to understanding the underlying abnormality of growth that is cancer than the architects of the NCP [National Cancer Program] could have imagined in 1971. . . . [Extending his discussion from the VCP to the entire NCP and the contract system of funding, DeVita states that] we have often been asked if the NCP has been a success. While I acknowledge a bias, my answer is an unqualified "yes." The success of the Virus Cancer Program which prompted this essay is a good example. Since its inception, this Program has cost almost $1 billion. If asked what I would pay now for the information generated by that Program, I would say that the extraordinarily powerful new knowledge available to us as a result of this investment would make the entire budget allocated to the NCP since the passage of the Cancer Act worthwhile. There may well be practical applications of this work in the prevention, diagnosis, and treatment of cancer that constitute a significant paradigm change. The work in viral oncology has indeed yielded a trust fund of information, the dividend of which defies the imagination.

Both oncogene researchers and cancer research administrators argued then that the "new" oncogene research would be based on the "extraordinarily powerful new knowledge" produced by past investments. The viral cancer genes constructed from the investments of the NCI in the Viral Cancer Program during the 1960s and 1970s have in the 1980s and early 1990s become human cancer genes through the proto-oncogene theory and recombinant DNA technologies. Viral cancer genes with no previous connection to human cancer have now become human cancer genes. And many tumor virologists, cut off from NCI's viral research funds, became cellular oncogene researchers.

Indeed, NCI director DeVita used the proto-oncogene theory to justify the entire National Cancer Program in 1984. DeVita told me that he used oncogenes to sell their general future program of molecular genetic research on cancer to Congress:

> Molecular genetics is a term nobody in Congress understands. Oncogenes they know. How do they know? I tell them. I can explain oncogenes to them much better than I can explain molecular genetics. When I point my finger at a Congressman, I say, "Mr. So-and-So, you and I both have genes in us, which we believe are the genes that are responsible for causing cancer." It gets their attention. They say, "My God! What do you mean I have genes in me. . . . ?" I have to explain it to them. If I tried to explain molecular genetics, they'd fall asleep on me.

The statement that we all have genes that can be triggered to cause cancer can engender great fear in Congressional members and their constituents. This fear, however, was laid to rest by the claim that a unifying pathway to all cancers may exist and that there may be ways to intervene in this pathway.

The National Cancer Institute supported and promoted the proto-oncogene theory and molecular genetic approaches to cancer in general. DeVita summarized for me NCI's investments in molecular genetic cancer research for 1984.

> [In 1984] we had $198 million in molecular genetics. . . . [That figure] includes oncogenes, but it also includes people who are walking up and down the genome, tripping on oncogenes but looking for something else. And they're going to find the regulatory elements that control the oncogenes, [which is] really the major step. Oncogenes have told us something very important, but now what you want to find out is what regulates these genes so that you can use this information to turn them on and off.

There is also other evidence of NCI's commitment to molecular genetic approaches. In 1981, while NCI leaders were in the process of reorganizing research at the Frederick Cancer Research

Facility, they decided to similarly shift the facility's emphasis. "We put three or four crackerjack oncogene scientists up there, and they're up there cranking out the data and having a fun old time," said one administrator. NCI also appointed a viral carcinogenecist who had worked with George Todaro, one of the originators of an early version of the proto-oncogene theory, to the position of associate director of the entire NCI. He oversaw the Frederick Cancer Research Facility and specifically kept track of oncogene research progress (Shapley 1983:5). Finally, NCI also committed funds to a supercomputer to facilitate oncogene and other molecular biological research.

To summarize, while proto-oncogenes were used to justify the NCI's previous commitments, they also gained a major portion of NCI's current commitments—a mutually beneficial state of affairs.

Linking Proto-Oncogenes to Applications: Clinical and Epidemiological Genetics

Oncogene researchers also proposed potential applications via clinical and epidemiological genetics to treatments and possibly to cures. For example, Bishop (1982:91) states:

> Medical geneticists may have detected the effects of cancer genes years ago, when they first identified families whose members inherit a predisposition to some particular form of cancer. Now, it appears, tumor virologists may have come on cancer genes directly in the form of cellular oncogenes.

In 1982 Bishop (1982:92) also proposed that the proto-oncogene might lead to "a final common pathway" for all cancers that would provide a common explanation for chemical, viral, and radiation carcinogenesis and for normal growth and development.

> Normal cells may bear the seeds of their own destruction in the form of cancer genes. The activities of these genes may represent the final common pathway by which many carcinogens act. Cancer genes may be not unwanted guests but essential constituents of the cell's genetic apparatus, betraying the cell only when their structure or control is disturbed by carcinogens.

Viruses, he continues, were merely the tool that made visible this underlying mechanism. "At least some of these genes may have appeared in retroviruses, where they are exposed to easy identification, manipulation and characterization" (Bishop 1982:92). Bishop (1989:7) reiterated this view in his Nobel Lecture: "By means of accidental molecular piracy, retroviruses may have brought to view the genetic keyboard on which many different causes of cancer can play, a *final common pathway* to the neoplastic phenotype" (emphasis added).

Weinberg (1983:134) similarly speculated broadly that the proto-oncogene theory accounted for findings in many lines of cancer research.

> What is most heartening is that the confluence of evidence from a number of lines of research is beginning to make sense of a disease that only five years ago seemed incomprehensible. The recent findings at the level of the gene are consistent with earlier insights into carcinogenesis based on epidemiological data and on laboratory studies of transformation.

In a volume entitled *RNA Tumor Viruses, Oncogenes, Human Cancer and AIDS: On the Frontiers of Understanding*, the editors Furmanski, Hager, and Rich (1985:20) proclaimed that

> we must turn these same tools of molecular biology and tumor virology, so valuable in dissecting and analyzing the causes of cancer, to the task of understanding other equally critical aspects of the cancer problem: progression, heterogeneity, and the metastatic process. These are absolutely crucial to our solving the clinical difficulties of cancer: detection, diagnosis and effective treatment.

Indeed, oncogenes have already moved into the arena of medical applications via diagnostics. For example, N-*myc* is an oncogene that is used as a tool in prognosis to supplement the other signals of the virulence of cancer cases. Indeed, according to Bishop (1989:14), "The New England Journal of Medicine has now provided its imprimatur by arguing that 'in neuroblastoma, amplification of the N-myc gene is of greater prognostic value than the clinical stage of the disease. Thirty years after deserting the bedside, I have found clinical relevance in my research." Additionally, cellular oncogenes have moved into cardiovascular specialties and are being studied for their possible action on cardiac muscle cells (Mulvagh, Roberts, and Schneider 1988).

My argument is that J. Michael Bishop and Harold T. Varmus,[19] working in the late 1970s in a microbiological laboratory at the University of California, San Francisco, Medical Center, constructed a theory that, they proposed, mapped onto the intellectual problems of many different scientific social worlds. As one measure of their success, Bishop and Varmus were awarded prizes for their research. In 1983 they won the Albert Lasker Medical Research Award for Basic Research, "the next best thing to a Nobel prize in medicine according to many Americans" (Newmark 1983:470) and the Armand Hammer Prize for cancer research. In 1989 they were awarded the Nobel Prize in Physiology or Medicine. As they and their colleagues crafted and recrafted the theory, they constructed equivalences between previously unequivalent units of analysis, for example, between genes controlling cancer and genes controlling normal growth and development. For other lines of research like those under the auspices of the Virus Cancer Program, they constructed continuities through time and locales while introducing novelty into the existing lines of research. The theory provided alternative ways of studying and explaining biological activities (for example, of carcinogenesis) at the molecular level and gave form to other researchers' efforts to recraft their existing lines of research using this new unit of analysis. However, proto-oncogene theorists did not do all of this by themselves. Working relationships were created between the oncogene theorists and biologists in evolutionary biology and population genetics, medical genetics, tumor virology, molecular biology, cell biology, developmental biology, and carcinogenesis. While early oncogene proponents suggested potential connections between their theory and questions in other fields of biology, researchers in these other fields took up the lines and drew on their work to support and extend their own lines of research. I have sketched a few of these interactions and discussed their importance to the development of a robust proto-oncogene theory and to the development of the oncogene line of research. However, these theoretical constructions alone do not explain the growth of this research. Of parallel significance were innovative recombinant DNA technologies.

RECOMBINANT DNA TECHNOLOGIES

As stated earlier, recombinant DNA techniques provided researchers access to the eukaryotic cell nucleus where they could actually *intervene* and *manipulate* nucleic acids to study their activities in ways never before possible. That is, it provided a new space of representation. Researchers used (and refined and extended) these technologies to frame new molecular questions (and answers) about problems previously approached at the cellular and organismal levels. However, equally important for the purposes of understanding the extraordinary growth of this line of research was the standardization of these techniques. Standardization created continuity in representations across laboratories and science production worlds.

Before 1981 the tools were limited in three ways. Despite the successful recombination of genes of different species of organisms in August 1973 (Morrow et al. 1974), several difficulties still limited the basic techniques to the exploratory, cutting-edge research of methodologists until 1976. First, according to Wright (1986:320), improvements required to make the procedures more easily reproducible in other laboratories included: (1) "the means to define and control precisely the piece of DNA to be inserted into a living cell and the manner of its insertion"; (2) "the means to increase the efficiency of replication of a piece of foreign DNA"; and (3) "the provision of a wide variety of sources of pure DNA so that gene-splicing techniques might be applied to any gene from any source."[20] Second, although recombinant DNA methodologists had resolved these technical difficulties by 1976, it was not until fall 1977 that they managed to refine a technique for "expressing" the genes of "higher" organisms by bacteria (Itakura et al. 1977). As finalized in their techniques, expression took the form of bacteria first incorporating a gene from another organism (such as humans) and then producing the proteins of that human gene. According to molecular biologists, this step was necessary before they could apply the set of techniques to problems involving the genes of "higher" organisms. Third, the techniques were still not refined and standardized enough for nonrecombinant DNA methodologists to efficiently (in terms of time, skills, materials, and money) use in substantive experiments. Methodologists achieved this final stage of refinement by 1980, as indicated by a course entitled "Molecular Cloning and Eukaryotic Genes" at the Cold Spring Harbor Laboratory's annual meeting in that year. Despite advertisements by recombinant DNA methodologists of the promised capabilities of their tools beginning in the early 1970s, they did not complete their package for "export" until sometime between 1980 and 1981. Until 1980 recombinant DNA technologies were the research objects of "high-tech" methodologists, rather than the applied methods of biological researchers. It was not until 1980 that recombinant DNA technologies were available to non-molecular methodologists.

By 1980–1981 the researchers who developed recombinant DNA techniques had refined a set of technologies that they exported from their laboratories. Despite claims of "state-of-the-art" methods, molecular biologists had transformed a set of artful and complex tasks and knowledges for manipulating DNA, RNA, and proteins into a set of relatively routine tools and standard practices. The term *state of the art* confounded technical and substantive capabilities. The definition of *state of the art* often refers to situations where tacit knowledge and trial-and-error procedures are still very much a part of daily operations in technical work. In contrast, recombinant DNA techniques were procedurally relatively routine by the early 1980s.

Rather than credit the transformation of recombinant DNA technologies only to technical improvements, I propose that the transformation was also the outcome of the gradual building of a network of laboratories, private enterprises, government funding agencies, practitioners (scientists, technicians, students), and material products. By adopting these tools, researchers and research laboratories have transformed their practices and their skills into molecular knowledges and technologies. By funding and producing these tools, federal and state funding agencies and private industry have contributed to building a network of support for molecular genetic technologies. The private and public commitments to these technologies reduced the cost of the materials and instruments and provided access to these technologies to laboratories with limited financial and technological resources. (In our capitalist economy, science too is subject to commodification.) Together they have transformed a set of novel techniques into accessible and productive tools that have in turn transformed the practices, problems, representations, and working tools of other biological subdisciplines to produce "the new [molecular] biology." Through the transformation of diverse research worlds, these technologies have been further standardized. Thus, recombinant DNA and other molecular genetic technologies have also reconstructed practices and representations in other worlds which in turn furthered their own standardization.

In my terminology, a technology is standardized when it has become collective conventional action, through either explicit, tacit, or coerced agreement. These commitments can become relatively solidified in experimental protocols, research materials, and instruments that regiment work, as they have in many molecular genetic technologies. Scientists spoke of standard technologies, in one sense, as "low tech" in that most "bugs" had been worked out such that they worked "well enough."

Standardized tools are relatively stabilized ways of acting that reduce the uncertainties of daily bench work. Scientists often assume that they will use a particular tool or technique in a particular context to accomplish a particular task. By the time tools are standardized, the number of unknown factors researchers must consider while conducting their daily experimental tasks and the number of "moments" of question, articulation, and discretion have been reduced. Researchers use these tools in a taken-for-granted manner in their work processes—that is, they regard them as relatively unproblematic—and concentrate on their use for solving new problems. Their focus is on making the problematic unproblematic, rather than vice versa. In contrast to luck or serendipity and intensive trial and error, researchers rely on these tools for the *routine* production of data.

In this case, routinization does not preclude the production of novelty. Indeed, recombinant DNA technologies continue to produce novel *representations* of nature and, perhaps even more compelling, novel *phenomena* (qua *realities*) including recombinant genes, proteins, and organisms.[21] These new natural phenomena are created through the "state of the art" capabilities of recombinant DNA technologies for genetic manipulations—for example, previously unfathomable hybridization experiments—that could not have been done using older biochemical techniques.

Working with standardized or taken-for-granted techniques also does not preclude the development of new techniques. The latter is methodology, the development or modification of techniques or methods for constructing representations and phenomena. Methodologists still tinker with recombinant DNA technologies to attempt new ways of doing the same thing or to develop new ways to do new things. The early 1980s saw the invention of new technologies such as the polymerase chain reaction (PCR) to speed up the amplification of DNA. The various human genome projects around the world are currently adding many new materials and techniques (e.g., YACs or yeast artificial chromosomes, better automated DNA and RNA synthesizers, and bioinformatics software and databases) that increase the speed, specificity, versatility, and volume of production in biology laboratories. Some molecular biologists build careers by successfully developing faster or novel techniques. New techniques are marketable commodities in the quest for positions, research funds, laboratories, and students.

Neither is the performance of standardized techniques necessarily easy. As any molecular biologist will report, even the simplest technique often does not work and requires more time than expected. The techniques often are a set of sequential steps, each of which is required to work before the next step can be done. If one step does not work, researchers must repeat it until it does work. Novices learn to reproduce or reperform standardized techniques such as "plasmid preps" (cf. Jordan and Lynch 1992), cloning of DNA, and sequencing, often through painstaking processes. Each student or technician in a laboratory creates her or his own way of making the procedures work. Yet, with no or little apprenticeship guidance other than published manuals, even novices teach themselves how to perform these techniques in ways that work!

Standardized recombinant DNA technologies, the proto-oncogene theory and its historical pylons, including various previous and extant institutional, organizational, material, theoretical, and political commitments and contexts, together constituted a package of theory and techniques that served as a dynamic interface that linked different worlds across time and space.

THE RULER OVER A NEW REALM: THE THEORY-METHODS PACKAGE

By 1983 the new "unifying" proto-oncogene theory of cancer had been adopted and used as the basis of investigations in several new and established laboratories in several lines of biological and biomedical research as well as by the National Cancer Institute. Oncogene theorists had constructed several cancer-causing normal genes that they claimed mapped onto intellectual problems in many different scientific social worlds. They claimed that their cancer-causing genes accounted for findings in many other lines of cancer research and represented a unified pathway to cancer in humans and other "higher" organisms.

By 1984 the proto-oncogene theory linked retroviruses and viral oncogenes to such diverse elements as the course of evolution, *Drosophila* genetics, normal growth and development in developmental biology, and established lines of biomedical research on cancer. It linked new research on oncogenes to the Virus Cancer Program of the past, the future funding agendas of the NCI, and to the concerns of the United States Congress. These links were constructed through the reconstruction of the theory through time as different laboratories added their pieces of DNA sequences to the pot. In Callon's (1986a, 1986b) and Latour's (1987) terms, the proto-oncogene theory translated one laboratory's interests into the interests of many others, and vice versa, and connected laboratories in different lines of research and subdisciplines and even through time into a single network and a new world of scientific practice.

The oncogene theory was constructed and reconstructed in conjunction with recombinant DNA and other molecular genetic technologies as these were becoming standardized and attracting attention in most biological subdisciplines (and elsewhere). It was the combination of the theory and these methods, the *package*, that together attracted other researchers to engage in oncogene research. For "old-fashioned" cancer researchers incentives for adopting the package included a chance to incorporate "hot" new recombinant DNA technologies. Molecular biologists were offered a chance to attack the human cancer problem through the proto-oncogene theory. These incentives contributed to the development of oncogene research.

Oncogene theory proponents enrolled allies behind their theory not only by claiming to have accounted for findings in many other lines of cancer research, but also by framing and posing new doable problems on oncogenes for other researchers to investigate. That is, they posed questions which (1) scientists could experimentally investigate using recombinant DNA and other molecular biological technologies; (2) laboratories were already organized and equipped with resources to handle, or could relatively smoothly import the requisite resources; and (3) satisfied significant audiences.

The proposed problems were both specific and general. Researchers could immediately begin experimentation on specific problems, while thinking of possible ways to translate more general problems into specific experiments.

J. Michael Bishop's (1983:345–48) article on "Cellular Oncogenes and Retroviruses" in the 1983 *Annual Review of Biochemistry* is an excellent example of proposed problems that mapped onto established laboratory organizations and available technical skills. He first summarized work in several other lines of cancer research and then presented proposals for research on the role of oncogenes in cancer and in biological processes generally, including experimental carcinogenesis, evolution, normal growth and differentiation, medical genetics, and epidemiology.

At a more hands-on level, oncogene researchers distributed their probes for oncogenes to other laboratories and to suppliers, thus facilitating oncogene research in other labs by providing standardized tools. "We've had so many requests for our probes for [two cellular oncogenes] that we had one technician working full-time on making and sending them out. So we finally turned over the stocks to the American Type Culture Collection," said Bishop in 1984. While Bishop may be quite astute and enthusiastic about pushing his laboratory's research, it is also evident that other researchers are *interested*! Enrollment is explicitly mutual, clearly a two-way street.

The probes were more than physical materials. They were "naturalized" or materialized ideas, designed with reference to specific hypotheses about their involvement in cancer causation. They *embodied the specific approach (model/question and solution) to a problem of the laboratories in which they had been constructed.* Exporting probes is one attempt to reshape the world outside. With Bishop and Varmus's probe, one is more likely to find/create what Bishop and Varmus reported finding. Any researcher can call or write to ATCC to order the probes at the cost of maintenance and shipping. Oncogene researchers distributed the tools for testing and exploring their framing of reality to a host of other researchers.

Oncogene theory proponents also taught and talked about their work to students and researchers in other biological disciplines. A molecular biologist described the positive response of participants to an oncogene talk at a cell biology conference. Most of the conference participants, uninitiated in the complexities of oncogene research, were awed by the lecture and unable to evaluate the difficulties and complexities in the data.

Proponents also spoke about their work in the popular media. In 1984 *Newsweek* (Clark and Witherspoon 1984:67; see also Clark et al. 1984) acclaimed the new proto-oncogene research and even discussed its potential for producing diagnostic aids and treatments.

> Such discoveries shed important light on the fundamental processes of cancer as well as the growth and development of all forms of life. In the future, they will surely lead to better forms of diagnosis and treatment. The presence in cells of abnormal amounts of proteins caused by gene amplifications, for example, could lead to sensitive new tests for certain kinds of cancer. As for treatment, scientists envision the development of drugs designed to specifically inhibit oncogenes. These would be far better than anticancer drugs that indiscriminately kill normal cells along with cancerous ones. "We would," says Frank Rauscher of the American Cancer Society, "be using a rifle rather than a shotgun."

COMMITTING TO ONCOGENE RESEARCH

By 1984 oncogene research was a distinct and entrenched phenomenon. Researchers referred to the "oncogene bandwagon" in conversations. Scientists were acting on the basis of its existence. More generally, modern biology was molecular biology. Oncogene research had grown to the point where it was sustained by its own momentum. That is, many researchers joined the bandwagon because it *was* a "bandwagon," its continued growth produced by a "snowball" effect.

Participants in diverse social worlds had committed their resources to molecular biological cancer research. These commitments included: (1) very large increases in funding allocations;[22] (2) designated positions in academic departments, research institutes, and private industrial laboratories; (3) accessible training and tools, including knowledge, standardized technologies, materials, and instruments; and (4) a cadre of researchers training in molecular biological skills. That is, an infrastructure of skills, funding allocations, committed researchers and teachers, positions committed to molecular biologists, biological material suppliers and supplies, and even whole companies and research institutes committed to oncogene research problems was established by 1984. This infrastructure then constrained and influenced the decisions of new investigators. It served to maintain previous commitments as well as to gain new commitments.

Commitments by Students, New Investigators, Established Researchers

The growing commitments of tumor virologists, molecular biologists, and the NCI to oncogene and related molecular biological research, in turn, further provoked students, new investigators, and even researchers established in other lines of work to frame their theses and research problems in similar terms.

Besides considering interesting intellectual questions and the problem of curing cancer, new researchers had to attend to career development contingencies in making problem choices. The immediate foreground was filled with the exigencies of their daily work lives: researching and writing Ph.D. theses, establishing and maintaining laboratories and staff, publishing and gaining tenure, writing grant proposals, attracting and training students. In this situation, constructing doable problems that produce results that someone will publish is a practical and pressing concern. Thus, desirable "cancer research" became "productive research" became "doable research."

In this context, students and beginning researchers gained major advantages for establishing their careers and laboratories by choosing to investigate problems under the rubric of oncogene research. By 1983 these advantages included clearly articulated experiments, research funds, high credibility, short-term projects, increased job opportunities, and the promised generation of further doable problems.

By 1982 the combination of proto-oncogene theory and molecular biological technologies provided clear problem and experimental protocols. By the end of 1983, a graduate student in an oncogene laboratory explained his work as a set of logical steps. Students learn these logical steps from their laboratory directors. A senior oncogene investigator described his research problem formulation as a textbook reductionist, logical approach.

> It's always seemed to me, because I don't see any other way to go about it, that you have to employ a reductionist approach. You have to say, "I now know that out of the thou-

> sands of genes in a given cell, Genes A, B, and C are the genes that are screwed up in a cancer cell. And what's wrong with them is that this base has changed that way, and this gene has moved to that chromosome, and the effect of these changes—which you can spell out with nucleotide sequences—is to make a protein which has an altered kinase activity, or an altered location in a cell, or there is an abundance of the protein, or the protein's not properly regulated by some other factor which I now call "Y.". . ." It's the description of those changes [which] in the long run lead to, first of all, fundamental understanding of how cell growth is regulated, and secondly, to clinical insights . . . how to make diagnoses at the earliest possible time [and how] to think about the ultimate in strategies—blocking the activity of genes that are instrumental in cancer ("Rames" interview).

Obviously, research is rarely so neat and tidy as in this senior scientist's post hoc descriptions. However, the clarity of their descriptions of oncogene research trajectories speaks to the "logic" created by oncogene investigators as compared with many other, less focused lines of biological research.

Finally, new and more established investigators reported that they engaged in oncogene research because of promised intellectual payoff and new generations of downstream questions. Reports of novel oncogene findings provided glimpses of new ways to study difficult and challenging problems. For example, in 1983, *Nature* (Newmark, 1983:470) published an article in its "News and Views" entitled "Oncogenic Intelligence: The *Ras*matazz of Cancer Genes." The article was just one of many published between 1978 and 1985 announcing exciting new findings from oncogene research. Researchers constructed novel, intellectually exciting representations which they then used to construct further experimental questions.

The intellectual excitement was not limited to oncogene research but extended to all molecular biological research. Almost all respondents, independent of their political views about how the new molecular biological technologies should be used, echoed this excitement. A respondent, who had been conducting protein biochemical research, explained the excitement in terms of a whole new scale of analysis opening up to scientists. This new scale of analysis was the outcome of molecular biological technologies that allowed researchers not just to observe but to actually *change nature* in the laboratory.[23]

> You can ask certain sorts of questions which you can't really answer with just the biochemical methodology. . . . *Genetics essentially involves modifying what's already there, rather than simply describing what's going on.* It allows you to ask much more specific questions about which components of the system are necessary to do what. Recombinant DNA technology is starting to allow one to ask those sorts of questions in animal cells, tumor cells. . . . Questions which there is as yet no other way of approaching ("Grant" interview; emphasis added).

In this case, it is not only "young turks" involved in the excitement. Established researchers also found the possibilities for exploring new scales of analysis useful. An example was the senior investigator who had been studying the effect of radiation on transforming cells in culture. After much excitement about the oncogene theories of carcinogenesis, he sent his student to train in recombinant DNA techniques in a nearby laboratory in order to test two hypotheses: first, whether radiation played a role in the mutation or transposition of one or several proto-oncogenes and, second, whether radiation damage to cells made it easier for the viral oncogene to become integrated into the normal cellular genome. The graduate student gained the benefits enrolled behind the proto-oncogene theory, the senior investigator imported new skills and a new line of research into his laboratory, and radiation and oncogenes became associated.

Commitments by Private Industry

Despite the uncertain commercial payoff from oncogene research, several large pharmaceutical companies and major research and development (R&D) companies committed funds, researchers, and laboratories to oncogene research and recombinant DNA technologies (Koenig 1985). These pharmaceuticals included Hoffman-La Roche Inc., SmithKline Beckman Corporation, Merck and Co., and Abbot Laboratories. Investing R&D biotechnology companies included Genentech and Cetus[24] and especially smaller companies aimed specifically at oncogene products including Oncogene Science Inc., Oncogen, and Centocor Oncogene Research Partners.

One respondent regarded these commitments as efforts to "get in on the ground floor." Even if a particular company is not the home of the desired new discovery that leads to a patentable diagnostic or therapeutic product, it will have established the infrastructure for early entry into the race to produce the final commercial product(s). In 1985 a research director at Hoffman-La Roche stated, "If you're interested in [oncogene] products, you can't afford not to be in the race now" (Koenig 1985:25). Entering the oncogene market was regarded as both a race and a gamble.

These commitments to oncogene research on the part of private industry refer back to similar commitments of new researchers. These commercial investments provide both job opportunities and affordable research tools for new investigators. Oncogenes, academic oncogene research commitments, and private industrial commitments interactively produced each other!

THEORETICAL DISCUSSIONS

A Distributed Center of Authority

I have argued that the common element, the distributed center of authority, in the phenomenal growth of a line of research was a *package* of theory and methods. This new package consisted of the proto-oncogene theory of cancer and recombinant DNA and other molecular genetic technologies for testing and exploring the theory and for generating new problems for investigation in diverse worlds. However, by opening up this package, this paper tells a story of the mutual production of this package and the extended network of commitments made to oncogene research. That is, both what post hoc appears to be a package theory and methods became a package through its becoming a *distributed* center of authority. The proto-oncogene theory and molecular genetic technologies and the transformation of practices in many different situations (research laboratories and communities, sponsors, and DNA sequences) were coconstructed. It is because of this sense of continuous movement and transformation that scientific work should be understood as ecologies of action.

This paper examined the processes and situations through which the combination of theory and methods was constructed, disseminated, adopted, and modified. The proto-oncogene theory was constructed as an abstract notion, a hypothesis, using a new unit of analysis to study and conceptualize cancer. Researchers in many extant lines of research used this abstraction to interpret the theory to fit their separate concerns all under the rubric of oncogene research.[25] Researchers translated the general theory into concrete research problems in their laboratories without contradicting the existing basic framework. Further, the concrete expression of the theory was framed by oncogene researchers in the terms of recombinant DNA and other molecular biology technologies which by the early 1980s were accessible to other biology laboratories. This combination of the abstract, general proto-oncogene theory and the specific, standard technologies was used to generate novel doable problems. *By locally concretizing the abstraction in different practices,* researchers with ongoing enterprises (re)constructed the new idea and the new methods in new sites, thus further extending the network.

Indeed, the *growth of oncogene research* was both the cause and the consequence of this capacity for maintaining the integrity and continuity of the interests of the enrolled worlds while simultaneously providing them with new tools for doing new work. Laboratories in many different biological subdisciplines and medical specialties viewed the theory-method package as a means for constructing new doable problems and an opportunity to augment or replace their old, well-known routines with "sexy" new recombinant DNA techniques. At the same time, the proto-oncogene theory did not challenge the theories to which the researchers had made previous commitments. Indeed, the new research provided them with ways of triangulating, of providing new evidence using new methods to support their earlier ideas using a new unit of analysis. These views of oncogene research were *realized* through the efforts of these researchers and, in turn, this realization further extended the reach of oncogene research and the complexity of the theory. The increasingly complex theory is today taken to be the best representation of cancer as well as normal growth and development at the molecular and cellular levels. Yet, it is impossible to separate this "best" representation from those scientists who judge it to be so. These scientists are oncogene specialists whose work forms the basis of the complex version of the original proto-oncogene theory.

Similarly, the director of the National Cancer Institute (NCI) used this new research to justify their past investments in the Virus Cancer Program, whose legitimacy and productivity had been questioned, and to lobby Congress for increased appropriations to the NCI. He presented the new research to Congress as promising new hope for a possible cancer cure for their constituents. NCI as well as other funding agencies, public and private, provided ample support for the research.

University administrators used oncogene research to reorganize cancer research institutes now deemed "old fashioned" into "hot" molecular biology institutes.[26] Of course, the very label *old fashioned* was a consequence of the growing popularity and credibility of molecular biology. Biological supply and biotechnology R&D companies saw in this new research an opportunity to develop new products and markets in the then slow biotechnology business. These companies then promoted their products as the answer to the cancer problem. In the late 1980s, DuPont corporation advertised its transgenic "OncoMouse™" that physically incorporated a specific proto-oncogene in the laboratory animal itself.[27]

> The OncoMouse™/*ras* transgenic animal is the first *in vivo* model to contain an activated oncogene. Each OncoMouse carries the *ras* oncogene in all germ and somatic cells. This transgenic model, available commercially for the first time, predictably undergoes carcinogenesis. OncoMouse reliably develops neoplasms within months . . . and offers you a shorter path to new answers about cancer. Available to researchers only from DuPont where better things for better living come to life. (DuPont advertisement in the journal *Science*)

All of these different actors used proto-oncogene research to maintain and extend their lines of work and simultaneously extended the reach of proto-oncogene research and molecular genetic technologies.

I have discussed the importance of standardization for making the package accessible to more than the elite laboratories. By the early 1980s, recombinant DNA technologies were organized into a set of protocols encompassing specific tasks, procedures, materials, and a few instruments (with more instruments created in the mid-1980s). What was to be done to which material for what reason or purpose and with what outcome were partially built into the technologies and protocols. Molecular biologists had constructed tools for manipulating DNA in eukaryotic organisms (including humans) which then became adopted by other laboratories. Through this process, "state of the art" tools became conventional tools used by many laboratories. These conventional tools were more accessible to researchers in other biological specialties, to new investigators, and to researchers far from the laboratories where the technologies were first created (University of California, San Francisco, Stanford University, California Institute of Technology, Massachusetts Institute of Technology, and Harvard University).[28]

The tools, available funds, a host of researchers and laboratories and institutes, specialized biotechnology companies, the proto-oncogene theory, and the oncogenes were thus coproduced.[29] Each buttressed, strengthened, and sustained the other to create doable oncogene problems (Fujimura 1987). The theory-method package provided procedures for a relatively straightforward construction of doable problems, or what Kuhn (1970) would call "normal science."

Through the extension of oncogene research from a few sites to a host of laboratories, private companies, and government supporting agencies, it has come to represent and facilitate collective work by members of different social worlds and also created oncogenes as stable *facts*. The package and the collective work *behind* it define a conceptual and technical work space that was less abstract, more structured, less ambiguous, and more concrete than that defined by the original conceptual model alone. Standardized molecular genetic technologies further define the conceptual model, and the conceptual model frames the use of the standardized technologies. Such codefinition of the range of possible actions and practices has created a greater degree of stability. The package has served as a dynamic *interface* between multiple social worlds and concurrently represents the contingent articulations of oncogene research at different sites.

The package then represents a transformation of practices in multiple social worlds and the emergence of a new definition of cancer. This new molecular genetic definition of cancer is maintained by the reorganization of commitments and practices of cancer research as well as other biological research worlds. The linkages thus created, in turn, shape subsequent commitments and work organizations made by researchers, laboratories, research institutions, biotechnology companies, and even diagnostic clinics. Twenty years ago the word/concept *oncogene* did not exist. Today oncogenes are facts in undergraduate biology textbooks and the building blocks of new research programs, theories, and diagnostics. The world has been reconstructed to include proto-oncogenes.

Coconstructing Practices, Problems, Theories, Laboratories, and the World

The account presented here shows researchers crafting science by shaping and adjusting materials, instruments, problems, theories and other representations, and social worlds as well as themselves and their laboratories in a process I call "coconstruction." I used the concepts of packages and the linking of practices and interests to show how different social worlds interacted through time and space to collectively craft science. Each world changed in some manner, yet each also maintained its uniqueness and integrity, in the construction and adoption of the proto-oncogene theory and recombinant DNA technologies. The theory and methods provided both dynamic opportunities for divergent meanings and uses as well as stability. In crafting and recrafting the theory, Bishop and Varmus and their collaborators constructed associations between oncogene research, on the one hand, and evolutionary biology, developmental biology, cell biology, carcinogenesis research, and more, on the other hand. As they drew on concepts and arguments from these lines of research, they were installing their theory, practices, inscriptions, and materials into these ongoing lines of research. The theory and methods were used to construct "homologies" between laboratories as well as between representations of phenomena. That is, the theory and technologies were being constructed, and they were also used to reconstruct laboratory work organizations as well as experimentally produced representations.

Crafting science is a historically located activity. The construction of new research programs is situated action, built on past practices, concepts and representations. In this case, scientific actors considered new proposals, theories, and methods in light of earlier routines and representations and in light of the conditions of the new situation. For years, as part of the war on cancer, tumor virologists had studied particular sequences of RNA in retroviruses, a class of viruses whose genetic material is RNA, not DNA.[30] Their research showed that these sequences caused cells in culture to turn cancerous and sometimes caused tumors in laboratory animals. Viral strains were developed and stored in order to study whether viruses or at least these viral oncogenes were possibly involved in human cancers. However, large investments of effort, time, and finances provided no indications of a significant involvement in human cancers. Instead they provided grounds for criticism of this research from many quarters. Into this historical scenario, Bishop and Varmus introduced a new hypothesis that *reversed the direction* of the relationship from gene insertion to gene pilfering by the virus. They claimed that their research found DNA sequences "homologous" (similar) to the RNA tumor-causing sequences in normal cells of life forms from yeast to humans. They argued that, contrary to the existing hypothesis that viruses had transferred their cancer-causing sequences to humans, their research established that the viruses had gained their cancer-causing sequences from normal cellular DNA.[31] The significance of this reversal, they argued, is that cancer-causing genetic elements are conserved through many species and thus must be related to some essential function in all organisms. Their new research program thus aimed at finding and studying the properties and expression of these normal DNA, called "proto-oncogene," sequences in normal and cancerous cells of different species, but especially in humans.

> If cells contain genes capable of becoming oncogenes by transduction into retroviruses, perhaps the same genes might also become oncogenes within the cell, without ever encountering a virus. By means of accidental molecular piracy, retroviruses may have brought to view the genetic keyboard on which many different causes of cancer can play, a final common pathway to the neoplastic phenotype (Bishop 1989:7).

When Bishop and Varmus won the Nobel Prize for their work, newspapers reported their views:

> "It has been known for a long time that cancer was in some sense a genetic disease. The importance of our finding is that one can identify explicitly the genes that play a role in cancer," said Varmus. As a result of their contribution, more

> than 50 different genes involved in cancer have been identified, he said. (Dolores Kong, *Boston Globe,* 10 October 1989).

> Bishop and Varmus discovered that the source of the gene [in a cancer-causing virus] had been the chickens themselves. An apparently normal gene was pilfered by the virus and was now able to cause cancer when it infected the birds (Richard Saltus, *Boston Globe,* 10 October 1989).

Bishop and Varmus had turned the earlier virus studies "on their heads." The movement of genetic elements reversed directions literally and figuratively, but the same or homologous gene sequence continued through time and space. The vehicles of this "gene conservation" were molecular probes created using molecular hybridization techniques. Critically, however, their work implicated the same (or similar) DNA sequences that previously had been studied extensively and that had gained enormous material and sentimental support from many researchers, organizations, and institutions. The construction of the proto-oncogene theory was situated within a network of historical and contemporary commitments to particular representations and representational practices in cancer research, such as Temin's proto-virus ideas, the provirus theory, viral cancer research and collected materials, new molecular genetic methods, and prior commitments of funding agencies.[32]

Although one could assume that the evidence proved that Bishop and Varmus were correct in their claims about proto-oncogenes, a few dissenting voices have challenged the hegemony of the proto-oncogene theory. While there has been much support for the theory in the late 1980s and work continues within this frame in the early 1990s, a few cancer researchers and molecular biologists have openly criticized the proto-oncogene theory.[33] One set of criticisms launched at the proto-oncogene theory of cancer causation is based on the theme "If you have a hammer, every problem is a nail." That is, the introduction of new technology (including knowledge) in a particular line of work results in a corresponding theory of cancer causation.[34]

The commonalities in representations and rerepresentations—for example, from cancer as caused by viruses to cancer as caused by proto-oncogenes—do not demonstrate any ultimate progression in the history of science and scientific development. Neither do they demonstrate any necessary connections among the pasts, presents, and futures, or between contemporary situations (cf. Becker 1993). Instead, in each situation actors construct new hybrid representations, theories, and methods, by constructing networks of common practices. My description of historical continuity does not infer causality. It is instead an account of association between the past and the future. Histories and social studies of science have painstakingly shown that inference and prediction are not the stuff of writing histories. In this case, the success of a particular representation and explanation of cancer etiology is part of a particular construction of history and local practice. The Nobel Prize Committee's decision to award its 1989 prize in Physiology and Medicine to Bishop and Varmus is part of this historical construction and construction of history.

Change and Continuity in Representations and Technologies of Representation

This paper has provided a story of creation, change, and continuity in scientific representations. I treated conceptual change and continuity in science as embedded in individual and collective changes in scientific work organization and not only in their representations. I examined the connection between the process of theoretical or conceptual shifts and both the local and broad-scale organization of work and technical infrastructures of science.

The proto-oncogene theory was framed in "molecular genetic language," and new molecular genetic methods were used to probe questions originally framed in the "language" of other methods. While the new theory provides a *metaphoric* tying together of different lines of research, the work was done by many heterogeneous actors and objects using recombinant DNA and other molecular genetic technologies.

Representational tools create our visions of nature.[35] By adopting molecular genetic technologies, scientists are reconstructing the ways in which they organize nature. Given this frame, new molecular genetic technologies and the bandwagon of molecular genetic studies of cancer appear to have tremendous power to reconstruct representations of cancerous/abnormal and *normal* growth and development across laboratories and throughout multiple social worlds. Across laboratories, this power takes the form of generatively entrenched or embedded representations and materializations.

However, again, scientific ideas are not fixed. After a few years, the proto-oncogene theory and molecular genetic protocols themselves had been *reconstructed* by proto-oncogene researchers. The proto-oncogene theory has been adjusted many times over to fit with new experimental findings. Single point mutations are no longer considered to be sufficient to cause a cell to become cancerous. "Catalogues of genetic damage within individual tumors are taking shape, showing us how the malfunction of several different genes might combine to produce the malignant phenotype: for example, carcinomas of the colon contain no less than five different yet prevalent lesions—some genetically dominant, others recessive; carcinoma of the breast, at least five lesions; carcinoma of the lung, at least four; and neuroblastoma, at least three" (Bishop 1989:15). Another modification of the original proto-oncogene theory is the addition of "anti-oncogenes" or tumor suppressor genes introduced by Robert Weinberg.[36] For example, the retinoblastoma and p53 anti-oncogenes, that are thought to cause retinoblastomas and colon cancers if they are inactivated by point mutations, truncations, or deletions (Cooper 1990; Stanbridge 1990).[37] Inactivation of these anti-oncogenes is a proposed mechanism by which normal genes can become cancer-causing genes.

Similarly, the relationship between a representation and its interpretation or use, and therefore its consequences, is not fixed. Representations change depending on who is using and interpreting them, when, and for which purposes.[38] Indeed, the same person may interpret the notion of the gene in one way on one day and in another way on another day. As such, the study of representations should incorporate its diverse interpretations, renderings, and consequences. This highlights the active role of audiences. The consequences of representations—that is, realities—are the products of cooperation, conflict, negotiation, and sometimes power struggles between audiences and producers, between readers and authors. To understand science as a way of producing realities, we also have to understand this interaction.

There is one point that sets molecular biological research apart from previous biological research. Molecular genetic technologies can produce novel "natural" phenomena, including recombinant genes, proteins, and organisms. Previous generations of biologists have constructed novel research materials as laboratories/sites for housing experiments. These include inbred strains of animals, "nude" mice (with no thymus), tissue culture for experiments in petri dishes, and other research materials, most of which cannot survive outside of contrived laboratory conditions. Similarly, agricultural research has produced new "recombinants" or hybrids through crossbreeding strains and sometimes even species.[39] The difference here is that molecular genetics has produced a set of technologies that has changed the level and degree of the experimental production of nature. Molecular geneticists can create recombinants, first, that could not have been produced through "clumsy" crossbreeding methods and, second, that can survive outside of experimental laboratory conditions. It is the combination that makes the molecular genetic production of nature historically novel and creative of a new ecology of action.

NOTES

1. In another paper, I write about how scientists in controversies construct and employ histories of medicine, technology, and science to support their arguments or to deconstruct others' arguments (see Fujimura and Chou 1994). As an academic writing this (hi)story, I participate in a similar enterprise when I make my arguments. For discussions of the constructing of histories and other narratives, see, e.g., Becker 1986, Borofsky 1987, Clifford and Marcus 1986, Dening 1988, Haraway 1991a, Rosaldo 1989, and Traweek in press.

2. See also Star (this volume) on ecologies of knowledge and Strauss (1993) on continual permutations of action .

3. See Thomas and Znaniecki (1918), Hughes (1971a), and Shalin (1986) on indeterminacy and the definition of the situation. For other views of situated action in science and technology studies, see Haraway (1988), Lynch (1985a), and Suchman (1988).

4. See, for example, Callon (1986a,1986b), Callon and Latour (1992), Latour (1987, 1988c, 1989, 1990b, 1992a), and Law (1986a).

5. Latour (1992a) has summarized his model/ontology/theory and methodology for talking about science, society, history. Using the sociology of associations or actor-networks, he extends his frame beyond science to the historical development of our "society(ies)."

6. It should be noted that oncogene research is not all of cancer research. Clinical and epidemiological research with no ties to molecular biological methods continue. Molecular biological cancer research occupies only part of the vast cancer research world.

7. In my approach, which is part of the symbolic interactionist tradition in sociology, there are no a priori "conditions" for work or *a priori*

sets of constant variables. Instead, a situation (including what many sociologists might call its "context") is a contingent and interactional *achievement.*

8. Compare the proto-oncogene theory with Duesberg and Schwartz's (1992) "rare recombinants" theory of cancer causation.

9. RNA tumor viruses are retroviruses which have genes constituted of RNA sequences rather than DNA. They replicate by producing a strand of DNA sequences through the activities of an enzyme called reverse transcriptase. See Watson et al. (1987).

10. However, researchers did report suspected links between some human cancers and retroviruses. See especially Gallo (1986).

11. According to the researchers involved, the work that led to this announcement began much earlier. "The oncogene we studied was not discovered until 1970. By 1972 I had several people in the lab thinking about the oncogene itself and what it does. Our first and still most important experiments were done within that year. By 1974 we had found that that gene was a normal cellular gene as well. And that is the work for which we are still best known. That work was conceived of and done within that first year of beginning to think about oncogenes" (Bishop interview, 1984).

12. Evolutionary biology, and especially evolutionary genetics, is so embroiled in debates that oncogene researchers may succeed in this effort to propose a role for oncogenes in evolutionary biology. The units-of-selection debates so closely studied by philosophers of science are just one indication of the lack of consensus about the unit, levels, and processes by which selection and evolution occur. See, for example, Lloyd (1988) and Brandon and Birian (1984) for overviews and analyses of the units-of-selection debates.

13. Other suggestions of oncogenes as a source of genetic variation and as an indication of the course of evolution were made by Temin (1971, 1980, 1983) and by Walter Gilbert's research group (Schwartz et al. 1983), respectively.

14. It is often difficult to distinguish between molecular biologists and tumor virologists because, by now, their technologies are very similar and overlapping, if not identical. These researchers often call themselves by several field or disciplinary titles. I make the distinction here based on the techniques used and problems framed by the researchers at the beginning of their research programs. I base my distinction on the early training of the researchers studied. . . . As a caveat, I do not assume that disciplines are stable, bounded entities in nature or society. I agree with Keating et al. (1992), for example, that disciplinary boundaries are also constructed and

therefore can be destabilized. What molecular biology is, for instance, has changed from its "birth" through its imperialism into other realms of biological research and biological institutions. For example, in the late 1980s the University of California, Berkeley, reorganized its many biological subdisciplines into two general "divisions," "molecular and cell biology" and "integrated biology," in part because of the general molecularization of biology. Keating et al.'s (1992:315) view of disciplines as "dynamic, shifting stakes and not as purely static institutions" is similar to the definition of Strauss and colleagues (Bucher and Strauss 1961; Strauss 1978a, 1978b; Strauss et al. 1964) of social worlds as "negotiated orders." Indeed, the social world is defined as *activities and processes.*

15. These "normal" cells, called NIH 3T3 cells, are somewhat ambiguous cells. They are not entirely normal, since they have been "passaged" so many times in the laboratory. That is, the original cells taken from normal mouse tissue in the early 1960s have by now adapted to the artificial conditions of cell cultures (plates of agar filled with nutrients to feed them and antibiotics to prevent them from being infected with bacteria) and are no longer entirely normal. They are referred to as "immortalized cells" that might already be partially transformed. The scientists I interviewed acknowledged that cells from NIH 3T3 cell lines have already undergone the first step in transformation from normal cells to cancer cells. However, they continued to use these cells as the "normal" base from which to examine the processes of "transformation."

16. After these events had occurred, a science journalist on leave from the *New York Times* observed Weinberg's laboratory for six months in 1986. From interviews with Weinberg and his postdocs and graduate students, Natalie Angier (1988) reconstructed the events of 1981 and 1982 when a few laboratories raced for priority in cloning transfected human cancer genes. Although Angier is an obviously sympathetic observer in that laboratory, she presents another useful perspective of those events.

17. Weinberg's claims have since been modified. Current views are that at least two events, and perhaps up to eight events, are necessary to transform "truly normal" cells into cancer cells.

18. However, see DeVita (1984) and Gallo (1986).

19. Varmus later established his own laboratory also at the University of California, San Francisco, Medical Center, and is currently the director of the National Institutes of Health.

20. Wright (1986) gives a summary of technical developments that solved these difficulties between 1974 and 1976. See especially pages 320–24.

21. These new genes, proteins, and organisms are regarded by some as "beneficial," by others as potentially dangerous to the environment and human evolution, by even others as both. For discussions of the controversies about physical, social, and ethical hazards of recombinant DNA research, see, for example, Billings (1991), Davis (1991), Duster (1990), Jackson and Stich (1979), Haraway (1992a), Holtzman (1989), Hubbard (1990), Hubbard and Wald (1993), Juma (1989), Keller (1992), Kevles (1985), Kevles and Hood (1992), Kenney (1986, 1987), Lappe (1984), Lewontin (1991), Lippman (1988), Nelkin and Tancredi (1989), Suzuki and Knudtson (1990), Teitelman (1989), and Wright (1986). Since the inception of the Human Genome Initiative and its ELSI programs administered by the National Institutes of Health (NIH) and the Department of Energy (DOE), the human genome projects around the world have generated a new industry of genomics critical studies (conferences, articles, books, research projects) and their counterparts refuting these studies. This includes a conference with the amazing title of "Genes R Us: So Who is That?" in which I participated (held at the University of California Humanities Research Institute, Irvine, California, 2–3 May 1991).

22. NCI commitments to oncogenes and other molecular biological cancer research studies meant cutbacks in other basic research lines, as a 1983 article (Shapley 1983:5) in the journal *Nature* indicates. If NCI allowed spending on oncogene research to expand naturally, did this mean less prominence for important traditional fields such as chemotherapy? At the time, DeVita said that some other work must obviously go, given the fact that NCI was unlikely to receive any budget increases in the next few years. He noted that chemotherapy had been cut by about 30 percent in the past six years on scientific grounds: "some things we didn't need to do any more." As viewed by Alan Rabson, director of NCI's Division of Cancer Biology and Diagnosis, "If you understand oncogenes you may learn where to go in chemotherapy. It may open up whole new areas of chemotherapy." Although chemotherapy was the only area mentioned in the article, other areas of research were also neglected just by the fact that Congress did not increase their budget while oncogene research expanded. It also hints that chemotherapists would benefit by taking oncogenes as their starting point.

23. Hacking (1983) calls these kinds of changes in nature "intervening." See the section on probes and the last paragraph of this paper for further discussion on changing nature.

24. Cetus has since been acquired by Roche multinational corporation and was formally dissolved (except for a group incorporated into Chiron, another biotechnology firm in the San Francisco Bay Area) as of December 12, 1991.

25. See also Hacking (1992), Pickering (1990, 1994), and Star and Griesemer (1989) for discussions of malleable concepts.

26. These comments are based on interviews with many scientists, administrators, and students. See also Boffey (1986) and Moss (1989) on Sloan-Kettering's more general shift in research from traditional immunological to molecular immunological approaches to understanding cancer during that period.

27. See also Haraway (1992a) on the OncoMouse as cultural actor.

28. I do not imply that innovation occurs only at universities such as those named. In fact, innovation occurs everywhere, often not in universities. However, the finances and cultural capital (Bourdieu 1977) critical to building a network, to convincing others that one's innovation is worth pursuing, are often located in these "big name" universities.

29. For further discussions of tools and their ecologies, see Clarke and Fujimura (1992a). See also Rheinberger (1992) on experimental systems.

30. Another large part of the viral research funded by the national Virus Cancer Program focused on DNA tumor viruses.

31. Temin (1971) had earlier suggested a similar mechanism by which "proto-viruses" were formed as an intermediate stage in the process of the cell's normal genetic activities.

32. I do not regard the theory-method package as constituting a necessary connection. The coupling of the oncogene theory and recombinant DNA with other molecular biology technologies is constructed and not born in nature. The theory may in the future continue to exist as an entity separate from these techniques or coupled to another set of techniques. Similarly, the technologies are coupled with quite different theories in other lines of biological research.

33. See, for example, Duesberg (1983, 1985, 1987), Habeshaw (1983), Rubin (1983, 1985), Shubik (1983), and Teitelman (1985) for early arguments against these claims. See Temin (1983) for an early review of claims and counterclaims. See Duesberg and Schwartz (1992) for a recent critique of current data. See also Finch (1990: 446) for an example of other scientists' use of Duesberg's critique.

34. The following quote is an example of this type of criticism: "I believe the current rush to accept cellular oncogenes as the origin of human cancer . . . is at best premature. I have discussed previously some of the problems inherent in a simple genetic interpretation of cancer . . . , and others . . . have pointed out flaws in experimental design which raise serious questions of interpretation of the results that engendered the pre-

sent excitement. Further detailed criticism is unlikely to have much effect. It should be pointed out, however, that explanations for the origin of cancer have been varied and plentiful in this century. A limited list would include early theories of chromosomal alterations, virus infection, high glycolytic rates, and damaged grana (mitochondria), enzyme deletion, and reduced immunological surveillance. Each of these was carried to the fore by developments in a corresponding area of basic biology or biochemistry, and each time many were convinced that a final answer had been found. In retrospect, the supporting evidence always was strong, but it later turned out to be inadequate to establish causality. I believe we have confused advances in molecular biology and its attendant technology with deepened understanding of the nature of malignancy" (Rubin 1983:1170).

35. Donna Haraway (1991b) calls this process the "reinvention of nature."

36. See Weinberg (1988) for a nontechnical description of this concept.

37. See Duesberg and Schwartz (1992) for a critical view of these adjustments.

38. See, e.g., Rabinow (1986), Duster (1990), Fleck (1979 [1935]), Knorr-Cetina (1981), Latour (1988a), Lynch and Woolgar (1990), Star (1989a).

39. However, what constitutes a species and the boundaries of a species is still under debate by philosophers of science and by scientists.

Bibliography

Abbott, Andrew. 1988. *The System of Professions: An Essay on the Division of Expert Labor.* Chicago: University of Chicago Press.

Adas, Michael. 1989. *Machines as the Measure of Men: Science, Technology, and Ideologies of Western Dominance.* Ithaca, NY: Cornell University Press.

Addelson, Kathryn Pyne. 1991. *Impure Thoughts: Essays on Philosophy, Feminism and Ethics.* Phildelphia: Temple University Press.

Agate, Frederick J., Jr. 1973. "Philip Edward Smith." Pp. 472–78 in Charles Coulston Gillispie, ed. *Dictionary of Scientific Biography* 12. New York: Scribner.

Akrich, Madeleine. 1987. "Comment décrite les objects techniques." *Culture et Technique* 5:49–63.

Alcoff, Linda. 1988. "Cultural Feminism versus Post-Structuralism: The Identity Crisis in Feminist Theory." *Signs* 133:405–36.

Allen, Edgar. 1922. "The Oestrus Cycle in the Mouse." *American Journal Anat.* 30:297-372.

Allen, Edgar. 1926. "The Time of Ovulation in the Menstrual Cycle of the Monkey, M. Rhesus." *Proceedings of the Society for Experimental Biology and Medicine* 23:281–383.

Allen, Edgar. 1928. "Reactions of Immature Monkeys (M. Rhesus) to Injections of Ovarian Hormone." *Journal of Morphology and Physiology* 46:279–519.

Allen, Edgar, J. P. Pratt, Q. U. Newell and L. J. Bland. 1930. "Human Tubal Ova: Related Early Corpora Lutea and Uterine Tubes." *Carnegie Contributions in Embryology* 27:45-75.

Allen, Garland. 1975. "The Introduction of *Drosophila* into the Study of Heredity and Evolution, 1900–1910." *Isis* 66:322–33.

Allen, Garland. 1978. *Thomas Hunt Morgan: The Man and His Science.* Princeton, NJ: Princeton University Press.

Allen, Garland. 1979. "The Transformation of a Science: T. H. Morgan and the Emergence of a New American Biology," in Alexandra Oleson and John Voss, eds. *The Organization of Knowledge in Modern America,* 1869-1920. Baltimore, MD: Johns Hopkins University Press.

Allen, Garland. 1981a. "Morphology and Twentieth Century Biology: A Response." *Journal of the History of Biol.* 14:159–76.

Allen, Garland. 1981b. "Naturalists and Experimentalists: The Genotype and Phenotype." Pp. 179–209 in William Coleman and Camille Limoges, eds. *Studies in the History of Biology 3* . Baltimore, MD: Johns Hopkins University Press.

Alpers, Svetlana. 1983. *The Art of Describing: Dutch Art in the Seventeenth Century.* Chicago: University of Chicago Press.

Amann, Klaus and Karin Knorr-Cetina. 1988. "The Fixation of Visual Evidence." *Human Studies* 11:133–69.

Amoroso, E. C. and George Corner. 1975. "Herbert McLean Evans, 1882–1971. "*Biographical Memoires of the Fellows of the Royal Society of London* 21: 83–186.

Angier, N. 1988. *Natural Obsessions: The Search for the Oncogene.* Boston: Houghton Mifflin.

Annas, George J. and Sherman Elias. 1989. "The Politics of Transplantation of Human Fetal Tissue." *New England Journal of Medicine* 321:1609.

Anzaldúa, Gloria. 1987. *Borderlands/La Frontera: The New Mestiza.* San Francisco: Aunt Lute.

Applegate, J. and K. Day. August 7, 1984. "Computer Firm's Aim: Artificial Intelligence. *Los Angeles Times,* Part IV:1, 17.

Arditti, Rita, Pat Brennan and Steve Cavrak, eds. 1980. *Science and Liberation.* Boston: South End Press.

Arendt, Hannah. 1963. *Eichmann in Jerusalem: A Report on the Banality of Evil.* New York: Viking Press.

Arnold, Erik. 1984. *Computer-Aided Design in Europe.* Sussex: Sussex European Research Centre.

Aronowitz, Stanley. 1988. *Science as Power: Discourse and Ideology in Modern Society.* Minneapolis, MN: University of Minnesota Press.

Aronson, Naomi. 1984. "Science as a Claims-making Activity: Implications for Social Problems Research." Pp. 1–30 in Joseph Schneider, and John Kitsuse, eds. *Studies in the Sociology of Social Problems.* Norwood, NJ: Ablex Publishing Co.

"Artificial Intelligence Is Here: Computers That Mimic Human Reasoning Are Already at Work." July 9, 1984. *Business Week*: 54–57, 60–62.

Ashmore, Malcolm. 1985. "A Question of Reflexivity: Wrighting the Sociology of Scientific Knowledge." D.Phil. Dissertation, University of York.

Attenborough, David. 1980. *The Zoo Quest Expeditions.* London: Lutterworth.

Attewell, Paul and James Rule. 1984. "Computing and Organizations: What We Know and What We Don't Know." *Communications of the ACM* 27:1184–92.

Bachelard, Gaston. 1953. *Le materialism rationnel* . Paris: P. U. F.

Backman, Carl. 1982. "Institutional History." Unpublished manuscript.

Backman, Carl. 1983. "Resource Utilization in Biomedical Science: Patterns of Research on Nonhuman Primate Reproductive Physiology." Ph.D. Dissertation, Cornell University.

Bacon, Francis. 1960 [1620]. *The New Organon and Related Writings,* edited by F. H. Anderson. New York: Bobbs-Merrill.

Bakunin, M. 1970/1916. *God and the State.* Dover: New York.

Bakunin, M. 1980. *Bakunin on Anarchism.* Montréal: Black Rose Books.

Bang, Frederick B. 1977. "History of Tissue Culture at Johns Hopkins." *Bulletin of the History of Medicine* 51:516–37.

Barber, Bernard. 1952. *Science and the Social Order.* New York: The Free Press.

Barnes, Barry. 1977. *Interests and the Growth of Knowledge.* London: Routledge and Kegan Paul.

Barnes, Barry and David Bloor. 1982. "Relativism, Rationalism and the Sociology of Knowledge." Pp. 21–47 in Martin Hollis and Steven Lukes, eds. *Rationality and Relativism.* Oxford: Blackwell.

Barnes, Barry. 1982. *T. S. Kuhn and Social Science.* New York: Columbia University Press.

Bartelmez, George W. 1933. "Histological Studies on the Menstruation Mucous Membrane of the Human Uterus." *Carnegie Contributions in Embryology* 142:131–86.

Bartelmez, George W. 1935. "The Circulation in the Intervillous Space of the Macaque Placenta." *Anatomical Record* 61 Suppl. A:4.

Bartelmez, George W. 1937. "Menstruation." *Physiol. Rev.* 17:28–72.

Barthes, Roland. 1981. *Camera Lucida: Reflections on Photography*, trans. Richard Howard. New York: Hill and Wang.

Bastide, Francoise. 1990. "The Iconography of Scientific Texts: Principles of Analysis," trans. G. Myers. Pp. 187–230 in Michael Lynch and Steve Woolgar, eds. *Representation in Scientific Practice.* Cambridge, MA: MIT Press.

Baudrillard, Jean. 1983. *Simulations,* trans. Paul Foss, Paul Patton and Philip Beitchman. New York: Semiotext(e).

Baxandall, Michael. 1985. *Patterns of Invention: On the Historical Explanation of Pictures.* New Haven, CT: Yale University Press.

Beamont, Roland. 1968. *Phoenix into Ashes.* London: William Kimber.

Beamont, Roland. 1980. *Testing Years.* London: Ian Allen.

Becker, Howard. 1967. "Whose Side Are We On?" *Social Problems* 14: 239–47.

Becker, Howard S. 1984. *Art Worlds.* Berkeley: University of California Press.

Becker, Howard S. 1986. "Telling about Society." Pp. 121–35 in his *Doing Things Together: Selected Papers.* Evanston, IL: Northwestern University Press.

Becker, Howard S. 1993. " 'Foi por Acaso': Conceptualizing Coincidence." Unpublished paper, University of Washington, Seattle.

Becker, Howard and James Carper. 1956. "The Development of Identification with an Occupation." *American Journal of Sociology* 61: 289–98.

Bell, Daniel. 1979. "The Social Framework of the Information Society." Pp. 163–211 in Michael Dertouzos and Joel Moses, eds. *The Computer Age: A Twenty-Year View.* Cambridge, MA: MIT Press.

Ben-David, Joseph. 1965. "The Scientific Role: The Conditions of its Establishment in Europe." *Minerva* 4:15–54.

Bendifallah, Sallah and Walt Scacchi. 1987. "Understanding Software Maintenance Work." *IEEE Transactions on Software Engineering*. SE–13 3:311–23.

Bendiner, Robert. 1981. *The Fall of the Wild and the Rise of the Zoo*. New York: Dutton.

Beninger, James. 1986. *The Control Revolution: Technological and Economic Origins of the Information Society*. Cambridge, MA: Harvard University Press.

Benjamin, Walter. 1969. *Illuminations*. New York: Schocken.

Benson, Keith R. 1981. "Problems of Individual Development: Descriptive Embryological Morphology in America at the Turn of the Century." *Journal of the History of Biology* 14: 115–28.

Benson, Keith R. 1985. "American Morphology in the Late Nineteenth Century: The Biology Department at Johns Hopkins University." *Journal of the History of Biology* 18:163–205.

Bentley, Arthur. 1975. [1954]. "The Human Skin: Philosophy's Last Line of Defense," Pp. 195–211 in his *Inquiry into Inquiries: Essays in Social Theory*. S. Ratner, ed. Westport, CT: Greenwood Press.

Berman, Morris. 1984. *The Reenchantment of the World*. New York: Bantam Books.

Best, Joel. 1987. "Rhetoric in Claims-making: Constructing the Missing Children Problem." *Social Problems* 34:101–21.

Bijker, Wiebe, Thomas Hughes and Trevor Pinch, eds. 1986. *New Directions in the Social Study of Technology*. Cambridge, MA: MIT Press.

Billings, P. 1991. "How Many Genetic Diseases?" *Lancet* 338:1603–4.

Bishop, J.M. 1982. "Oncogenes." *Scientific American* 246:80–92.

Bishop, J.M. 1983. "Cellular Oncogenes and Retroviruses." *Annual Review of Biochemistry* 52: 301–54.

Bishop, J.M. December 8, 1989. "Retroviruses and Oncogenes II." Nobel Lecture, Stockholm, Sweden.

Bishop, J.M. and Varmus H.E. 1982. "Functions and Origins of Retroviral Transforming Genes." Pp. 999–1108 in R. Weiss et al., eds., *Molecu-*

lar Biology of Tumor Viruses, 2d ed.: RNA Tumor Viruses. Cold Spring Harbor, NY: Cold Spring Harbor Laboratory.

Bjelic, Dusan and Michael Lynch. 1992. "The Work of a Scientific Demonstration." Pp. 50–70 in Graham Watson and Robert Seiler, eds. in *Text in Context: Contributions to Ethnomethodology*. London: Sage.

Blake, John B. 1980. "Anatomy." P. 43 in Ronald L. Numbers, ed. *The Education of American Physicians: Historical Essays*. Berkeley: University of California Press.

Bledstein, Burton J. 1978. *The Culture of Professionalism*. New York: W. W. Norton.

Bloor, David. 1976. *Knowledge and Social Imagery*. London: Routledge and Kegan Paul.

Blumer, Herbert. 1969. "Social Movements." Pp. 8–29 in B. McLaughlin, ed. *Studies in Social Movements: A Social Psychological Perspective*. New York: Free Press.

Boffey, P. M. 1986. "Dr. Marks' Crusade: Shaking up Sloan-Kettering for a New Assault on Cancer." *New York Times Magazine* April 26:25–31, 60–67.

Bogen, J.E. and G.M. Bogen. 1976. "Wernicke's Region—Where Is It?"*Annals of the New York Academy of Sciences* 280:834–43.

Borell, Merriley. 1986. "Extending the Senses: The Graphic Method." *Medical Heritage* 2:114-21.

Borofsky, R. 1987. *Making History: Pukapukan and Anthropological Constructions of Knowledge*. New York: Cambridge University Press.

Borell, Merriley. 1987. "Instruments and an Independent Physiology: The Harvard Physiology Laboratory, 1871–1906." Pp. 350–371 in Gerald Geison ed. *Physiology in the American Context, 1850–1940*. Bethesda, MD: American Physiological Society.

Bourdieu, Pierre. 1977. *Outline of a Theory of Practice*. Cambridge: Cambridge University Press.

Bourne, Geoffrey, ed. 1973. *Nonhuman Primates and Medical Research*. New York: Academic.

Bowers, John. in press. "The Politics of Formalism," in Martin Lea, ed. *Issues in CSCW*.

Bowker, Geoffrey. 1987. "A Well-Ordered Reality: Aspects of the Development of Schlumberger, 1920–1939." *Social Studies of Science* 17: 611–55.

Bowker, Geoffrey. 1988. "Not Hung—Drawn and Quartered; Pictures from the Subsoil, 1939." Pp. 221–54 in Gordon Fyfe and J. Law, eds. *Picturing Power: Visual Depiction and Social Relations*. Sociological Review Monograph 35. London: Routledge.

Bowker, Geoffrey. 1994. *Science on the Run: Information Management and Industrial Geophysics at Schlumberger, 1920–1940*. Cambridge, MA: MIT Press.

Bowker, Geof and Bruno Latour. 1987. "A Booming Discipline Short of Discipline: (Social) Studies of Science in France." *Social Studies of Science* 17:715–48.

Brante, Thomas, Steve Fuller, and William Lynch. 1993. *Controversial Science: From Content to Contention*. Albany, NY: State University of New York Press.

Braverman, Harry. 1975. *Labor and Monopoly Capital: The Degradation of Work in the Twentieth Century*. New York: Monthly Review Press.

Brewer, Thomas H. 1971. "Disease and Social Class." Pp. 143–62 in Martin Brown, ed. *The Social Responsibility of the Scientist*. New York: The Free Press.

Bridges, William. 1966. *Gathering of Animals: An Unconventional History of the New York Zoological Society*. New York: Harper and Row.

Brighton Women and Science Group. 1980. *Alice Through the Microscope: The Power of Science over Women's Lives*. Linda Birke, Wendy Faulkner, Sandy Best, Deirdre Janson-Smith, Kathy Overfield, eds. London: Virago.

Broad, William J. 1987. "The Men Who Made the Sun Rise." Review of Richard Rhodes, *The Making of the Atomic Bomb*, New York: Simon & Schuster, 1987. *New York Times Book Review* February 8: 1,39.

Brosco, Jeffrey P. 1991. "Anatomy and Ambition: the Evolution of a Research Institute." *Transactions and Studies of the College of Physicians of Philadelphia* Ser. 5, 13:1–28.

Bucher, Rue. 1962. "Pathology: A Study of Social Movements Within a Profession." *Social Problems* 10:40–51.

Bucher, Rue and Anselm Strauss. 1961. "Professions in Process." *American Journal of Sociology* 66:325–34.

Buck, Frank, with Edward Anthony. 1930. *Bring 'Em Back Alive*. New York: Simon and Schuster.

Burian, Richard M. 1993. "How the Choice of Experimental Organism Matters: Epistemological Reflections on an Aspect of Biological Practice." *Journal of the History of Biology* 26:351–68.

Burnham, David. 1983. *The Rise of the Computer State*. New York: Random House.

Butler, Judith. 1990. *Gender Trouble : Feminism and the Subversion of Identity*. New York: Routledge.

Butler, Samuel. 1970 [1872]. *Erewhon*. Harmondsworth: Penguin.

Callon, Michel. 1986a. "The Sociology of an Actor-Network." Pp. 19–34 in Michel Callon, John Law and Arie Rip, eds. *Mapping the Dynamics of Science and Technology*. London: Macmillan.

Callon, Michel. 1986b. "Some Elements of a Sociology of Translation: Domestication of the Scallops and the Fishermen of St. Brieue Bay." Pp. 196–223 in John Law, ed. *Power, Action, and Belief: A New Sociology of Knowledge?* London: Routledge and Kegan Paul.

Callon, Michel. 1987b. "Society in the Making: The Study of Technology as a Tool for Sociological Analysis." Pp. 83–103 in Wiebe E. Bijker, Thomas P. Hughes, and Trevor Pinch, eds. *The Social Construction of Technological Systems*. Cambridge, MA: MIT Press.

Callon, Michel and John Law. 1989. "On the Construction of Sociotechnical Networks: Content and Context Revisited." *Knowledge and Society* 9:57–83.

Callon, Michel and Bruno Latour. 1992. "Don't Throw the Baby Out with the Bath School!" Pp. 343–68 in Andrew Pickering ed. *Science as Practice and Culture*. Chicago: University of Chicago Press.

Cambrosio, Alberto and Peter Keating. 1988. "'Going Monoclonal': Art, Science and Magic in the Day-To-Day Use of Hybridoma Technology." *Social Problems* 35: 244–60.

Camilleri, Joseph A. 1976. *Civilization in Crisis*. Cambridge: Cambridge University Press.

Campbell, Donald T. 1987. "Guidelines for Monitoring the Scientific Competence of Preventive Intervention Research Centers: An Exercise in the Sociology of Scientific Validity." *Knowledge* 8:389–430.

Campbell, John. 1983. *Roy Jenkins, a Biography*. London: Weidenfeld and Nicolson.

Carroll , John M. and Robert L. Campbell. 1986. "Softening Up Hard Science: Reply to Newell and Card."*Human-Computer Interaction* 2:227–49.

Carroll, P. Thomas. 1986. "American Science Transformed." *American Scientist* 74:466–85.

Cartwright, Lisa. 1992. " 'Experiments of Destruction': Cinematic Inscriptions of Physiology." *Representations* 40:129–52.

Casper, Monica. 1994a. "At the Margins of Humanity: Fetal Positions in Science and Medicine." *Science, Technology and Human Values* 19:307–23.

Casper, Monica. 1994b. "Reframing and Grounding Nonhuman Agency: What Makes a Fetus an Agent." *American Behavioral Scientist* 37:839–56.

Casper, Monica. In press. "Fetal Cyborgs and Technomoms on the Reproductive Frontier, or Which Way to the Carnival?" In Chris Hables Gray, Heidi Figueroa-Sarriera and Steven Mentor, eds. *The Cyborg Handbook*. New York: Routledge.

Cawelti, John. 1976. *Adventure, Mystery, and Romance. Formula Stories as Art and Popular Culture*. Chicago: University of Chicago Press.

Chubin, Daryl and Ellen Chu. 1989. *Science Off the Pedestal : Social Perspectives on Science and Technology*. Belmont, CA: Wadsworth Publishing.

Churchill, Frederick B. 1981. "In Search of the New Biology: An Epilogue." *Journal of the History of Biology* 14:177–91.

Clark, M. and D. Witherspoon. 1984. "Cancer: The Enemy Within." *Newsweek*, March 5:66–67.

Clark, M., M. Gosnell, D. Shapiro and M. Hager. 1984. "Medicine: A Brave New World." *Newsweek*, March 5:64–70.

Clarke, Basil. 1965.*Supersonic Flight*. London: Frederick Muller.

Clarke, Adele E. 1985. "Emergence of the Reproductive Research Enterprise: A Sociology of Biological, Medical and Agricultural Science in the United States, 1910–1940." Ph.D. Dissertation, University of California, San Francisco.

Clarke, Adele E. 1987. "Research Materials and Reproductive Science in the United States, 1910–1940." Pp. 323–50 in Gerald L. Geison ed.

Physiology in the American Context, 1850–1940. Bethesda, MD: American Physiological Society.

Clarke, Adele E. 1990. "Controversy and the Development of Reproductive Sciences." *Social Problems* 37:18–37.

Clarke, Adele E. 1991. "Social Worlds/Arenas Theory as Organizational Theory." Pp. 119–158 in David Maines, ed. *Social Organization and Social Process: Essays in Honor of Anselm Strauss.* Hawthorne, NY: Aldine de Gruyter.

Clarke, Adele E. In press a. *Disciplining Reproduction: Modernity, the American Life Sciences and the 'Problem of Sex.'* Berkeley, CA: University of California Press.

Clarke, Adele E. In press b. "Modernity, Postmodernity and Reproduction, 1890–1993, or 'Mommy, Where Do Cyborgs Come from Anyway?' " In Gray, Chris Hables, Heidi Figueroa-Sarriera and Steven Mentor, eds. *The Cyborg Handbook.* NY: Routledge.

Clarke, Adele E. and Joan H. Fujimura, eds. 1992b. "What Tools? Which Jobs? Why Right?" Pp. 3-44 in Adele E. Clarke and Joan H. Fujimura, eds. *The Right Tools For the Job: At Work in Twentieth Century Life Sciences.* Princeton: Princeton University Press.

Clarke, Adele and Joan H. Fujimura, eds. 1992a. *The Right Tools for the Job: At Work in 20th Century Life Sciences.* Princeton: Princeton University Press.

Clarke, Adele and Theresa Montini. 1993. "The Many Faces of RU486: Tales of Situated Knowledges and Technological Contestations." *Science, Technology and Human Values* 18:42–78.

Clause, Bonnie Tocher. 1993. "The Wistar Rat as a Right Choice: Establishing Mammalian Standards and the Ideal of a Standardized Mammal." *Journal of the History of Biology* 26:329–50.

Clifford, James and George E. Marcus, eds. 1986. *Writing Culture: The Poetics and Politics of Ethnography.* Berkeley: University of California Press.

Coalition Working Group on Teaching and Learning. 1993. "Call for Project Descriptions." (April). Available by ftp from cni.org in /cniftp/calls/netteach/netteach.txt'.

Code, Lorraine. 1991. *What Can She Know? Feminist Theory and the Construction of Knowledge.* Ithaca, NY: Cornell University Press.

Coghill, G. E. 1939. "Studies on Rearing the Opossum Delphys Virginiana." *Ohio Journal of Science* 39:329–349.

Cole, H. H. and G. H. Hart. 1930. "The Potency of Blood Serum of Mares in Progressive Stages of Pregnancy in Effecting the Sexual Maturity of the Immature Rat." *American Journal of Physiology* 94:37–68.

Cole, Ralph I. 1972. "Some Reflections Concerning the Future of Society, Computers and Education." Pp. 135–45 in Robert Lee Chartrand, ed. *Computers in the Service of Society*. New York, Pergamon Press.

Coleman, William. 1977. *Biology in the Nineteenth Century: Problems of Form, Function and Transformation*. New York: Cambridge University Press.

Coleman, William. 1985. "The Cognitive Basis of the Discipline: Claude Bernard on Physiology." *Isis* 76:49–70.

Collins, Harry. 1981. "Introduction," to *Knowledge and Controversy: Studies in Modern Natural Science*. Special issue of *Social Studies of Science* 11:1.

Collins, Harry. 1985. *Changing Order: Replication and Induction in Scientific Practice*. London: Sage.

Collins, Harry M. 1987a. "Expert Systems and the Science of Knowledge." Pp. 329–48 in Wiebe Bijker, Thomas P. Hughes and Trevor J. Pinch, eds. *The Social Construction of Technological Systems: New Directions in the Sociology and History of Technology*. Cambridge, MA: MIT Press.

Collins, Harry M. 1987b. "Expert Systems, Artificial Intelligences, and the Co-ordinates of Action."Pp. 258–82 in Brian Bloomfield, ed. *The Question of Artificial Intelligence: Philosophical and Sociological Perspectives*. London: Croom Helm.

Collins, Harry M. and Trevor Pinch. 1982. *Frames of Meaning: The Social Construction of Extraordinary Science*. London: Routledge and Kegan Paul.

Collins, Patricia Hill. 1986. "Learning from the Outsider Within: The Sociological Significance of Black Feminist Thought." *Social Problems* 33: 514–32.

Collins, Randall and Sal Restivo. 1983. "Development, Diversity, and Conflict in the Sociology of Science." *Sociological Quarterly* 2:185–200.

Cooley, M. J. E. 1980. "Some Social Implications of CAD." Pp. 97–116 in J. Mermont, ed. *CAD in Medium-sized and Small Industries*. Amsterdam: North-Holland.

Cooper, G. M. 1982. "Cellular Transforming Genes." *Science* 217:801–6.

Corner, George. 1923. "Ovulation and Menstruation in Macacus Rhesus." *Carnegie Contributions to Embryology* 73:73–110.

Corner, George W. 1981. *Seven Ages of a Medical Scientist: An Autobiography.* Philadelphia, PA: University of Pennsylvania Press.

Coulam, R. F. 1977. *Illusions of Choice: The F 111 and the Problem of Weapons Acquisition Reform.* Princeton: Princeton University Press.

Coulter, Jeff. 1979. *The Social Construction of Mind.* London: Macmillan.

Coulter, Jeff. 1983. *Rethinking Cognitive Theory.* London and New York: Macmillan/St. Martin's Press.

Coulter, Jeff. 1989. *Mind in Action.* Oxford: Polity Press.

Cowan, Ruth S. 1983. *More Work for Mother.* New York: Basic Books.

Cozzens, Susan E. and Thomas F. Gieryn, eds. 1990. *Theories of Science in Society.* Bloomington, IN: Indiana University Press.

Crandall, Lee S., in collaboration with William Bridges. 1966. *A Zoo Man's Notebook.* Chicago: University of Chicago Press.

Crane, Diana. 1972. *Invisible Colleges: The Diffusion of Knowledge in Scientific Communities.* Chicago: University of Chicago Press.

Crichton, Michael. 1990. *Jurassic Park.* New York: Alfred A. Knopf.

Crossman, Richard. 1975. *Minister of Housing 1964–66: The Diaries of a Cabinet Minister,* vol. 1. London: Hamish Hamilton and Jonathan Cape.

Curtis, Bill, Herb Krasner and Neil Iscoe. 1988. "A Field Study of the Software Design Process for Large Systems." *Communications of the ACM* 31:1268–87.

Cyert, Richard. 1984. "New Teacher's Pet: The Computer." *IEEE Spectrum* 21:120–22.

Daly, Mary. 1978. *Gyn/Ecology: The Metaethics of Radical Feminism.* Boston: Beacon Press.

Daly, Mary. 1985. *Beyond God the Father: Toward a Philosophy of Women's Liberation.* Boston: Beacon Press.

Danziger, James. 1977. "Computers, Local Government, and The Litany to EDP." *Public Administration Review* 37:28–37.

Danziger, James, William Dutton, Rob Kling, and Kenneth Kraemer. 1982. *Computers and Politics: High Technology in American Local Governments.* New York: Columbia University Press.

Davis, J. 1991. *Mapping the Code: The Human Genome Project and the Choices of Modern Science.* New York: Wiley.

Dawson, Alden B. and Harry B. Friedgood. 1940. "The Time and Sequence of Preovulatory Changes in the Cat Ovary after Mating or Mechanical Stimulation of the Cervix Uter," *Anatomical Record* 76:411–29.

DeMillo, Richard A., Richard J. Lipton and Alan J. Perlis. 1979. "Social Processes and Proofs of Theorems and Programs." *Communications of the ACM* 22 :271–80.

Dening, G. 1988. *History's Anthropology: The Death of William Gooch.* (ASAO special publications; number 2). Lanham, MD: University Press of America.

Denning, Peter, Douglas Comer, David Gries, Michael Mulder, Allen Tucker, Joe Turner and Paul Young. 1989. "Computing as a Discipline." *Communications of the ACM* 32:9–23.

Denzin, Norman. 1989. *Interpretive Interactionism.* Newbury Park, CA: Sage.

Der, C. J., T. G. Krontiris and G. M. Cooper. 1982. "Transforming Genes of Human Bladder and Lung Carcinoma Cell Lines Are Homologous to the Ras Genes of Harvey and Kirsten Sarcoma Viruses." *Proceedings of the National Academy of Sciences* 79:3637–40.

DeVita, V. T. 1984. "The Governance of Science at the National Cancer Institute: A Perspective on Misperceptions." Pp. 1–5 in *Management Operations of the National Cancer Institute That Influence the Governance of Science, National Cancer Institute Monograph 64.* Bethesda, MD: U.S. Department of Health and Human Services (NIH Publication No. 84–2651).

Dewey, John. 1981. [1896] "The Reflex Arc Concept in Psychology." Pp. 136–48 in J. J. McDermott, ed. *The Philosophy of John Dewey.* Chicago: University of Chicago Press.

Dickson, David. 1979. "Science and Political Hegemony in the 17th Century." *Radical Science Journal* 8:7–37.

Dickson, David. 1984. *The New Politics of Science.* New York: Pantheon Books.

Dickson, David. 1988. *The New Politics of Science.* Chicago, IL: The University of Chicago Press.

Diddle, A. W. and T. H. Burford. 1935. "A Study of a Set of Quadruplets." *Anatomical Record* 61:282.

di Leonardo, Micaela. 1991. "Introduction: Gender, Culture, and Political Economy: Feminist Anthropology in Historical Perspective." Pp.

1–48 in M. di Leonardo, ed. *Gender at the Crossroads of Knowledge: Feminist Anthropology in the Postmodern Era*. Berkeley: University of California Press.

Dizard, Wilson. 1982. *The Coming Information Age*. New York: Longman.

Doolittle, R. F., M. W. Hunkapiller, L. E. Hood, S. G. DeVare, K. C. Robbins, et al. 1983. "Simian Sacroma Virus *Onc* Gene, V-*Sis*, Is Derived from the Gene (Or Genes) Encoding a Platelet-Derived Growth Factor." *Science* 221:275–76.

Downey, Gary. 1986. "Ideology and the Clamshell Identity: Organizational Dilemmas in the Anti-nuclear Power Movement." *Social Problems* 33:357–73.

Downey, Gary Lee, Joseph Dumit, and Sarah Williams. In press. "Granting Membership to the Cyborg Image." In Chris Hables Gray, Heidi Figueroa-Sarriera and Steven Mentor, eds. *The Cyborg Handbook*. New York: Routledge.

Dreyfus, Hubert. 1965. *Alchemy and Artificial Intelligence*. Santa Monica, CA: Rand Corporation.

Dreyfus, Hubert. 1979. *What Computers Can't Do: The Limits of Artificial Intelligence*. 2d ed. New York: Basic Books.

Dreyfus, Hubert and Stuart E. Dreyfus. 1986. *Mind Over Machine: The Power of Human Intuition and Expertise in the Era of the Computer*. Oxford: Blackwell.

Duden, Barbara. 1993. *Disembodying Women: Perspectives on Pregnancy and the Unborn*. Trans. Lee Hoinacki. Cambridge, MA: Harvard University Press.

Duesberg, P. H. 1983. "Retroviral Transforming Genes in Normal Cells?" *Nature* 304:219–25.

Duesberg, P. H. 1985. "Activated Proto-onc Genes: Sufficient or Necessary for Cancer?" *Science* 228:669–77.

Duesberg, Peter. 1987. "Retroviruses as Carcinogens and Pathogens: Expectations and Reality." *Cancer Research* 47:1199–2200.

Duesberg, P. H. and J. Schwartz. 1992. "Latent Viruses and Mutated Oncogenes: No Evidence for Pathogenicity." *Progress in Nucleic Acid Research and Molecular Biology* 43:135–204.

Dugdale, Anni and Joan Fujimura, eds. *Making Sex, Fabricating Bodies: Gender and the Construction of Knowledge in the Biomedical Sciences,* in prep.

Dunlop, Charles and Rob Kling, eds. 1991. *Computerization and Controversy: Value Conflicts and Social Choices.* San Diego: Academic Press.

Durkheim, Émile. 1961 [1912] *The Elementary Forms of the Religious Life.* New York: Collier Books.

Durrell, Gerald. 1964. *A Zoo in My Luggage* . Middlesex, UK: Penguin.

Duster, Troy. 1990. *Eugenics Through the Back Door.* Berkeley: University of California Press.

Easlea, Brian. 1983. *Fathering the Unthinkable: Masculinity, Scientsts, and the Nuclear Arms Race.* London: Pluto Press.

Edge, David and Michael Mulkay. 1976. *Astronomy Transformed: The Emergence of Radio Astronomy in Britain.* New York: Wiley.

Edgerton, Samuel. 1975. *The Renaissance Rediscovery of Linear Perspective.* New York: Harper & Row.

Elzinga, Aant. 1993. "Science as the Continuation of Politics by Other Means." Pp. 127–52 in T. Brante, et al., eds. *Controversial Science: From Content to Contention.* Albany, NY: State University of New York Press.

Elzinga, Aant and Ingemar Bohlin. 1989. "The Politics of Science in Polar Regions." *Ambio* 181:71–77.

Engeström, Yrjö. 1990. "When Is a Tool? Multiple Meanings of Artifacts in Human Activity." Pp. 171–95 in his *Learning, Working and Imagining.* Helsinki: Orienta-Konsultit Oy.

Evans, John. W. and Richard A. Knisely, 1970. *Integrated Municipal Information Systems: Some Potential Impacts.* Washington, D.C.: U.S. Department of Housing and Urban Development.

Eyestone, Willard H. 1966. "Scientific and Administrative Concepts Behind the Establishment of the U.S. Primate Center." Pp. 1–9 in R. N. Fienes, ed. *Some Recent Developments in Comparative Medicine. Symposium of the Zoological Society of London* . Number 17. London: Academic Press for Zoological Society.

Farber, Paul L. 1982a. *The Emergence of Ornithology as a Scientific Discipline.* Boston: D. Reidel.

Farber, Paul L. 1982b. "The Transformation of Natural History in the Nineteenth Century." *Journal of the History of Biology* 15:145–52.

Fausto-Sterling, Anne. 1992. *Myths of Gender: Biological Theories of Men and Women*. New York: Basic Books.

Fee, Elizabeth. 1981. "Is There a Feminist Science?" *Science and Nature* 4:46–57.

Fee, Elizabeth. 1983. "Women's Nature and Scientific Objectivity." Pp. 9–27 in Marian Lowe and Ruth Hubbard, eds. *Woman's Nature*. New York: Pergamon Press.

Feigenbaum, Edward and Pamela McCorduck. 1983. *Fifth Generation: Artificial Intelligence and Japan 's Challenge to the World*. Reading MA: Addison-Wesley.

Fetzer, James. 1988. "Program Verification: The Very Idea." *Communications of the ACM* 31:1048–63.

Feyerabend, Paul K. 1975. *Against Method*. London:Verso.

Feyerabend, Paul K. 1978. *Science in a Free Society*. London:Verso.

Finch, C. E. 1990. *Longevity, Senescence, and the Genome*. Chicago: University of Chicago Press.

Fleck, Ludwig. 1979 [1935]. *Genesis and Development of a Scientific Fact*. Chicago: University of Chicago Press.

Forsythe, Diana. 1992. "Blaming the User in Medical Informatics: The Cultural Nature of Scientific Practice." *Knowledge and Society* 9: 95–111.

Forsythe, Diana. 1993. "Engineering Knowledge: The Construction of Knowledge in Artificial Intelligence." *Social Studies of Science* 23:445–77.

Foss, Daniel A. and Ralph Larkin. 1986. *Beyond Revolution: A New Theory of Social Movements*. South Hadley, MA: Bergen and Garvey Publishers.

Foucault, Michel. 1972. *The Archeology of Knowledge*, trans. Alan Sheridan. New York: Random House.

Foucault, Michel. 1979. *Discipline and Punish: The Birth of the Prison*, trans. Alan Sheridan. New York: Random House.

Foucault, Michel. 1984. *The Foucault Reader*, ed. Paul Rabinow. New York: Pantheon.

Foulk, P. W. 1984. "CAD in Electronics." *Computer Aided Design* 16:166–71.

Fox, Daniel M. and Christopher Lawrence. 1989. *Photographing Medicine: Images and Power in Britain and America Since 1840*. New York: Greenwood Press.

Fox, Renee and Judith Swazey, with Judith Watkins. 1992. *Spare Parts: Organ Replacement in American Society*. New York: Oxford University Press.

Freeman, E. M. 1984. "Computer Aided Engineering For Electrical Engineers." *Proceedings of the International Conference on Computer Aided Engineering*. Conference Publication Number 243. Coventry, UK: Venue.

Freeman, Jo. 1975. *The Politics of Women's Liberation: A Case Study of an Emerging Social Movement and Its Relation to the Policy Process.* New York : McKay.

Friedson, Eliot. 1970. *Profession of Medicine*. New York: Dodd, Mead and Co.

Freidson, Eliot. 1976. "The Division of Labor at Social Interaction." *Social Problems* 23:304–13.

Friedson, Eliot. 1986. *Professional Power: A Study of the Institutionalization of Formal Knowledge.* Chicago: University of Chicago Press.

Fujimura, Joan. 1986. "Bandwagons in Science: Doable Problems and Transportable Packages as Factors in the Development of the Molecular Genetic Bandwagon in Cancer Research." Ph.D. Dissertation, University of California, Berkeley.

Fujimura, Joan H. 1987. "Constructing 'Do-able' Problems in Cancer Research: Articulating Alignment." *Social Studies of Science* 17:257–93.

Fujimura, Joan H. 1992a. "Crafting Science: Standardized Packages, Boundary Objects, and 'Translation.'" Pp. 168–214 in Andew Pickering, ed. *Science as Practice and Culture*. Chicago: University of Chicago Press.

Fujimura, Joan H. 1992b. "On Methods, Ontologies, and Representation in the Sociology of Science: Where Do We Stand?" In David Maines, ed. *Social Organization and Social Process: Essays in Honor of Anselm L. Strauss*. Hawthorne, NY: Aldine de Gruyter.

Fujimura, Joan H. and D. Chou. 1994. "Dissent in Science: Styles of Scientific Practice and the Controversy over the Cause of AIDS." *Social Science and Medicine* 38:1017–36.

Fuller, Steven, M. de Mey, T. Shinn and S. Woolgar, eds. 1989. *The Cognitive Turn: Sociological and Psychological Perspectives on Science.* Dordrecht: D. Reidel.

Fulton, John F. and Leonard G. Wilson. 1966. *Selected Readings in the History of Physiology*, 2d ed. Springfield, IL: Thomas.

Furmanski, P., J. C. Hager and M. A. Rich, eds. 1985. *RNA Tumor Viruses, Oncogenes, Human Cancer and AIDS: On the Frontiers of Understanding. Proceedings of the International Conference on RNA Tumor Viruses in Human Cancer (Denver, Colorado, June 10–14, 1984).* Boston: Martinus Nijhoff.

Gadamer, Hans-Georg. 1984. *Truth and Method.* New York: Crossroad.

Galison, Peter. 1987. *How Experiments End.* Chicago: University of Chicago Press.

Galison, Peter. 1989. "The Trading Zone: Coordination Between Experiment and Theory in the Modern Laboratory." Paper presented at the International Workshop in the Place of Knowledge, Tel Aviv and Jerusalem, May 15–18.

Gallo, Robert C. 1986. "The First Human Retrovirus." *Scientific American* 255:88–98.

Gardner, Richard. 1981. *British Aircraft Corporation: A History.* London: Batsford.

Garfinkel, Harold, ed. 1986. *Ethnomethodological Studies of Work.* London: Routledge and Kegan Paul.

Garfinkel, Harold, L. Livingston, M. Lynch, D. MacBeth, and A. B. Robillard. 1989. "Respecifying the Natural Sciences as Discovering Sciences of Practical Action, I and II: Doing So Ethnographically by Administering a Schedule of Contingencies in Discussions with Laboratory Scientists and by Hanging Around Their Laboratories." Unpublished paper, Department of Sociology, University of California, Los Angeles.

Garfinkel, Harold, Michael Lynch, and Eric Livingston. 1981. "The Work of a Discovering Science Construed with Materials from the Optically Discovered Pulsar." *Philosophy of the Social Sciences* 11:131–58.

Garry, Ann and Marilyn Pearsall, eds. 1989. *Women, Knowledge, and Reality: Explorations in Feminist Philosophy.* Boston, MA: Unwin Hyman.

Gasser, Les. 1986. "The Integration of Computing and Routine Work." *ACM Transactions on Office Information Systems* 4:205–25.

Geison, Gerald L. 1978. *Michael Foster and the Cambridge School of Physiology: The Scientific Enterprise in Late Victorian Society.* Princeton, NJ: Princeton University Press.

Geison, Gerald L. 1979. "Divided We Stand: Physiologists and Clinicians in the American Context." Pp. 67–90 in Morris J. Vogel and Charles Rosenberg, eds. *The Therapeutic Revolution: Essays in the Social History of American Medicine.* Philadelphia, PA: University of Pennsylvania Press.

Geison, Gerald. 1981. "Scientific Change, Emerging Specialities and Research Schools." *History of Science* 19:20–40.

Geison, Gerald, ed. 1983. *Professions and Professional Idelogy in America.* Chapel Hill: University of North Carolina Press.

Geller, E. 1964. *Thought and Change.* Chicago: University of Chicago Press.

Gerson, E. 1976. "On Quality of Life." *American Sociological Review* 41:793–806.

Gerson, E. 1982. "The Realignment of Population Biology, 1880–1925." Paper presented at the Society for Social Studies of Science, Philadelphia, PA.

Gerson, E. 1983a."Scientific Work and Social Worlds." *Knowledge* 4:357–77.

Gerson, E. 1983b. "Styles of Scientific Work and the Population Realignment in Biology, 1880–1925." Paper presented at the Conference on History and Philosophy of Biology, Denison University, Granville, OH.

Gerson, E. and Susan Leigh Star. 1986. "Analyzing Due Process in the Workplace."*ACM Transactions on Office Information Systems* 4:257–70.

Gerson, E., and Susan Leigh Star. 1988. "Representations and Re-representations in Scientific Work." Unpublished paper, Tremont Research Institute, San Francisco.

Gibson, James J. 1986. *The Ecological Approach to Visual Perception.* London and Hillsdale, NJ: Lawrence Erlbaum.

Gieryn, Thomas. 1992. "Science as a Social Problem." Pp. 21–33 in Craig Calhoun and George Ritzer, eds. *Social Problems.* New York: McGraw Hill.

Ginzberg, Ruth. 1989. "Uncovering Gynocentric Science." Pp. 64–84 in Nancy Tuana ed. *Feminism and Science.* Bloomington: Indiana University Press.

Giuliano, Vincent. 1982. "The Mechanization of Office Work." *Scientific American* 247:148–64.

Glaser, Barney. 1964. *Organizational Scientists.* Indianapolis: Bobbs- Merrill.

Glaser, Barney. 1978. *Theoretical Sensitivity: Advances in the Methodology of Grounded Theory.* Mill Valley, CA: Sociology Press.

Glaser, Barney and Anselm Strauss. 1967. *The Discovery of Grounded Theory.* Chicago: Aldine.

Glass, Fred. 1989. "The 'New Bad Future': *Robocop* and 1980s Sci-Fi Films." *Science as Culture* 5:7–49.

Godwin, William. 1971. *Enquiry Concerning Political Justice.* New York: Oxford University Press.

Goffman, Erving. 1963. *Stigma: Notes on the Management of Spoiled Identity.* Englewood Cliffs, NJ: Prentice-Hall.

Goffman, Erving. 1967. "Where the Action Is." Pp. 149–270 in his *Interactional Ritual: Essays on Face-to-Face Behavior.* Garden City, NY: Doubleday.

Goguen, Joseph. 1992. "The Dry and the Wet." Pp. 1-17 in Eckhard Falkenberg, Rolland Colette, and El-Sayed El-Sayed Nasr-El-Dein, eds. *Information Systems Concepts.* Amsterdam: Elsevier North-Holland.

Goldfarb, M., K. Shimizu, M. Perucho and M. Wigler. 1982. "Isolation and Preliminary Characterization of a Human Transforming Gene from T24 Bladder Carcinoma Cells." *Nature* 296:404–9.

Gombrich, E. H. 1960. *Art and Illusion.* Princeton, NJ: Princeton University Press.

Gooding, David. 1986. "How Do Scientists Reach Agreement about Novel Observations?" *Studies in History and Philosophy of Science* 17:205–30.

Gooding, David. 1988. "Mapping Experiment as a Learning Process." Paper presented at the Annual Meeting of the Society for Social Studies of Science, Amsterdam.

Goodwin, Michael. 1988. "Wild-man Warren." *PC World* 6 (January):108–9, 114.

Graham, Loren, Wolf Lepenies, and Peter Weingart, eds. 1983. *Functions and Uses of Disciplinary Histories*. Boston: Kluwer.

Gray, Chris Hables, Heidi Figueroa-Sarriera and Steven Mentor, eds. In press. *The Cyborg Handbook*. New York: Routledge.

Green, Jerry E. 1985. *The Planning and Management of Zoological Parks: A Selected Annotated Bibliography*. Monticello, IL: Vance Bibliographies.

Griffin, Susan. 1978. *Woman and Nature: The Roaring Inside Her*. New York: Harper & Row.

Griffith, John Q. 1942. *The Rat in Lab Investigations*. Philadelphia, PA: Lippincott.

Guerrini, Anita. 1993. "Animal Tragedies: The Moral Theater of Anatomy, 1660–1750." Paper presented at the History of Science Society Meetings, Santa Fe.

Gumplowicz, Ludwig. 1905. *Grundrisse der Soziologie*. Vienna: Manz.

Gunston, Bill. 1974. *Attack Aircraft of the West*. London: Ian Allen.

Gurwitsch, Aron. 1964. *The Field of Consciousness*. Pittsburgh: Duquesne University Press.

Gusfield, Joseph. 1981. *The Culture of Public Problems*. Chicago: University of Chicago Press.

Habeshaw, J. A. 1983. "Letter to Nature." *Nature* 301:652.

Hacker, P. M. S. 1987. *Appearance and Reality*. Oxford: Blackwell.

Hacker, Sally. 1990. *Doing It the Hard Way: Investigations of Gender and Technology*. Dorothy Smith and Susan Turner, eds. Boston, MA: Unwin Hyman.

Hacking, Ian. 1983. *Representing and Intervening: Introductory Topics in the Philosophy of Natural Science*. Cambridge: Cambridge University Press.

Hacking, Ian. 1992. "The Self-Vindication of the Laboratory Sciences." Pp. 29–64 in Andrew Pickering, ed. *Science as Practice and Culture.* Chicago: University of Chicago Press.

Hagen, Joel B. 1990. "Problems in the Institutionalization of Tropical Biology: The Case of the Barro Colorado Island Biological Laboratory." *History and Philosophy of the Life Sciences* 12:225–37.

Hales, Mike. 1974. "Management Science and the 'Second Industrial Revolution.'" *Radical Science Journal* 1:5-28.

Hall, Rogers. 1989. "Computational Approaches to Analogical Reasoning: A Comparative Analysis." *Artificial Intelligence* 39: 39–120.

Hall, Rogers. 1990. "Making Math on Paper: Constructing Representations of Stories about Related Linear Functions." Ph.D. Dissertation, University of California, Irvine.

Hamlett, G. W. D. 1935. "Primordial Germ Cells in a 4.3-M. M. Human Embryo." *Anatomical Record* 61:273.

Haraway, Donna. 1984. "Signs of Dominance: From a Physiology to a Cybernetics of Primate Society: C. R. Carpenter, 1930–1970." *Studies in the History of Biology* 6:129–219.

Haraway, Donna. 1985. "Manifesto for Cyborgs: Science, Technology and Socialist Feminism in the 1980s." *Socialist Review* 152:65–107.

Haraway, Donna. 1988. "Situated Knowledges: The Science Question in Feminism and the Privilege of Partial Perspective." *Feminist Studies* 14:575–99.

Haraway, Donna. 1989. *Primate Visions: Gender, Race and Nature in the World of Modern Science.* New York: Routledge.

Haraway, Donna. 1991a. "The Past is a Contested Zone: Human Nature and Theories of Production and Reproduction in Primate Behavior Studies." Pp. 21–42 in her *Simians, Cyborgs and Women: The Reinvention of Nature.* New York: Routledge.

Haraway, Donna. 1991b. *Simians, Cyborgs and Women: The Reinvention of Nature.* New York: Routledge.

Haraway, Donna. 1992a. "When Man Is on the Menu." Pp. 38–43 in J. Crary and S. Kwinter, eds. *Incorporations.* New York: Zone Press.

Haraway, Donna. 1992b. "The Promises of Monsters: A Regenerative Politics for Inappropriate/d Others." Pp. 295–337 in P. Treichler, C. Nelson, and L. Grossberg, eds. *Cultural Studies Now and in the Future.* New York: Rutledge.

Harding, Sandra. 1986. *The Science Question in Feminism*. Ithaca, NY: Cornell University Press.

Harding, Sandra. 1991. *Whose Science? Whose Knowledge?: Thinking from Women's Lives*. Ithaca, NY: Cornell University Press.

Harris, Marvin. 1987. *Why Nothing Works*. 2d ed. New York: Simon and Schuster; orig. title *America Now*.

Hartman, Carl. 1924. "Observation on the Viability of the Mammalian Ovum." *American Journal of Obstetrics and Gynecology* 7:40–43.

Hartman, Carl. 1930. "Bimanual Rectal Palpation as Applied to the Female Rhesus Monkey." *Anatomical Record* 45:263.

Hartman, Carl. 1932. "Ovulation and the Transport and Viability of Ova and Sperm in the Female Genital Tract." In Edger Allen, ed. *Sex and Internal Secretions*. Baltimore: Williams and Wilkins.

Hartman, Carl G. 1939. "Studies on Reproduction in the Monkey and Their Bearing in Gynecology and Anthropology." *Endocrinology* 25:676.

Hartman, Carl. 1945. "The Mating of Mammals [Including Monkeys]."*Annals of the New York Academy of Sciences* 46:23–44.

Hartman, Carl G. 1956. "The Scientific Achievements of George W. Corner." *American Journal of Anatomy* 98:8.

Hartsock, Nancy M. 1983. *Money, Sex, and Power: Toward a Feminist Historical Materialism*. Boston: Northeastern University Press.

Hartsock, Nancy M. 1987. "Rethinking Modernism: Minority vs. Majority Theories." *Cultural Critique* 7:187–206.

Harvey, A. McGehee. 1976. *Adventures in Medical Research: A Century of Discovery at Johns Hopkins*. Baltimore, MD: Johns Hopkins University Press.

Harvey, A. McGehee. 1983. *Research and Discovery in Medicine: Contributions from Johns Hopkins*. Baltimore: Johns Hopkins University Press.

Hastings, Stephen. 1966. *The Murder of TSR 2*. London: Macdonald.

Hauser, Arnold. 1974. *The Sociology of Art*. Chicago: University of Chicago Press.

Hayes-Roth, Frederick. 1984. "The Machine as Partner of the New Professional." *IEEE Spectrum* 21:28–31.

Heelan, Patrick A. 1983. *Space Perception and the Philosophy of Science.* Berkeley: University of California Press.

Hekman, Susan J. 1990. *Gender and Knowledge: Elements of a Postmodern Feminism.* Boston: Northeastern University Press.

Helsinger, Elizabeth K., Robin Lauterbach Sheets, and William Veeder. 1989. *The Woman Question: Society and Literature in England and America 1837–1883.* Chicago: University of Chicago Press.

Hendrickson, Robert. 1988. *More Cunning than Man: A Social History of Rats and Men.* New York: Dorset Press.

Herman, Harry A. 1981. *Improving Cattle by the Millions.* Columbia, MO: University of Missouri Press.

Hertwig, Arthur T. and John Rock. 1939. "On a Complete Normal 12-day Human Ovum of the Pre-villous Stage." *Anatomical Record* 73 Supplement: 26–27.

Hesse, Mary. 1980. *Revolutions and Reconstructions in the Philosophy of Science.* Bloomington: Indiana University Press.

Heuser, G. F., and G. Streeter. 1941. "Development of the Macaque Embryo." *Carnegie Contributions in Embryology* 181:15–55.

Hill, Dwight D. and David R. Coelho. 1987. *Multi-Level Simulation for VLSI Design.* Boston: Kluwer.

Hiltz, Starr Roxanne and Murray Turoff. 1978. *The Network Nation: Human Communication via Computer.* Reading, MA: Addison-Wesley.

Hogle, Linda. 1993. "Margins of Life: Boundaries of the Body." Paper presented at meetings of the American Association for Applied Anthropology, San Antonio.

Holden, Constance. 1985. "An Omnifarious Data Bank for Biology?" *Science* 228:1412–13.

Holmes, F. L. 1993. "The Old Martyr of Science: The Frog in Experimental Physiology." *Journal of the History of Biology* 26:311–28.

Hogle, Linda. 1994. "Dead, Double-dead, Triple-dead: Technoscientific, Legal and Economic Definitions of 'Life' and 'Human.'" Paper presented at the meetings of the Society for Social Studies of Science, New Orleans.

Hogle, Linda. In press a. "Breadboarding, Finetuning and Interpreting: 'Standard' Medical Protocols at the Level of Everyday Practice." *Science, Technology and Human Values.*

Hogle, Linda. In press b. "Tales from the Cryptic: Technology Meets Organism in the Living Cadaver." In Chris Hables Gray, Heidi Figueroa-Sarriera and Steven Mentor, eds. *The Cyborg Handbook. New York: Routledge.*

Holstein, Jean. 1979. *The First Fifty Years at the Jackson Laboratory, 1929–1979.* Bar Harbor, ME: The Jackson Laboratory.

Holtzman, N. 1989. *Proceed with Caution: Predicting Genetic Risks in the Recombinant DNA Era.* Baltimore, MD: Johns Hopkins University Press.

Hooker, Clifford. 1987. *A Realistic Theory of Science.* Albany, NY: State University of New York Press.

hooks, bell. 1990. *Yearning: Race, Gender, and Cultural Politics.* Boston: South End Press.

Hornstein, Gail. 1988. "Quantifying Psychological Phenomena: Debates, Dilemmas, and Implications." In Jill Morawski, ed. *The Rise of Experimentation in American Psychology.* New Haven: Yale University Press.

Horowitz, I. L., ed. 1964. "Introduction." Pp. 15–64 in *The Anarchists.* New York: Dell Publishing.

Horstmann, Paul W. 1983. "Expert Systems and Logic Programming for CAD." *VLSI Design* 4:34–46.

Hubbard, Ruth. 1990. *The Politics of Women's Biology.* New Brunswick: Rutgers University Press.

Hubbard, Ruth and M. Lowe. 1983. *Women's Nature: Rationalization of Inequality.* New York: Pergamon.

Hubbard, Ruth and Elijah Wald. 1993. *Exploding the Gene Myth.* Boston: Beacon Press.

Huebner, R. J. and G. J. Todaro. 1969. "Oncogenes of RNA Tumor Viruses as Determinants of Cancer." *Proceedings of the National Academy of Sciences of the USA* 64:1087–94.

Hughes, Everett C. 1971a. *The Sociological Eye.* Chicago: Aldine.

Hughes, Everett C. 1971b. "Good People and Dirty Work." Pp. 87–97 in *The Sociological Eye: Selected Papers.* Chicago: Aldine.

Hughes, Thomas P. 1986. "The Seamless Web: Technology, Science, Etcetera, Etcetera." *Social Studies of Science* 16:281–92.

Hughes, Thomas P. 1987. "The Evolution of Large Scale Technical Systems." Pp. 51–82 in Wiebe Bijker, Thomas Hugues and Trevor Pinch, eds. *New Developments in the Social Studies of Technology.* Cambridge, MA: MIT Press.

Iacono, Suzanne and Rob Kling. 1987. "Changing Office Technologies and Transformations of Clerical Work: A Historical Perspective." Pp. 53–75 in Robert Kraut, ed. *Technology and the Transformation of White Collar Work.* Hillsdale, NJ: Lawrence Erlbaum.

Itakura, K., T. Hirose, R. Crea, A. D. Riggs, H. L. Heyneker, F. F. Bolivar and H. W. Boyer. 1977. "Expression in *Escherichia Coli* of a Chemically Synthesized Gene for the Hormone Somatostatin." *Science* 198:1056–63.

Jackson, D. A. and S. P. Stich, eds. 1979. *The Recombinant DNA Debate.* Englewood Cliffs, NJ: Prentice-Hall.

Jacob, Margaret. 1988. *The Cultural Roots of Science.* Philadelphia: Temple University Press.

Jameson, Frederic. 1984. "Postmodernism: Or the Cultural Logic of Late Capitalism." *New Left Review* 146:53–93.

Janesick, S. and B. Blouke. 1987. "Sky on a Chip: The Fabulous CCD." *Sky and Telescope* 74:238–42.

Jewett, Tom and Rob Kling. 1990. "The Work Group Manager's Role in Developing Computing Infrastructure." *Proceedings of the 5th ACM Conference on Office Information Systems*, Boston, MA.

Johnston, Ron. 1976. "Contextual Knowledge: A Model for the Overthrow of the Internal/External Dichotomy in Science." *Australian and New Zealand Journal of Sociology* 12:193–203.

Joll, James. 1993. "Nietzsche vs. Nietzsche." *The New York Review of Books* 404:20–23.

Jones, Morton. 1979. "Integrated Circuits." *Materials Science and Engineering* 37:61–64.

Jordan, Kathleen and Michael Lynch. 1992. "The Sociology of a Genetic Engineering Technique: Ritual and Rationality in the Performance of the Plasmid Prep." Pp. 77–114 in Adele E. Clarke, and Joan H. Fujimura, eds. *The Right Tools for the Job: At Work in 20th Century Life Sciences.* Princeton: Princeton University Press.

Juma, C. 1989. *The Gene Hunters: Biotechnology and the Scramble for Seeds.* Princeton: Princeton University Press.

Kaplan, Bonnie. 1983. "Computers in Medicine, 1950–1980: The Relationship Between History and Policy." Ph.D. Dissertation, University of Chicago.

Karp, Herbert and Sal Restivo. 1974. "Ecological Factors in the Emergence of Modern Science." Pp. 123–43 in Sal Restivo, and Christopher K. Vanderpool, eds. *Comparative Studies in Science and Society*. Columbus: C. Merrill.

Katz, Randy. 1985. *Information Management for Engineering Design*. New York: Springer-Verlag.

Kay, Alan. 1977. "Microelectronics and the Personal Computer." *Scientific American* 237:230–244.

Keating, Peter, Alberto Cambrosio and Michael MacKenzie. 1992. "Tools of the Discipline? Standards, Models and Measures in the Affinity-Avidity Controversy in Immunology." Pp. 312–54 in Adele E. Clarke, and Joan H. Fujimura, eds. *The Right Tools for the Job: At Work in Twentieth Century Life Sciences*. Princeton: Princeton University Press.

Keller, Evelyn Fox. 1982. "Feminism and Science." *Signs* 73:589–602.

Keller, Evelyn Fox. 1985. *Reflections on Gender and Science*. New Haven: Yale University Press.

Keller, Evelyn Fox. 1992. "Nature, Nurture and the Human Genome Project." In Kevles, Daniel J. and Lee Hood, eds. 1992. *The Code of Codes: Scientific and Social Issues in the Human Genome Project*. Cambridge, MA: Harvard University Press.

Kenney, M. 1986. *Biotechnology: The University-Industry Complex*. New Haven: Yale University Press.

Kenney, M. 1987. *The Impact of the International Political Economy on National Biotechnology Programs*. National Symposium on the Role of Biotechnology in Crop Protection, Kalyani, West Bengal, India, January.

Kevles, Daniel. 1985. *In the Name of Eugenics: Genetics and the Uses of Human Heredity*. New York: Knopf.

Kevles, Daniel J. and Lee Hood, eds. 1992. *The Code of Codes: Scientific and Social Issues in the Human Genome Project*. Cambridge, MA: Harvard University Press.

Kimbrell, Andrew. 1993. *The Human Body Shop: The Engineering and Marketing of Life*. San Francisco: HarperSanFrancisco.

Kimmelman, Barbara A. 1992. "Organisms and Interests in Scientific Research: R. A. Emerson's Claims for the Unique Contribution of

Agricultural Genetics." Pp. 172–97 in Adele E. Clarke, and Joan Fujimura, eds. *The Right Tools for the Job: At Work in Twentieth Century Life Sciences.* Princeton: Princeton University Press.

King, Jessie L. 1926. "Menstrual Records and Vaginal Smears in a Selected Group of Normal Women." *Carnegie Contributions in Embryology* 95:79–94.

King, M. D. 1971. "Reason, Tradition, and the Progressivism of Science." *History and Theory* 10:3–32.

Kitsuse, John and Malcolm Spector. 1977. *Constructing Social Problems.* Menlo Park, CA: Cummings Publishing.

Kling, Rob. 1978a. "Automated Welfare Client-Tracking and Service Integration: The Political Economy of Computing." *Communications of the ACM* 21:484–93.

Kling, Rob. 1978b. "Value Conflicts and Social Choice in Electronic Funds Transfer Systems." *Communications of the ACM* 21:642–57.

Kling, Rob. 1980. "Computer Abuse and Computer Crime as Organizational Activities." *Computers and Law Journal* 2:403–27.

Kling, Rob. 1983. "Value Conflicts in the Deployment of Computing Applications: Cases in Developed and Developing Countries." *Telecommunications Policy* 7:12–34.

Kling, Rob. 1986. "The New Wave of Academic Computing in Colleges and Universities." *Outlook* 19:8–14.

Kling, Rob. 1987. "Defining the Boundaries of Computing Across Complex Organizations." Pp. 307–62 in Richard Boland, and Rudy Hirschheim, eds. *Critical Issues in Information Systems.* London: John Wiley.

Kling, Rob. 1990. "More Information, Better Jobs?: Occupational Stratification and Labor Market Segmentation in the United States' Information Labor Force." *The Information Society* 7:77–107.

Kling, Rob. 1991. "Cooperation, Coordination and Control in Computer-Supported Work." *Communications of the ACM* 34:83–88.

Kling, Rob. 1994. "Reading 'All About' Computerization: How Genre Conventions Shape Social Analyses." *The Information Society* 10:147–172.

Kling, Rob, ed. In press. *Computerization and Controversy: Value Conflicts and Social Choices.* (2nd edition.) San Diego: Academic Press.

Kling, Rob and Suzanne Iacono. 1984. "The Control of Information Systems Development after Implementation." *Communications of the ACM* 27:1218–26.

Kling, Rob and Suzanne Iacono. 1991. "Making the Computer Revolution" in Dunlop, Charles and Rob Kling, eds. *Computerization and Controversy: Value Conflicts and Social Choices*. San Diego: Academic Press.

Kling, Rob and Roberta Lamb. In press. "Conceptualizing Electronic Publishing and Digital Libraries." In Robin M. Peek, Lois Lunin, Gregory Newby, and Gerald Miller, eds. *Academia and Electronic Publishing: Confronting the Year 2000.* Cambridge, MA: MIT Press.

Kling, Rob and Walt Scacchi. 1982. "The Web of Computing: Computer Technology as Social Organization." *Advances in Computers* 21:1–90.

Knorr-Cetina, Karin. 1979. "Tinkering toward Success." *Theory and Society* 8:347–76.

Knorr-Cetina, Karin. 1981. *The Manufacture of Knowledge.* Oxford: Pergamon Press.

Knorr-Cetina, Karin and Michael Mulkay, eds. 1983. *Science Observed: Perspectives on the Social Study of Science.* Beverly Hills, CA: Sage.

Kochan, D. 1984. "Integrated Information Processing for Manufacturing—From CAD/CAM to CIM." *Computers in Industry* 5:311–18.

Koenig, R. 1985. "Technology: Product Payoffs Prove Elusive after a Cancer Research Gain." *Wall Street Journal* June 28:25.

Kohler, Robert. 1982. *From Medical Chemistry to Biochemistry: The Making of a Biomedical Discipline.* Cambridge: Cambridge University Press.

Kohler, Robert E. 1991a. "Drosophila and Evolutionary Genetics: The Moral Economy of Scientific Practice." *History of Science* 29:335–75.

Kohler, Robert E. 1991b. *Partners in Science.* Chicago: University of Chicago Press.

Kohler, Robert E. 1991c. "Systems of Production: Drosophila, Neurospora and Biochemical Genetics." *Historical Studies in the Physical and Biological Sciences* 22:87–130.

Kohler, Robert E. 1993a. "*Drosophila*: A Life in the Laboratory." *Journal of the History of Biology* 26:281–310.

Kohler, Robert E. 1993b. *Lords of the Fly: Drosophila and the Experimental Life.* Chicago : University of Chicago Press.

Kornhauser, William. 1962. *Scientists in Industry.* Berkeley: University of California Press.

Kowalski, Thaddeus J. 1986. *An Artificial Intelligence Approach to VLSI Design.* Boston: Kluwer.

Kraemer, Kenneth, Siegfried Dickhoven, Susan Fallows Tierney and John L. King. 1987. *Datawars: The Politics of Modelling in Federal Policymaking.* New York: Columbia University Press.

Kraemer, Kenneth and John L. King. 1978. "Requiem for USAC." *Policy Analysis* 5:313–49.

Kropotkin, P. 1970. [1927] *Kropotkin's Revolutionary Pamphlets.* Dover: New York.

Kuhn, Thomas S. 1970 [1962]. *The Structure of Scientific Revolutions.* 2nd Edition. Chicago: University of Chicago Press.

Kuhn, Thomas S. 1983. "Reflections on Receiving the [Bernal] Award." *4S Review* 1:26–30.

Ladd, John. 1970. "Morality and the Ideal of Rationality in Formal Organizations." *The Monist* 54 :488–516.

Ladd, John. 1986. "Computers and Moral Responsibility: A Framework for an Ethical Analysis." Paper presented at conference on Ethical and Social Implications of Computer Networking, Stevens Institute, April 11–12, 1986.

Land, H., L. F. Parada and R. A. Weinberg. 1983. "Cellular Oncogenes and Multistep Carcinogenesis." *Science* 222:771–78.

Lane-Petter, W. 1963. "The Experimental Animal in Research." In Peter Eckstein and Francis Knowles, eds. *Techniques in Endocrine Research.* London: Academic.

Lappe, M. 1984. *Broken Code: The Exploitation of DNA.* San Francisco: Sierra Club Books.

Latour, Bruno. 1984. *Les microbes: Guerre et paix suivi de irréductions.* Paris; Metaillie.

Latour, Bruno. 1986. "Visualization and Cognition: Thinking with Eyes and Hands." *Knowledge and Society: Studies in the Sociology of Culture Past and Present* 6:1–40.

Latour, Bruno. 1987. *Science in Action*. Cambridge, MA: Harvard University Press.

Latour, Bruno. 1988a. "How to Write *The Prince* for Machines as Well as for Machinations." Pp. 20–63 in Brian Elliot, ed. *Technology and Social Change*. Edinburgh: Edinburgh University Press.

Latour, Bruno. 1988b. "Mixing Humans and Nonhumans Together: The Sociology of a Door-Closer." *Social Problems* 35:298–310. Reprinted in this volume.

Latour, Bruno. 1988c. *The Pasteurization of France*, trans. Alan Sheridan and John Law. Cambridge. MA: Harvard University Press.

Latour, Bruno. 1988d. "A Relativistic Account of Einstein's Relativity." *Social Studies of Science* 18:3–44.

Latour, Bruno. 1989. "Do We Really Need the Notion of Ideology? A Case to Get Rid of the Notion by Using Pasteur's Historiography." Paper presented at the Conference on Ideology in the Life Sciences, Harvard University, April.

Latour, Bruno. 1990a. "Drawing Things Together." Pp. 19–68 in Michael Lynch and Steve Woolgar, eds. *Representation in Scientific Practice*. Cambridge, MA: MIT Press.

Latour, B. 1990b. "Postmodern? No, Simply Amodern! Steps towards an Anthropology of Science." *Studies in the History and Philosophy of Science* 21:145–71.

Latour, Bruno. 1992a. *Nous n'avons jamais été modernes*. Paris: La Decouverte.

Latour, Bruno. 1992b. "Where are the Missing Masses? The Sociology of a Few Mundane Artefacts" Pp. 25–58 inWiebe E. Bijker and John Law, eds. *Shaping Technology/Building Society*. Cambridge, MA: MIT Press.

Latour, Bruno, and Steve Woolgar. 1986 [1979]. *Laboratory Life: The Social Construction of Scientific Facts*. London: Sage. Revised and reprinted by Princeton University Press.

Laudon, Kenneth C. 1974. *Computers and Bureaucratic Reform*. New York: John Wiley and Sons.

Laudon, Kenneth C. 1986. *Dossier Society: Value Choices in the Design of National Information Systems*. New York: Columbia University Press.

Lave, Jean. 1988a. *Cognition in Practice.* Cambridge: Cambridge University Press.

Lave, Jean. 1988b. "The Values of Quantification." Pp. 88–111 in John Law ed. *Power, Action and Belief: A New Sociology of Knowledge?* Sociological Review Monograph 32. London: Routledge.

Law, John. 1974. "Theories and Methods in the Sociology of Science: An Interpretative Approach." *Social Science Information* 13: 163–172.

Law, John. 1985. "Les textes et leurs alliés," *Culture Technique* 14:59–69.

Law, John. 1986a. "On the Methods of Long-distance Control: Vessels, Navigation and the Portuguese Route to India." Pp. 234–63 in John Law, ed. *Power, Action and Belief: A New Sociology of Knowledge?* Sociological Review Monograph. London: Routledge.

Law, John, ed. 1986b. *Power, Action and Belief: A New Sociology of Knowledge?* London: Routledge.

Law, John, 1987. "Technology and Heterogeneous Engineering: The Case of Portugese Expansion." Pp. 111–34 in W. Bijker, T. P. Hughes and T. J. Pinch, eds. *The Social Construction of Technological Systems: New Directions in the Sociology and History of Technology.* Cambridge, MA: MIT Press.

Law, John. 1988a."The Anatomy of a Sociotechnical Struggle: The Design of the TSR 2." Pp. 44–69 in Brian Elliott, ed. *Technology and Social Process.* Edinburgh: Edinburgh University Press.

Law, John. 1988b. "Purity, Production and Power: A Note on the Organisation of Sociotechnical Control." Unpublished manuscript, University of Keele, November.

Law, John, ed. 1991. *A Sociology of Monsters? Power, Technology and the Modern World.* Sociological Review Monograph. No. 38. London: Routledge.

Law, John. 1992a. "Notes on the Theory of the Actor-Network: Ordering, Strategy and Heterogeneity." *Systems Practice* 5:379–93.

Law, John. 1992b. "The Olympus 320 Engine: A Case Study in Design, Autonomy and Organisational Control." *Technology and Culture* 33: 409–40.

Lederer, Susan E. 1984. "The Rights and Wrongs of Making Experiments on Human Beings." *Bulletin of the History of Medicine* 58:380–98.

Lederer, Susan E. 1985. "Hideyo Noguchi's Luetin Experiment and the Antivivisectionists." *Isis* 76:31–48.

Lederer, Susan E. 1987. "The Controversy Over Animal Experimentation in America, 1880–1914." Pp. 235–58 in Nicolaas A. Rupke, ed.*Vivisection in Historical Perspective*. London: Croom Helm.

Lederer, Susan E. 1991a. "Dogfights: The Use of Pet Animals in Biomedical Research, 1938–1966." Paper presented at the meetings of the American Association for the History of Medicine, Cleveland, Ohio.

Lederer, Susan E. 1991b. "Political Animals: The Shaping of Biomedical Research Literature in Twentieth-Century America." *Isis* 83:61–79.

Lederer, Susan E. 1993. "Laboratory Life on the Silver Screen: Animal Experimentation and the Film Industry in the 1930s." Paper presented at the History of Science Society Meetings, Santa Fe.

Lederman, Muriel and Richard M. Burian. 1993. "The Right Organism for the Job: Introduction." *Journal of the History of Biology* 26:235–38.

Lederman, Muriel and Sue A. Tolin. 1993. "OVATOOMB: Other Viruses and the Origins of Molecular Biology." *Journal of the History of Biology* 26:239–54.

Lemaine, Gerard, Roy MacLeod, Michael Mulkay, and Peter Weingart, eds. 1976. *Perspectives on the Emergence of Scientific Disciplines*. Chicago: Aldine.

Lenoir, Timothy. 1989. "The Politics of Vision: Optics, Painting and Ideology in Germany, 1845–1880." Paper presented at the International Conference on Ideology in the Life Sciences, Harvard-Fidia Lecture and Conference Series, Harvard University, April.

Levy, Steven. 1984. *Hackers: Heroes of the Computer Revolution*. Garden City, NY: Anchor/Doubleday.

Lewontin, Richard C. 1991. *Biology as Ideology: The Doctrine of DNA*. New York: HarperPerennial.

Lewontin, Richard. 1992. "Doubts about the Human Genome Project." *The New York Review of Books* 92:31–40.

Liberman, Ken. 1985. *Understanding Interaction in Central Australia: An Ethnomethodological Study of Australian Aboriginal People*. London: Routledge and Kegan Paul.

Lillie, Frank. 1917. "The Free-Martin: A Study of the Action of Sex Hormones in the Foetal Life of Cattle." *Journal of Experimental Zoology* 23:371–452.

Lindsey, J. Russell. 1979. "Historical Foundations." Pp. 1–36 in Henry J. Baker, J. Russell Lindsey and Steven A. Weisbroth, eds. *The Laboratory Rat*, vol. I, *Biology and Diseases.* New York: Academic Press.

Lippman, A. 1988. "Prenatal Genetic Screening and Testing: Constructing Needs and Reinforcing Inequities. *American Journal of Law and Medicine* 17:15–50.

Livingston, Eric. 1986. *The Ethnomethodological Foundations of Mathematics*. London: Routledge and Kegan Paul.

Lloyd, E. A. 1988. *The Structure and Confirmation of Evolutionary Theory*. Westport, CT: Greenwood Press.

Loeb, Leo. 1911. "The Cyclic Changes in the Ovary of the Guinea Pig." *Journal of Morphology* 23:37–70.

Long, Diana E. 1987. "Physiological Identity of American Sex Researchers between the Two World Wars." Pp. 263–78 in Gerald Geison, ed. *Physiology in the American Context, 1850–1940.* Bethesda, MD: American Physiological Society.

Long, J. A. and H. M. Evans. 1922. "The Oestrus Cycle in the Rat and Its Associated Phenomena." *Memoirs of the University of California* 6.

Longino, Helen. 1990. *Science as Social Knowledge: Values and Objectivity in Scientific Inquiry.* Princeton, NJ: Princeton University Press.

Lorde, Audre. 1981. "The Master's Tools Will Never Dismantle the Master's House." Pp. 98–101 in Cher'rie Moraga and Gloria Anzaldúa, eds. *This Bridge Called My Back: Writings by Radical Women of Color*. Watertown, MA: Persephone Press.

Loughlin, Julia. 1993. "The Feminist Challenge to Social Studies of Science." Pp. 3–20 in T. Brante, S. Fuller and W. Lynch, eds. *Controversial Science: From Content to Contention*. Albany, NY: State University of New York Press.

Lowy, Ilana. 1993. "On Mice, Men and Tumors: The Development of Medical Oncology in the U.S. after the Second War." Paper presented at the History of Science Society Meetings, Santa Fe, 1993.

Lynch, Michael. 1985a. *Art and Artefact in the Laboratory: A Study of Shop Work and Shop Talk in a Research Laboratory*. London: Routledge and Kegan Paul.

Lynch, Michael. 1985b. "Discipline and the Material Form of Images: An Analysis of Scientific Visibility." *Social Studies of Science* 15:37–66.

Lynch, Michael. 1988a. "The Externalized Retina: Selection and Mathematization in the Visual Documentation of Objects in the Life Sciences." Pp. 201–34 in Michael Lynch and Steve Woolgar, eds. *Representations in Scientific Practice*. Cambridge, MA: MIT Press.

Lynch, Michael. 1988b. "Sacrifice and the Transformation of Animal Body into Scientific Object: Laboratory Culture and Ritual Practice in the Neurosciences." *Social Studies of Science* 18:265–89.

Lynch, Michael and Samuel Y. Edgerton. 1988. "Aesthetics and Digital Image Processing." Pp. 184–220 in Gordon Fyfe and John Law, eds. *Picturing Power: Visual Depiction and Social Relations*. Sociological Review Monograph 35. London: Routledge and Kegan Paul.

Lynch, Michael, Eric Livingston, and Harold Garfinkel. 1983. "Temporal Order in Laboratory Work." Pp. 205–80 in Karin Knorr-Cetina and Michael Mulkay, eds. *Science Observed*. London: Sage.

Lynch, Michael and Steve Woolgar, eds. 1990. *Representation in Scientific Practice*. Cambridge, MA: MIT Press.

"Machine of the Year." January 3, 1982. *Time*: 13–39.

McAdam, Doug. 1988. *Freedom Summer*. New York: Oxford University Press.

Macbeth, Douglas. 1992. "Classroom 'Floors': Material Organizations as a Course of Affairs." *Qualitative Sociology* 15:123–50.

McCarthy, John and Mayer Zald. 1977. "Resource Mobilization and Social Movements: A Partial Theory." *American Journal of Sociology* 82:1212–41.

McCorduck, Pamela. 1979. *Machines Who Think*. San Francisco: Freeman.

McCrady, Edward. 1938. "The Embryology of the Opossum." *American Anatomical Memoirs*. 16:9–14.

MacDowell, E. C. and E. M. Lord. 1926. "The Relative Viability of Male and Female Mouse Embryos." *American Journal of Anatomy* 37:127–40.

MacKenzie, Donald A. 1981. *Statistics in Britain, 1865–1930: The Social Construction of Scientific Knowledge*. Edinburgh: Edinburgh University Press.

MacKenzie, Donald. 1986. "Science and Technology Studies and the Question of the Military." *Social Studies of Science* 16:361–71.

MacKenzie, Donald and Judy Wajcman. 1985a. "Introductory Essay: The Social Shaping of Technology." Pp. 2–25 in Donald MacKenzie and Judy Wajcman, eds. *The Social Shaping of Technology: How the Refrigerator Got Its Hum.* Philadelphia: Milton Keynes and Open University Press.

MacKenzie, Donald and Judy Wacjman, eds. 1985b. *The Social Shaping of Technology: How the Refrigerator Got Its Hum.* Philadelphia: Milton Keynes and Open University Press.

McLaughlin, Loretta. 1982. *The Pill, John Rock, and the Church.* Boston: Little, Brown.

McNeill, William. 1963. *The Rise of the West.* Chicago: University of Chicago Press.

McNeill, William. 1982. *The Pursuit of Power: Technology, Armed Force, and Society.* Chicago: University of Chicago Press.

Maienschein, Jane. 1981. "Shifting Assumptions in American Biology; Embryology, 1890–1910." *Journal of the History of Biology* 14:89–113.

Maienschein, Jane. 1983. "Experimental Biology in Transition: Harrison's Embryology, 1895–1910." Pp. 107–127 in William Coleman and Camille Limoges, eds. *Studies in the History of Biology* 6. Baltimore, MD: Johns Hopkins University Press.

Maienschein, Jane. 1985a. "Agassiz, Hyatt, Whitman and the Birth of the Marine Biological Laboratory." *Biological Bulletin of Woods Hole* 168 Supplement:26–34.

Maienschein, Jane. 1985b. "Early Struggles at the Marine, Biological Laboratory." *Biological Bulletin of Woods Hole* 168 Supplement:192–96.

Maienschein, Jane. 1985c. "First Impressions: American Biologists at Naples." *Biological Bulletin of Woods Hole* 168 Supplement:187–91.

Maienschein, Jane, ed. 1987. *Defining Biology: Lectures from the 1890's.* Cambridge, MA: Harvard University Press.

Maienschein, Jane, Ronald Rainger and Keith Benson. 1981. "Introduction: Were American Morphologists in Revolt?" *Journal of the History of Biology* 14: 83–87.

Mall, Franklin. 1893. "Early Human Embryos and the Mode of Their Preservation." *Bulletin of the Johns Hopkins Hospital* 36:115.

Mall, Franklin. 1903. "Note on the Collection of Human Embryos in the Anatomical Laboratory of Johns Hopkins University." *Bulletin of the Johns Hopkins Hospital* 14:39–33.

Mander, Jerry. 1991. *In the Absence of the Sacred: The Failure of Technology and the Survival of the Indian Nations.* San Francisco: Sierra Club Books.

Mankoff, Milton. 1972. *The Poverty of Progress: The Political Economy of American Social Problems.* New York: Holt, Rinehart, and Winston.

Marcson, Simon. 1960. *The Scientist in American Industry.* New York: Harper.

Martin, Brian. 1979. *The Bias of Science.* Canberra, Australia: Society for Social Responsibility in Science.

Martin, Emily. 1987. *The Woman in the Body: A Cultural Analysis of Reproduction.* Boston: Beacon Press.

Marx, Karl. 1956. *Economic and Philosophic Manuscripts of 1844,* trans. Martin Milligan. Moscow: Foreign Languages Publishing House.

Marx, Karl. 1973. *Grundrisse,* trans. Martin Nicolaus. New York: Vintage Press.

Mascia-Lees, Frances, Patricia Sharpe and Colleen Ballerino Cohen. 1988. "The Postmodernist Turn in Anthropology: Cautions from a Feminist Perspective." *Signs* 151:7–33.

Maynard-Moody, Steven. 1984. [1979]. "The Fetal Research Dispute." Pp. 213–32 in Dorothy Nelkin, ed. *Controversy: The Politics of Technical Decisions.* Newbury Park, CA: Sage.

Meiksins, Peter F. 1982. "Science in the Labor Process: Engineers as Workers." Pp. 121–40 in Charles Derber, ed. *Professionals as Workers.* New York: G. K. Hall & Co.

Merchant, Carolyn. 1980. *The Death of Nature: Women, Ecology, and the Scientific Revolution.* Harper & Row.

Merleau-Ponty, Maurice. 1962. *Phenomenology of Perception,* trans. Colin Smith. London: Routledge and Kegan Paul.

Merton, Robert K. 1968. *Social Theory and Social Structure,* enlarged ed. New York: The Free Press.

Merton, Robert K. 1970. [1938]. *Science, Technology and Society in Seventeenth-Century England.* New York: Harper Torchbooks.

Merton, R. K. 1973. *The Sociology of Science: Theoretical and Empirical Investigations.* N. W. Storer, ed. Chicago: University of Chicago Press.

La Mettrie. 1784. *L'homme machine.* Leiden.

Michie, D. and R. Johnson. 1985. *The Creative Computer: Machine Intelligence and Human Knowledge.* Harmondsworth: Penguin.

Middleton, David and Derek Edwards, eds. 1990. *Collective Remembering.* London: Sage.

Mies, Maria. 1990. "Science, Violence, and Responsibility." *Women's Studies International Forum* 13:443–41.

Mills, C.Wright. 1961. *The Sociological Imagination.* New York: Grove Press.

Mills, C. Wright. 1963. *Power, Politics and People.* New York: Ballantine Books.

Ministry of Defence. 1957. *Defence: Outline of Future Policy.* Cmnd. 124. London: Her Majesty's Stationer Office. Ministry of Defence and Ministry of Supply.

Ministry of Defence and Ministry of Supply. 1955. *The Supply of Military Aircraft.* Cmnd. 9388. London: Her Majesty's Stationery Office.

Mitman, Gregg. 1993a. "Cinematic Nature: Hollywood Technology, Popular Culture, and the Science of Animal Behavior, 1925–1940." *Isis* 84:637–61.

Mitman, Gregg. 1993b. "The Private Life of the Dolphin: What Enquiring Minds Never Knew." Paper presented at the History of Science Society Meetings, Santa Fe.

Mitman, Gregg and Anne Fausto-Sterling. 1992. "Whatever Happened to *Planaria*? C. M. Child and the Physiology of Inheritance." Pp. 172–97 in Adele Clarke and Joan Fujimura, eds. *The Right Tools for the Job: At Work in Twentieth Century Life Sciences.* Princeton: Princeton University Press.

Montgomery, Scott L. 1991. "Science as Kitsch: The Dinosaur and Other Cultural Icons." *Science and Culture* 21:7–58.

Moore, Carl R. and David Bodian. 1940. "Opossum Pouch Young as Experimental Material." *Anatomical Record* 76:319.

Morrell, Virginia. 1992. "30-Million-Year-Old DNA Boosts an Emerging Field." *Science* 257:1860–62.

Morrison, K. L. 1990. "Some Researchable Recurrences in Disciplinary-Specific Inquiry." Pp. 141–57 in D. T. Helm, W. T. Anderson, A. J. Mechan, and A. W. Rawls, eds. *The Instructional Order: New Directions in the Study of Social Order*. New York: Irvington Publishers.

Morrow, J. F., S. N. Cohen, A. C. Y. Chang, H. W. Boyer, H. M. Goodman and R. B. Helling. 1974. "Replication and Transcription of Eucaryotic DNA in *Escherichia Coli." Proceedings of the National Academy of Sciences, U.S.A.* 71:1743–47.

Moss, R. W. 1989. *The Cancer Industry: Unraveling the Politics*. New York: Paragon House.

Mowshowitz, Abbe. 1976. *The Conquest of Will: Information Processing in Human Affairs*. Reading, MA: Addison-Wesley.

Mulkay, Michael 1979. *Science and the Sociology of Knowledge*. London: George Allen and Unwin.

Mulkay, Michael J. 1985. *The Word and The World: Exploration in the Form of Sociological Analysis*. London: George Allen and Unwin.

Mullins, Nicholas. 1973. *Theory and Theory Groups in Contemporary American Sociology*. New York: Harper & Row.

Mulvagh, S. L., Roberts, R., Schneider, M. D. 1988. "Cellular Oncogenes in Cardiovascular Disease." *Journal of Molecular and Cellular Cardiology* 20:657–62.

Mumford, Lewis. 1966. *The Myth of the Machine*. New York: Harcourt.

Musson, Albert E. and Eric Robinson. 1969. *Science and Technology in the Industrial Revolution*. Manchester: Manchester University Press.

Myers, Greg. 1988. "Every Picture Tells a Story: Illustrations in E. O. Wilson's *Sociobiology*." Pp. 235–70 in Michael Lynch and Steve Woolgar, eds. *Representations in Scientific Practice*. Cambridge, MA: MIT Press.

Myers, W. 1985. "CAD/CAM: The Need for a Broader Focus." *Computer* 15:105–16.

Naisbitt, John. 1984. *Megatrends*. New York: Warner Books.

Nandy, Ashis. 1988. *Science, Hegemony, and Violence*. New Delhi: Oxford University Press.

Needham, Joseph. 1928. *Man a Machine, in Answer to a Romantical and Unscientific Treatise Written by Sig. Eugenio Rignano and Entitled "Man Not A Machine."* New York.

Nelkin, Dorothy and Tancredi, L. 1989. *Dangerous Diagnostics: The Social Power of Biological Information.* New York: BasicBooks.

Newell, Allen and Stuart Card. 1985. "The Prospects for Psychological Science in Human-Computer Interaction." Part 1 *Human-Computer Interaction* 1:209–42.

Newell, Allen and Stuart Card. 1987. "Straightening Out Softening Up: Response to Carroll and Campbell." Part 2 *Human-Computer Interaction* 2:251–67.

Newhall, Beaumont. 1964. *The History of Photography.* New York: Museum of Modern Art.

Newman, H. H. 1948. "History of the Department of Zoology in The University of Chicago." *Bios* 19:215–39.

Newmark, P. 1983. "Oncogenic Intelligence: The *Ras*matazz of Cancer Genes." *Nature* 305:470–71.

Nietzsche, F. 1974. *The Gay Science* trans. Walter Kaufman. Vintage: New York.

Noble, David F. 1977. *America by Design: Science and Technology and the Rise of Corporate Capitalism.* Oxford: Oxford University Press.

Noble, David F. 1985. *Forces of Production: Science, Technology, and the Rise of Corporate Capitalism.* New York: Oxford University Press.

Osborne, Anthony. 1979. *Running Wild: The Next Industrial Revolution.* Berkeley, CA: Osborne-McGraw Hill.

Oudshoorn, Nelly. 1990. "On the Making of Sex Hormones: Research Materials and the Production of New Knowledge." *Social Studies of Science* 20:5-33.

Ozonoff, David. 1979. "The Political Economy of Cancer Research." *Science and Nature* 2:13–16.

Papert, Seymour. 1979. "Computers and Learning." Pp. 73–86 in Michael L. Dertouzos and Joel Moses, eds. *The Computer Age: A Twenty-Year View.* Cambridge, MA: MIT Press.

Papert, Seymour. 1980. *Mindstorms: Children, Computers and Powerful Ideas.* New York: Basic Books.

Parada, L. F., C. J. Tabin, C. Shih and R. A. Weinberg. 1982. "Human EJ Bladder Carcinoma Oncogene Is Homologue of Harvey Sarcoma Virus *Ras* Gene." *Nature* 297:474–79.

Park, Robert Ezra. 1952. *Human Communities.* Glencoe, IL: The Free Press.

Paton, William. 1993. [1984]. *Man and Mouse: Animals in Medical Research.* London: Oxford University Press.

Pauly, Philip J. 1984. "The Appearance of Academic Biology in Late 19th Century America." *Journal of the History of Biology* 17: 369–97.

Perrolle, Judith. 1987. *Computers and Social Change: Information, Property, and Power.* Belmont, CA: Wadsworth.

Perrow, Charles. 1984. *Normal Accidents: Living with High-Risk Technologies.* New York: Basic Books.

Petchesky, Rosalind Pollack. 1987. "Fetal Images: The Power of Visual Culture in the Politics of Reproduction." Pp. 57–80 in Michelle Stanworth, ed. *Reproductive Technologies: Gender, Motherhood and Medicine.* Minneapolis: University of Minnesota Press.

Pickering, Andrew. 1988. "Big Science as a Form of Life." In M. de Maria and M. Grilli, eds. *The Restructuring of the Physical Sciences in Europe and the United States, 1945–1960.* Singapore: World Scientific Publishing.

Pickering, Andrew. 1990. "Knowledge, Practice and Mere Construction." *Social Studies of Science* 20:682–729.

Pickering, Andrew, ed. 1992. *Science as Practice and Culture.* Chicago: University of Chicago Press.

Pickering, Andrew. 1994. "The Mangle of Practice." *American Journal of Sociology* 99:559–89.

Pinch, Trevor and Wiebe E. Bijker. 1984. "The Social Construction of Facts and Artefacts: Or How the Sociology of Science and the Sociology of Technology Might Benefit Each Other." *Social Studies of Science* 14:399–441.

Pinch, Trevor and Wiebe E. Bijker. 1986. "Science, Relativism, and the New Sociology of Technology: Reply to Russell." *Social Studies of Science* 16:347–60.

Pocock, R. I. 1906. "Notes Upon Menstruation, Gestation and Parturition of Some Monkeys That Have Lived in the Society Gardens." *Proceedings of the Zoological Society of London*: 338–570.

Polanyi, Michael. 1958. *Personal Knowledge.* London: Routledge and Kegan Paul.

Poovey, Mary. 1988. *Uneven Developments: The Ideological Work of Gender in Mid-Victorian England.* Chicago: University of Chicago Press.

Postan, Michael M., Denys Hay and John D. Scott. 1964. *Design and Development of Weapons: Studies in Government and Industrial Organisation.* London: H.M. Stationery Office.

Potter, Jonathan. 1988. "What Is Reflexive about Discourse Analysis? The Case of Reading Readings." Pp. 37–54 in Steve Woolgar, ed. *Knowledge and Reflexivity: New Frontiers in the Sociology of Knowledge.* London: Sage.

Presthus, R. 1964. "The Social Dysfunction of Organization." Pp. 551-62 in I. L. Horowitz, ed. *The Anarchists.* New York: Dell Publishing.

Proudhon, P. 1977. "The Revolution and the Nation," Pp. 314–18 in G. Woodcock, ed. *The Anarchist Reader.* Glasgow: Fontana.

Pyenson, Lewis. 1983. *Neohumanism and the Persistence of Pure Mathematics in Wilhelmian Germany.* Philadelphia: American Philosophical Society.

Rabinow, Paul. 1986. "Representations Are Social Facts: Modernity and Post-Modernity in Anthropology." Pp. 234–61 in James Clifford and George E. Marcus, eds. *Writing Culture: The Poetics and Politics of Ethnography.* Berkeley: University of California Press.

Rader, Karen A. 1992. "Making Mice: Clarence Little, the Jackson Laboratory and the Standardization of *Mus Musculus* for Research." Dissertation proposal in History and Philosophy of Science Department, Indiana University.

Rader, Karen A. 1993a. "Animals as High-Performance Technologies: Laboratory Mice and Clarence Little's War on Cancer in 1930s America." Paper presented at the History of Science Society Meetings, Santa Fe.

Rader, Karen A. 1993b. "Turning Mice into Money: Clarence Little and Patronage for Cancer Research in 1930s America." Paper presented at "Science in a National Context: Style and Substance," Indiana University, Bloomington.

Rathmell, J. G. 1985. "Information flow in VLSI Design." *Integration* 4:185–91.

Reed, Bruce and Geoffrey Williams. 1971. *Denis Healey and the Policies of Power.* London: Sidgwick and Jackson.

Reich, Robert. 1991. *The Work of Nations.* New York: Knopf.

Reinecke, Ian. 1984. *Electronic Illusions.* New York: Penguin Books.

Reingold, Nathan, ed. 1979. *The Sciences in the American Context: New Perspectives.* Washington, DC: Smithsonian Institution.

Restivo, Sal 1983a. "The Myth of the Kuhnian Revolution." Pp. 293–305 in Randall Collins, ed. *Sociological Theory.* San Francisco: Jossey-Bass.

Restivo, Sal. 1983b. *The Social Relations of Physics, Mysticism, and Mathematics.* Dordrecht, Holland: D. Reidel.

Restivo, Sal 1988. "Modern Science as a Social Problem." *Social Problems* 35: 206–25.

Restivo, Sal. 1989. "In the Clutches of Daedalus: Science, Society, and Progress." Pp. 145–76 in S. Goldman, ed. *Science, Technology and Social Progress.* Bethlehem, PA: Lehigh University Press.

Restivo, Sal. 1991. *The Sociological Worldview.* Oxford: Blackwell.

Restivo, Sal. 1992. *Mathematics in Society and History.* Dordrecht: Kluwer.

Restivo, Sal. 1993a. *Science, Society, and Values: Toward a Sociology of Objectivity.* Bethlehem, PA: Lehigh University Press.

Restivo, Sal. 1993b. "Science, Sociology of Science, and the Anarchist Tradition." Pp. 21–40 in T. Brante, et al., eds. *Controversial Science: From Content to Contention.* Albany, NY: State University of New York Press.

Restivo, Sal and Julia Loughlin. 1987. "Critical Sociology of Science and Scientific Validity." *Knowledge* 3:486–508.

Restivo, Sal and Christopher K. Vanderpool. 1974. "Science: Social Activity and Social Process." Pp. 1–24 in Sal Restivo and Christopher K. Vanderpool, eds. *Comparative Studies in Science and Society.* Columbus, OH: C. E. Merrill.

Rettig, R. A. 1977. *Cancer Crusade: The Story of the National Cancer Act of 1971.* Princeton: Princeton University Press.

Rezac, R. R. 1984. "Driving Technical Forces and Model for Use of CAE Workstations." Pp. 221–23 in *IEEE International Conference on Computer-Aided Design Procedures.* Los Angeles, CA: IEEE Computer Society.

Rheinberger, H-J. 1992. "The Laboratory Production of Transfer RNA." *Studies in History and Philosophy of Science* 23:389–422.

Rich, Adrienne. 1978. "Hunger." In *The Dream of a Common Language: Poems 1974–77*. New York: Norton.

Rignano, E. 1926. *Man Not a Machine: A Study of the Finalistic Aspects of Life*. London.

Rosaldo, Renato. 1989. *Culture and Truth: The Remaking of Social Analysis*. Boston: Beacon Press.

Rose, Hilary and Steven Rose, eds. 1976. *The Political Economy of Science*. London: Macmillan.

Rosenberg, Charles. 1976. *No Other Gods: On Science and American Social Thought*. Baltimore, MD: Johns Hopkins University Press.

Rosenberg, Charles. 1979a. "Rationalization and Reality in Shaping American Agricultural Research, 1875–1914." Pp. 143–63 in Nathan Reingold, ed. *The Sciences in the American Context: New Perspectives*. Washington, D.C.: Smithsonian Institution.

Rosenberg, Charles. 1979b. "Toward an Ecology of Knowledge: On Discipline, Contexts and History." Pp. 440–55 in Alexandra Oleson and John Voss, eds. *The Organization of Knowledge in Modern America*, 1869–1920. Baltimore, MD: Johns Hopkins University Press.

Rosenfeld, L. C. 1941. *From Beast-Machine to Man-Machine: Animal Soul in French Letters from Descartes to La Mettrie*. New York: Oxford University Press.

Rossiter, Margaret. 1979. "The Organization of the Agricultural Sciences." Pp. 279–98 in Alexandra Oleson and John Voss, eds. *The Organization of Knowledge in Modern America*, 1869–1920. Baltimore, MD: Johns Hopkins University Press.

Roszak, Theodore. 1973. *Where the Wasteland Ends*. New York: Anchor Books.

Rothschild, Joan, ed. 1983. *Machina Ex Dea: Feminist Perspectives on Technology*. New York: Pergamon Press.

Rothschild, Miriam. 1983. *Dear Lord Rothschild: Birds, Butterflies and History*. Philadelphia, PA: Balaban.

Rowan, Andrew. 1984. *Of Mice, Models and Men*. Albany: State University of New York Press.

Rubin, H. 1983. "Letter to Science." *Science* 219:1170–71.

Rubin, H. 1985. "Cancer as a Dynamic Developmental Disorder." *Cancer Research* 45:2935–42.

Rudwick, Martin J. S. 1976. "The Emergence of a Visual Language for Geological Science, 1760–1840." *History of Science* 14:149–95.

Rupke, Nicolaas A. 1987. *Vivisection in Historical Perspective.* London: Croom Helm.

Ruse, Michael and Peter J. Taylor, eds. 1991. Special Issue on Pictorial Representation in Biology. *Biology and Philosophy* 6:125–294.

Russell, Stewart. 1986. "The Social Construction of Artefacts: A Response to Pinch and Bijker." *Social Studies of Science* 16:333–46.

Sabin, Florence Rena. 1934. *Franklin Paine Mall.* Baltimore: Johns Hopkins University Press.

Sacks, Harvey. 1984. "Notes on Methodology." Pp. 21–27 in J. M. Atkinson and J. Heritage, eds. *Structures of Social Action: Studies in Conversation Analysis.* Cambridge: Cambridge University Press.

Salerno, Lynne. 1985. "Whatever Happened to the Computer Revolution?" *Harvard Business Review* 85:129–38.

Santos, E., S. Tronick, S. Aaronson, S. Pulciani and M. Baracid. 1982. "T24 Human Bladder Carcinoma Oncogene Is an Activated Form of Normal Human Homologue of Balab-Harvey Msv Transforming Genes." *Nature* 298:343–47.

Saver, Robert M., ed. 1960. "Care and Diseases of the Research Monkey." *Annals of the New York Academy of Sciences* 85:735–992.

Scacchi, Walt. 1984. "Managing Software Engineering Projects: A Social Analysis." *IEEE Transactions on Software Engineering* SE–10(1):49–59.

Scacchi, Walt, S. Bendifallah, A. Bloch, S. Choi, P. K. Garg, A. Jazzar, A. Safavi, J. Skeer, and M. J. Turner. 1986. "Modeling System Development Work: A Knowledge-Based Approach." System factory working paper SF-86-05, Computer Science Department, University of Southern California, Los Angeles, June.

Schiebinger, Londa. 1989. *The Mind Has No Sex? Women in the Origins of Modern Science.* Cambridge, MA: Harvard University Press.

Schmidt, L. H. 1972. "Problems and Opportunities of Breeding Primates." Pp. 1–22 in W. I. B. Beveridge, ed. *Breeding Primates: Proceedings of the International Symposium on Breeding Non-Human Primates for Laboratory Use.* New York: Karger.

Schneider, Howard A. 1973. "Harry Steenbock, 1886–1967: A Biographical Sketch." *Journal of Nutrition* 103:1238.

Schneider, Joseph. 1984. "Morality, Social Problems, and Everyday Life." Pp. 80–205 in Joseph Schneider and John Kitsuse, eds. *Studies in the Sociology of Social Problems*. Norwood, NJ: Ablex Publishing Co.

Schneider, Joseph. 1985. "Social Problems Theory: The Constructionist View." *Annual Review of Sociology* 11:209–29.

Schneider, Joseph and John Kitsuse, eds. 1984. *Studies in the Sociology of Social Problems*. Norwood, NJ: Ablex.

Schultz, A. H. 1971. "The Rise of Primatology in the Twentieth Century." P. 4 in J. Biegert and W. Leutenegger, eds. *Proceedings of the Third International Congress of Primatology*. Vol. 1. Zurich, New York: S. Karger.

Schutz, Alfred. 1962. "On Multiple Realities." Pp. 207–59 in *The Problem of Social Reality*, vol. 1 of *Collected Papers*, ed. Maurice Natanson. The Hague: Nijhoff.

Schwartz, D. E., R. Tizard and W. Gilbert. 1983. "Nucleotide Sequence of Rous Sarcoma Virus." *Cell* 32:853–69.

Schwartz, Eugene. 1972. *Overkill*. New York: Ballentine Books.

Scipiades, Elemer. 1938. "Young Human Ovum Detected in Uterine Scraping." *Carnegie Contributions in Embryology*. 163:97.

Scott, W. Richard. 1966. "Professionals in Bureaucracy—Areas of Conflict." Pp. 265–75 in Howard M. Vollmer and Donald L. Mills, eds. *Professionalization*. Englewood Cliffs, NJ: Prentice-Hall.

Scott, W. Richard. 1992. *Organizations: Rational, Natural, and Open Systems*. 3rd edition. Englewood Cliffs, NJ: Prentice-Hall.

Searle, John. 1980. "Minds, Brains and Programs." *The Behavioral and Brain Sciences* 3:417–57.

Shaiken, Harlie. 1985. *Work Transformed: Automation and Labor in the Computer Age*. New York: Holt, Rinehart, and Winston.

Shalin, Dmitri. 1986. "Pragmatism and Social Interactionism." *American Sociological Review* 51:9–29.

Shapin, Steven. 1988. "The House of Experiment in Seventeenth-Century England." *Isis* 79:373–404.

Shapin, Steven and Simon Schaffer. 1985. *Leviathan and the Air-Pump: Hobbes, Boyle, and the Experimental Life*. Princeton University Press.

Shapin, Steven. 1989. "The Invisible Technician." *American Scientist* 77:553–63.

Shapley, D. 1983. "U.S. Cancer Research: Oncogenes Cause Cancer, Institute to Change Tack." *Nature* 301:5.

Shih, C., B. Shilo, M. Goldfarb, A. Dannenberg and R. Weinberg. 1979. "Passage of Phenotypes of Chemically Transformed Cells via Transfection of DNA and Chromantic." *Proceedings of the National Academy of Sciences* 76:5714–18.

Shimizu, K. , M. Goldfarb, Y. Suard, M. Perucho, Y. Li, T. Kamata, J. Feramiso, E. Starnezer, J. Fogh and M. Wigler. 1983. "Three Human Transforming Genes are Related to the Viral Ras Oncogenes." *Proceedings of the National Academy of Sciences* 80:2112–16.

Shubik, P. 1983. "Letter to Science." *Science* 220:1226–28.

Simon, Herbert. 1977. "What Computers Mean for Man and Society." *Science* 195:1186–90.

Simoudis, E. and S. Fickas. 1985. "The Application of Knowledge-Based Design Techniques to Circuit Design." Pp. 213–15 in *Proceedings of the IEEE Conference on Computer-Aided Design.* Washington, D.C.: IEEE Computer Society Press.

Singer, Peter. 1990. *Animal Liberation.* New York: Random House.

Singleton, Vicky and Mike Michael. 1993. "Actor-networks and Ambivalence: General Practitioners in the UK Cervical Screening Programme." *Social Studies of Science* 23: 227–64.

Smith, Dorothy. 1987. *The Everyday World as Problematic.* Boston: Northeastern University Press.

Smith, Dorothy. 1990. *The Conceptual Practices of Power: A Feminist Sociology of Knowledge.* Boston: Northeastern University Press.

Smith, Philip E. 1927. "The Disabilities Caused by Hypophysectomy in the Rat and Their Repair." *Journal American Medical Association* 88:158–61.

Sohn-Rethel, Alfred. 1979. "Science as Alientated Consciousness." *Radical Science Journal* 2/3:65–101.

Spector, D. H., H. E. Varmus and J. M. Bishop. 1978. "Nucleotide Sequences Related to the Transforming Gene of Avian Sarcoma Virus are Present in the DNA of Uninfected Vertebrates." *Proceedings of the National Academy of Sciences of the U.S.A.* 75:5023–27.

Spector, Malcolm and John Kitsuse. 1977. *Constructing Social Problems.* Menlo Park, CA: Cummings.

Spender, Dale. 1983. *Women of Ideas and What Men Have Done to Them.* London: Ark Paperbacks.

Star, Susan Leigh. 1983. "Simplification in Scientific Work: An Example from Neuroscience Research." *Social Studies of Science* 13:205–28.

Star, Susan Leigh. 1985. "Scientific Work and Uncertainty." *Social Studies of Science* 15:391–427.

Star, Susan Leigh. 1986. "Triangulating Clinical and Basic Research: British Localizationists, 1870–1906." *History of Science* 24:29–48.

Star, Susan Leigh. 1988. "Layered Representations and the Coordination of Information: Making Computer Chips and Mapping the Brain." Paper presented to the Society for the Social Study of Science/European Association for the Study of Science and Technology (4S/EASST), Amsterdam, November.

Star, Susan Leigh. 1989a. *Regions of the Mind: Brain Research and the Quest for Scientific Certainty.* Stanford, CA: Stanford University Press.

Star, Susan Leigh.1989b. "The Structure of Ill-Structured Solutions: Boundary Objects and Heterogeneous Distributed Problem Solving." Pp. 37–54 in M. Huhns and I. Gasser, eds. *Distributed Artificial Intelligence* 2. Menlo Park, CA: Morgan Kauffmann.

Star, Susan Leigh. 1991a. "Power, Technologies and the Phenomenology of Standards: On Being Allergic to Onions." Pp. 27–57 in John Law, ed. *A Sociology of Monsters? Power, Technology and the Modern World.* Sociological Review Monograph. No. 38. London: Routledge.

Star, Susan Leigh. 1991b. "The Sociology of the Invisible: The Primacy of Work in the Writings of Anselm Strauss." Pp. 265–83 in David Maines, ed. *Social Organization and Social Process: Essays in Honor of Anselm Strauss.* Hawthorne, NY: Aldine de Gruyter.

Star, Susan Leigh. 1992a. "Craft vs. Commodity, Mess vs. Transcendence: How the Right Tool Became the Wrong One in the Case of Taxidermy and Natural History." Pp. 257–86 in Adele E. Clarke and Joan Fujimura, eds. *The Right Tools for the Job: At Work in Twentieth Century Life Sciences.* Princeton: Princeton University Press.

Star, Susan Leigh. 1992b. "The Trojan Door: Organizations, Work, and the 'Open Black Box.'" *Systems/Practice* 5:395–410.

Star, Susan Leigh. 1993. "Thinking Paradoxically: Multiple Regimes in the Great Divide." Proceedings of Conference "Beyond the Great Divide: Social Science, Technical Systems, and Cooperative Work." Paris: CNRS, March.

Star, Susan Leigh. In press a. "Misplaced Concretism and Concrete Situations: Feminism, Method and Information Technology." Paper presented at the Gender, Culture, Nature Network Workshop. June 1, 1994. Århus University. Århus, Denmark. Gender, Nature, Culture Working Papers Series.

Star, Susan Leigh, ed. In press b. *The Cultures of Computing*. Sociological Review Monograph. Oxford: Basil Blackwell.

Star, Susan Leigh and E. Gerson. 1987. "The Management and Dynamics of Anomalies in Scientific Work." *Sociological Quarterly* 28:147–69.

Star, Susan Leigh and James R. Griesemer. 1989. "Institutional Ecology, 'Translations,' and Coherence: Amateurs and Professionals in Berkeley's Museum of Vertebrate Zoology, 1907–1939." *Social Studies of Science* 19: 387–420.

Stehelin, D., H. E. Varmus, J. M. Bishop and P. K. Vogt. 1976. "DNA Related to the Transforming Gene(s) of Avian Sarcoma Viruses Is Present in Normal Avian DNA." *Nature* 260:170–73.

Steinbock, Bonnie. 1992. *Life Before Birth: The Moral and Legal Status of Embryos and Fetuses*. New York: Oxford University Press.

Stockard, Charles R. and George Papanicolaous. 1917. "The Existence of a Typical Oestrus Cycle in the Guinea Pig, with a Study of Its Istological and Physiological Changes." *American Journal of Anatomy* 22:223–83.

Strassman, Paul. 1985. *Information Payoff : The Transformation of Work in the Electronic Age*. New York: Free Press.

Strauss, Anselm. 1971. *Professions, Work and Careers*. San Francisco: Sociology Press.

Strauss, Anselm. 1978a. *Negotiations: Varieties, Contexts, Processes, and Social Order*. San Francisco: Jossey-Bass.

Strauss, Anselm. 1978b. "A Social World Perspective." *Studies in Symbolic Interaction* 1:119–28.

Strauss, Anselm. 1985. "Work and the Division of Labor." *Sociological Quarterly* 26:1–19.

Strauss, Anselm. 1987. *Qualitative Analysis for Social Scientists.* Cambridge: Cambridge University Press.

Strauss, Anselm. 1993. *Continual Permutations of Action.* Hawthorne, NY: Aldine de Gruyter.

Strauss, Anselm and Lee Rainwater. 1962. *The Professional Scientist: A Study of American Chemists.* Chicago: Aldine.

Strauss, Anselm, Leonard Schatzman, Rue Bucher, Danuta Erlich and Melvin Sabshin. 1964. *Psychiatric Ideologies and Institutions.* New York: Free Press of Glencoe.

Strum, Shirley and Bruno Latour. 1987. "Redefining the Social Link: From Baboons to Humans." *Social Science Information* 26:783–802.

Sturdevant, A. H. 1971. "On the Choice of Materials for Genetic Studies." *Stadler Symposium* 1:51–57.

Suchman, Lucy. 1987. *Plans and Situated Actions: The Problem of Human-Machine Communication.* Cambridge: Cambridge University Press.

Suchman, Lucy. 1988. "Representing Practice in Cognitive Science." *Human Studies* 11:305.

Sudnow, David. 1979. *Talk's Body: A Mediation Between Two Keyboards.* Hammondsworth: Penguin.

Summers, William. 1993. "How and Why Bacteriophage Came to Be Used by the Phage Group." *Journal of the History of Biology* 26:255–68.

Suzuki, D. and P. Knudtson. 1990. *Genethics: The Ethics of Engineering Life.* Cambridge, MA: Harvard University Press.

Tabin, C. J., S. M. Bradley, C. I. Bargmann, R. A. Weinberg, A. G. Papageorge, E. M. Scolnick, R. Dhar, D. R. Lowy and E. H. Chang. 1982. "Mechanism of Activation of a Human Oncogene." *Nature* 300:143–49.

Tansey, E. M. 1994. "Protection Against Canine Distemper and Dogs Protection Bills: The Medical Research Council and Anti-Vivisectionist Protest, 1911–1933." *Medical History* 38: 1–19.

Taylor, H. Jeanie, Maureen Ebben and Cheris Kramarae, eds. 1993. *WITS: Women, Information Technology, and Scholarship Colloquium.* Urbana, IL: Center for Advanced Study.

Taylor, Peter J. and Ann S. Blum. 1991. "Pictorial Representation in Biology." *Biology and Philosophy* 6:125–34.

Taylor, Robert, ed. 1980. *The Computer in the School: Tutor, Tutee, Tool.* New York: Teachers College Press, Columbia University.

Teitelman, R. 1985. "The Baffling Standoff in Cancer Research." *Forbes,* July 15:110–14.

Teitelman, R. 1989. *Gene Dreams: Wall Street, Academia and the Rise of Biotechnology.* New York: Basic Books.

Temin, H. M. 1971. "The Protovirus Hypothesis: Speculations on the Significance of Rna Directed Dna Synthesis for Normal Development and Carcinogenesis." *Journal of the National Cancer Institute* 46:3–7.

Temin, H. M. 1980. "Origin of Retroviruses of Cellular Genetic Moveable Elements." *Cell* 21:599–600.

Temin, H. M. 1983. "We Still Don't Understand Cancer." 1983. *Nature* 302:656.

Thomas, W. I. and Dorothy Swaine Thomas. 1970 [1917]. "Situations Defined as Real are Real in their Consequences." Pp. 54–155 in Gregory P. Stone and Harvey A. Farberman, eds. *Social Psychology Through Symbolic Interaction.* Waltham, MA: Xerox Publishers.

Thomas, W. I. and Florian Znaniecki. 1918. *The Polish Peasant in Poland and America.* New York: Alfred A. Knopf.

Thompson, Edward P. 1980. *The Making of the English Working Class.* New York: Penguin Books.

Tibbetts, Paul. 1988. "Representation and the Realist-Constructivist Controversy." *Human Studies* 11:117–32.

Times (London). 1963. "Britain's Space Entry Nearer." December 4:7.

Times (London).1963b "Parliament." March 5:14.

Times (London). 1964a "TSR 2 makes its first flight." September 28:10.

Times (London). 1964b "Parliament." January 17:14.

Todaro, G. J., and Huebner, R. J. 1972. "The Viral Oncogene Hypothesis: New Evidence." *Proceedings of the National Academy of Sciences* 69:1009–15.

Toffler, Alvin. 1980. *The Third Wave.* New York: William Morrow.

Traugott, Mark. 1978. "Reconceiving Social Movements." *Social Problems* 26:38–49.

Traweek, Sharon. 1984. "Nature in the Age of Its Mechanical Reproduction: The Reproduction of Nature and Physicists in the High Energy Physics Community." Pp. 94–112 in Claire Belisle and Bernard Schiele, eds. *Les savoirs dans les pratiques quotidiennes*. Paris: Centre National de Recherche Scientifique.

Traweek, Sharon. 1989. *Beamtimes and Lifetimes*. Cambridge, MA: Harvard University Press.

Traweek, Sharon. In press. "Bodies of Evidence: Law and Order, Sexy Machines, and the Erotics of Fieldwork among Physicists." In S. Forster, ed. *Choreographing History*. Bloomington, IN: Indiana University Press.

Trescott, Martha M., ed. 1979. *Dynamos and Virgins Revisited: Women and Technological Change in History*. London: The Scarecrow Press.

Troyer, Ronald J. In press. "Are Social Problems and Social Movements the Same Thing?" In Gale Miller and James A. Holstein, eds. *Perspectives on Social Problems*. Greenwich, CT: JAI Press.

Tuana, Nancy, ed. 1989. *Feminism and Science*. Bloomington, IN: Indiana University Press.

Tucker, B. R. 1964. "State Socialism and Libertarianism." Pp. 169–82 in I. L. Horowitz, ed. *The Anarchists*. New York: Dell Publishing.

Turkle, Sherry. 1984. *The Second Self: Computers and the Human Spirit*. New York: Simon & Schuster.

Turner, Stephen. 1989. "Tacit Knowledge and the Project of Computer Modeling Cognitive Processes in Science." Pp. 83–94 in S. Fuller, M. deMey, T. Shinn and S. Woolgar, eds. *The Cognitive Turn*. Dordrecht: Kluwer.

Tyler, S. A. 1987. *The Unspeakable: Discourse, Dialogue and Rhetoric in the Postmodern World*. London: University of Wisconsin Press.

Uhlig, Ronald, David Farber and James Bair. 1979. *The Office of the Future: Communication and Computers*. New York: North-Holland.

Useem, Bert and Mayer Zald. 1982. "From Pressure Group to Social Movement: Organizational Dilemmas of the Effort to Promote Nuclear Power." *Social Problems* 30:144–56.

Valerio, D. A., et al. 1973. *Macaca mulatta: Management of a Laboratory Breeding Colony*. New York: Academic.

Varmus, H. T. 1989a. "An Historical Overview of Oncogenes." Pp. 3–44 in R. Weinberg, *Oncogenes and the Molecular Origins of Cancer*. Cold Spring Harbor, NY: Cold Spring Harbor Lab Press.

Varmus, H. T. 1989b. "Retroviruses and Oncogenes I." Nobel Lecture, Dec 8, Stockholm, Sweden.

Vartanian, A. 1960. *La Mettrie's* L'Homme Machine: *A Study in the Origins of an Idea.* Princeton: Princeton University Press.

Vaughan, Diane. 1989. "Regulating Risk: Implications of the Challenger Accident." *Law and Policy* 11:330.

Vaughan, Diane. 1990. "Autonomy, Interdependence, and Social Control: NASA and the Space Shuttle *Challenger*." *Administrative Science Quarterly* 35: 225–57.

Veblen, Thorstein. 1919. *The Place of Science in Modern Civilization and Other Essays.* New York: Viking Press.

Volberg, Rachel. 1983. "Constraints and Commitments in the Development of American Botany, 1880–1920." Ph.D. Dissertation, University of California, San Francisco.

Vollman, Rudolph P. 1965. "Fifty Years of Research on Mammalian Reproduction: Carl G. Hartman." USDHEW, PHS Publication 1281:5.

van Wagenen. 1950. "The Monkey." Pp. 1–42 in E. J. Farris, ed. *The Care and Breeding of Laboratory Animals*. New York: Wiley.

Wajcman, Judy. 1991. *Feminism Confronts Technology.* University Park, PA: Penn State Press.

Walsh, Edward. 1981. "Resource Mobilization and Citizen Protest in Communities Around Three Mile Island." *Social Problems* 29:1–21.

Waterfield, M. D., G. T. Scrace, N. Whittle, P. Stroobant, A. Johnson, A. Wasteson, B. Wetermark, J. Huang, and T. F. Deuel. 1983. "Platelet-derived Growth Factor Is Structurally Related to the Putative Transforming Protein P28 *Sis* of Simian Sarcoma Virus." *Nature* 304:35–39.

Watson, John D., et al. 1997. *Molecular Biology of the Gene,* Vol. 2. Menlo Park: W.A. Benjamin.

Webb, Cecil S. 1953. *A Wanderer in the Wild: The Odyssey of an Animal Collector*. London: Hutchinson.

Weber, Max. 1958. *The Protestant Ethic and the Spirit of Capitalism.* New York: Charles Scribner's Sons.

Webster, Frank and Kevin Robbins. 1986. *Information Technology: A Luddite Analysis*. Norwood, NJ: Ablex.

Weick, Karl. 1987. "Organizational Culture and High Reliability."*California Management Review* 29:112–27.

Weinberg, R. A. 1982. "Review: Oncogenes of Human Tumor Cells." In S. Prentis, ed. *Trends in Biochemical Sciences*, vol. 7. Amsterdam: Elsevier Biomedical Press BV.

Weinberg, R. A. 1983. "A Molecular Basis of Cancer." *Scientific American* 249:126–43.

Weinberg, R. 1988. "Finding the Anti-oncogene." *Scientific American* (September):44–51.

Weinberg, R. A., ed. 1989. *Oncogenes and the Molecular Origins of Cancer*. Cold Spring Harbor, NY: Cold Spring Harbor Press.

Weizenbaum, Joseph. 1976. *Computer Power and Human Reason: From Judgement to Calculation*. San Francisco: W. H. Freeman.

Werdinger, Jeffrey. 1980. "Embryology at Woods Hole: The Emergence of a New American Biology." Ph.D. Dissertation, Indiana University.

Whitman, Charles Otis. 1902a. "A Biological Farm." *Biological Bulletin of Woods Hole* 3:214–24.

Whitman, Charles Otis. 1902b. "A Biological Farm for the Experimental Investigation of Heredity, Variation and Evolution and for the Study of Life Histories, Habits, Instincts, and Intelligence." *Science* 16:504–10.

Wieckert, Kären. 1988. "The Impact of Organizational Setting on Expert Systems Development." *Proceedings of the 1988 Conference on the Impact of Artificial Intelligence*, Denton, TX.

Williams, Geoffrey, Frank Gregory, and John Simpson. 1969. *Crisis in Procurement: A Case Study of the TSR 2*. London: Royal United Service Institution.

Wilson, Harold. 1971. The Labour Government 1964–1970: A Personal Record. London: Weidenfeld and Nicolson and Michael Joseph.

Wilson, John. 1973. *Introduction to Social Movements*. New York: John Wiley.

Wilson, Karl M. 1926. "Histological Changes in the Vaginal Mucosa of the Sow." *American Journal of Anatomy* 37:418–32.

Wimsatt, William. 1980. "Reductionist Research Strategies and Their Biases in the Units of Selection Controversy." Pp. 213–59 in T. Nickles, ed. *Scientific Discovery: Case Studies*. Dordrecht, The Netherlands: Reidel.

Wimsatt, William C. 1986. "Developmental Constraints, Generative Entrenchment and the Innate-Acquired Distinction." Pp. 185–208 in P. W. Bechtel, ed. *Integrating Scientific Disciplines*. Dordrecht: Martinus-Nijhoff.

Winner, Langdon. 1977. *Autonomous Technology*. Cambridge, MA: MIT Press.

Winner, Langdon. 1984. "Mythinformation in the High-tech Era." *IEEE Spectrum* 21:90–96.

Winner, Langdon. 1985. "Do Artifacts Have Politics?" Pp. 26–38 in Donald MacKenzie and Judy Wajcman, eds. *The Social Shaping of Technology*. Philadelphia: Open University Press.

Winner, Langdon. 1986. *The Whale and the Reactor: A Search for the Limits in an Age of High Technology*. Chicago: University of Chicago Press.

Winston, Patrick and Karen Prendergast, eds. 1984. *The AI Business: Commercial Uses of Artificial Intelligence*. Cambridge, MA: MIT Press.

Witkowski, Jan A. 1983. "Experimental Pathology and the Origins of Tissue Culture: Leo Loeb's Contribution." *Medical History* 27:269–88.

Wittgenstein, Ludwig. 1953. *Philosophical Investigations*, trans. G. E. M. Anscombe. Oxford: Blackwell.

Wood, Derek. 1975. *Project Cancelled*. London: Macdonald and Jane's.

Woolgar, Steve. 1985. "Why Not a Sociology of Machines? The Case of Sociology and Artificial Intelligence." *Sociology* 19: 557–72

Woolgar, Steve. 1986. "The Chips Are Now Down?" *Nature* 324:182–83.

Woolgar, Steve. 1987. "Reconstructing Man and Machine: A Note on Sociological Critiques of Cognitivism." Pp. 31–32 in W. Bijker, T. P. Hughes and T. J. Pinch, eds. *The Social Construction of Technological Systems: New Directions in the Sociology and History of Technology*. Cambridge, MA: MIT Press.

Woolgar, Steve, ed. 1988a. *Knowledge and Reflexivity: New Frontiers in the Sociolgy of Knowledge*. London: Sage.

Woolgar, Steve. 1988b. *Science: The Very Idea.* London: Tavistock/Horwood.

Woolgar, Steve. 1988c "Time and Documents in Researcher Interaction: Some Ways of Making Out What Is Happening in Experimental Science." *Human Studies* 11:171–200.

Woolgar, Steve. 1988d. "The Turn to Technology in the Social Studies of Science." Working paper, presented to Edinburgh University Research Centre for the Social Sciences, 18 January.

Woolgar, Steve and Malcolm Ashmore. 1988. "The Next Step: An Introduction to the Reflexive Project." Pp. 1–13 in Steve Woolgar, ed. *Knowledge and Reflexivity: New Frontiers in the Sociolgy of Knowledge.* London: Sage.

Woolgar, Steve and Dorothy Pawluch. 1985. "Ontological Gerrymandering: The Anatomy of Social Problems Explanations." *Social Problems* 32:214–37.

Wright, Sewell. 1931. *Evolution in Mendelian Populations.* Chicago: University of Chicago Press.

Wright, S. 1986. "Recombinant DNA Technology and its Social Transformation, 1972–1982." *Osiris,* 2d Series 2:303–60.

Wright, Will. 1992. *Wild Knowledge: Science, Language, and Social Life in a Fragile Environment.* Minneapolis, MN: University of Minnesota Press.

Yerkes, Robert M. 1916. "Provision for the Study of Monkeys and Apes." *Science* 43:231.

Yerkes, Robert M. 1925. *Almost Human.* New York: Century.

Yerkes, Robert M. 1935. "Yale Laboratories of Primate Biology." *Science* 82:618–20.

Yerkes, Robert M. 1935–1936. "The Significance of Chimpanzee-Culture for Biological Research." *Harvey Lectures* 31:57–73.

Yerkes, Robert M. 1943. *Chimpanzees: A Laboratory Colony.* New Haven: Yale University Press.

Yerkes, Robert M. 1963. "Creating a Chimpanzee Community." Ed. Roberta W. Yerkes from archival materials. *Yale Journal of Biology and Medicine* 36:205–23.

Young, Andrew T. 1985. "What Color Is the Solar System?" *Sky and Telescope.* May: 399.

Yourdon, Edward. 1986. *Nations at Risk: The Impact of the Computer Revolution*. New York: Yourdon Press.

Yoxen, Edward. 1983. *The Gene Business: Who Should Control Biotechnology?* New York: Oxford University Press.

Yoxen, Edward. 1987. "Seeing with Sound: A Study of the Development of Medical Images." Pp. 281–303 in W. Bijker, T. P. Hughes, and T. J. Pinch, eds. *The Social Construction of Technological Systems: New Directions in the Sociology and History of Technology*. Cambridge, MA: MIT Press.

Zald, Mayer and Michel Berger. 1978. "Social Movements in Organizations: Coup d'Etat, Insurgency, and Mass Movements." *American Journal of Sociology* 83:823–61.

Zallen, Doris. 1993. "The 'Light' Organism for the Job: Green Algae and Photosynthesis Research." *Journal of the History of Biology* 26:269–80.

Zenzen, Michael and Sal Restivo 1982. "The Mysterious Morphology of Immiscible Liquids." *Social Science Information* 3:447–73.

Zondek, Bernhard and Michael Finkelstein. 1966. "Professor Bernhard Zondek: An Interview." *Journal of Reproduction and Fertility*. 12:3–19.

Zuckerman, Harriet. 1979. *Scientific Elite : Nobel Laureates in the United States*. New York : Free Press.

Zuckerman, Solly. 1930. "The Menstrual Cycle of the Primates, Part I: General Nature and Homology." *Proceedings of the Zoological Society of London*. 22: 691–734.

Zuckerman, Solly. 1970. "The History of Apes and Monkeys as Objects of Enquiry." Pp. 205–14 in Solly Zuckerman, ed. Appendix 1. *Beyond the Ivory Tower: Frontiers of Public and Private Science*. New York: Taplinger.

Contributors

Michel Callon is professor of Sociology and director of the Centre de Sociologie de l'Innovation at the École des mines de Paris. He has published widely in the sociology of science and technology, the economics of research and development, the sociology of translation, and scientometrics. Together with Bruno Latour, John Law and other colleagues, he contributed to the development of actor-network theory. He has pionereed a range of techniques for describing scientific and technical change, including the co-word method. He was directly involved in the definition and implementation of policy on research in France in the early 1980s. He is the coeditor of *Mapping the Dynamics of Science and Technology*.

Adele E. Clarke is associate professor of Sociology, Department of Social and Behavioral Sciences, co-director of the Women, Health and Healing Program, and adjunct associate professor of History of Health Sciences at the University of California, San Francisco. A recent paper is "Money, Sex and Legitimacy at Chicago, 1900–1940: Lillie's Center of Reproductive Biology" (Fall 1993 Special Issue of *Perspectives on Science* on Biology at the University of Chicago, coedited with Gregg Mitman and Jane Maienschein). She is currently finishing a book, *Disciplining Reproduction: Modernity, American Life Sciences and the "Problem of Sex"* (University of California Press, 1995). Her paper "Modernity, Postmodernity and Human Reproductive Processes, c.1890–Present, or 'Mommy, Where Do Cyborgs Come from Anyway?'" will appear in Chris Hables Gray, Heido Figueroa-Sarriera and Steven Mentor, eds. *The Cyborg Handbook* (Routledge, 1995).

Jennifer Croissant is instructor of Sociology in the Department of Science and Technology Studies at Rensselaer Polytechnic Institute in Troy, New York, where she recently completed her Ph.D. in Science and Technology Studies. Her dissertation, "Bodies, Movements, Representations: Elements for a Feminist Theory of Knowledge," was supported by NSF and AAUW Fellowships.

Joan Fujimura is the Henry R. Luce Professor of Biotechnology and Society at Stanford University. She is associate professor in the Department of Anthropology and in the History and Philosophy of Science Program. Fujimura works on the sociology and ethnography of recent biomedicine and molecular biology. She has written on the development of oncogene research, on recent debates in AIDS research, and on theoretical issues in science studies. Her current research is on bioinformatics, the standardization of biomedical knowledge and technologies, and other aspects of the uses of computers in the international human genome projects as well as on the standardization of disease classifications in databases used internationally.

C. Suzanne Iacono is assistant professor of Management Information Systems, School of Management, Boston University. She earned her Ph.D. in Management Information Systems from the University of Arizona and her B.A. and M.A. from the University of California, Irvine. Before joining Boston University, she held a research position at the Public Policy Research Organization at the University of California, Irvine, where she studied the social dynamics of the development and use of computer-based information systems in manufacturing and white-collar work organizations. Current research focuses on the role of emerging technologies in organizational innovation and change. In particular, she is investigating the links between electronic work groups and political processes in organizational settings. A current project focuses on cultural differences and the effectiveness of distance teams in global organizations. Her articles have appeared in a variety of publications, including *Communications of the ACM, Journal of Social Issues, Social Problems, Journal of Information, Technology and Management,* and *Revue International de Sociologie.*

Rob Kling completed his graduate studies, specializing in Artificial Intelligence, at Stanford University. He has been on the faculty of the University of California, Irvine, since 1973. In 1987 he was awarded an honorary doctorate in Social Sciences by the Free University of Brussels. Since the early 1970s he has studied the social opportunities and dilemmas of computerization for managers, professionals, workers, and the public. His current research focuses on the ways that computerization is a social process with technical elements, how intensive computerization transforms work, and how computerization entails many social choices. He has also studied the ways that complex information systems and expert systems are integrated into the social life of organizations. He is coauthor of *Computers and Politics: High Technology in American Local Governments* (Columbia University Press) and coeditor of two recent books, *PostSuburban California: The Transformation of Postwar Orange County* (University of California Press) and *Computerization and Controversy: Value Conflict & Social Choices* (Academic Press).

Bruno Latour was trained as a philosopher and an anthropologist. After field studies in Africa and California he specialized in the analysis of scientists and engineers at work. He has written *Laboratory Life: The Construction of Scientific Facts* (with Steve Woolgar, Princeton University Press), *Science in Action*, and *The Pasteurization of France* (both at Harvard University Press). He recently published a new field study on an automatic subway system, *Aramis or the Love of Technology* (to be published by Harvard), and an essay on symmetric anthropology, *We Have Never Been Modern* (also with Harvard). He has just published *La clef de Berlin*, a series of essays on science and techniques. He is professor at the Centre de Sociologie de l'Innovation in the *École nationale superieure des mines* in Paris.

John Law is Professor of Sociology at Keele University. He writes on technology and organization, and is particularly concerned with materiality, spatiality, and heterogeneity. His recent book, *Organizing Modernity* (Blackwell, 1994), is a theoretical study of social ordering in a laboratory.

Michael Lynch is senior lecturer in the Department of Human Sciences at Brunel University. His publications include *Art and Artifact in Laboratory Science* (London, 1985), an ethnographic study of practical activities in a neurosciences laboratory, and *Scientific Practice and Ordinary Action* (Cambridge, UK, 1993), a theoretical study of the relationships between sociology of scientific knowledge and ethnomethodology. He also has worked on visual representations in science (see his coedited volume with Steve Woolgar, *Representation in Scientific Practice* (Cambridge, MA, 1990).

Sal Restivo is professor of Sociology and Science Studies in the Department of Science and Technology Studies at Rensselaer Polytechnic Institute in Troy, New York. He is president of the Society for Social Studies of Science (1993–1995). His latest book is

Science, Society and Values: Toward a Sociology of Objectivity (Lehigh, 1994).

Susan Leigh Star is associate professor of Sociology and faculty affiliate in Women's Studies, Library and Information Science, and Computer Science at the University of Illinois, Urbana-Champaign. She is also affiliate research scientist, Institute for Research on Learning, Palo Alto. She is author of *Regions of the Mind: Brain Research and the Quest for Scientific Certainty* (Stanford, 1989) and editor of *The Cultures of Computing* (Blackwell, 1995). She does ethnographic and historical research on information systems and medical classification and writes feminist theory.

Steve Woolgar is professor of Sociology and director of CRICT (Centre for Research into Innovation, Culture and Technology) at Brunel, the University of West London. He is author of *Science: The Very Idea* (Routledge, 1988) and *Laboratory Life: The Construction of Scientific Facts* (with Bruno Latour, Princeton, 1986) and editor of *Knowledge and Reflexivity* (Sage, 1988) and *Representation in Scientific Practice* (with Michael Lynch, MIT, 1990). He is currently engaged in several projects exploring the textual and reflexive dimensions of new information technologies, and is researching the theoretical bases of technology transfer.

Index

F

G

H

I

J

K

N

O

P

R

S

T

U

V

W

Y

Z